Computational Fluid and Solid Mechanics

Series Editor:

Klaus-Jürgen Bathe
Massachusetts Institute of Technology
Cambridge, MA, USA

Advisors:

Franco Brezzi
University of Pavia
Pavia, Italy

Olivier Pironneau
Université Pierre et Marie Curie
Paris, France

Miloš Kojić · Klaus-Jürgen Bathe

Inelastic Analysis of Solids and Structures

With 136 Figures and 33 Tables

 Springer

Authors

Miloš Kojić
Department of Mechanical Engineering
University of Kragujevac, Serbia and Montenegro;
Harvard School of Public Health, Boston, MA, USA

Klaus-Jürgen Bathe
Massachusetts Institute of Technology
Cambridge, MA, USA

ISBN 3-540-22793-8 **Springer Berlin Heidelberg New York**

Library of Congress Control Number: 2004112098

Springer is a part of Springer Science+Business Media

springeronline.com

© Springer-Verlag Berlin Heidelberg 2005
Printeu ın Germany

Tvnesettino: Data conversion by the authors.
Final processing by PTP-Berlin Protago-TEX
Cover-Design: deblik, Berlin
Printed on acid-free paper   62/3020Yu - 5 4 3 2 1 0

Preface

This book is a result of the authors' research and teaching in the field of finite element nonlinear analysis over many years.

In the research related to this book, our objective has been to develop robust, accurate and efficient computational procedures to calculate stresses from given strains (strain-driven problems) within incremental (finite element) inelastic analyses. To this aim, M. Kojić worked at ADINA R&D, Inc., intermittently, for a few years about 18 years ago. During that time we published together four valuable papers (Bathe et. al. 1984, Kojić and Bathe 1987a, 1987b, 1987c). In these papers we introduced the "effective-stress-function" method for the implicit integration of thermo-elasto-plastic and creep material models. We found that the function of the effective stress (a single parameter which governs the inelastic deformations in the time step) is monotonic, and that the zero provides the solution for the stresses, creep and plastic strains at the end of the time step. The simple algorithm was surprisingly robust, accurate and efficient, and suitable for general applications. Based on the exciting research we had conducted, we formally agreed at that time to write a book together on the inelastic analysis of solids and structures.

Since 1992, M. Kojić no longer contributed at ADINA R&D, but we both continued to work independently on the development of inelastic analysis procedures. M. Kojić published a series of papers in which he called a generalization of the effective-stress-function method the "governing parameter method". Independently, K.J. Bathe continued his research on inelastic analysis procedures and used the effective-stress-function method and generalizations thereof, and ADINA R&D continued to develop material models in ADINA. Of course, some of these material models are the subject of this book.

In teaching, our objective has been to present in a unified manner the physical and theoretical background of inelastic material models and computational methods, and to illustrate the behavior of the models in typical engineering conditions.

With the above objectives we started about five years ago to work on this text. We prepared this book to give the fundamentals of inelastic material models based on experimental observations and principles of mechanics, to describe computational algorithms for stress calculation (stress integration within a time step), and to present solved examples. We give the theoretical background to an extent necessary to describe the commonly employed material models in metal isotropic and orthotropic plasticity, thermoplasticity and viscoplasticity, and the plasticity of geological materials. The computational algorithms are developed in a unified manner with some detailed derivations of the algorithmic relations. The solved examples are designed to give insight into the material behavior in various engineering conditions (general three-dimensional deformations, plane strain, axisymmetric, plane stress, shell, beam, pipe conditions), and to demonstrate the application of the computational algorithms.

In the book we do not focus on some of the current research areas in computational inelasticity, as, for example, non-local models, gradient plasticity theory, damage models, fracture and the dynamics of inelasticity. But the presented computational methods can, of course, be used for the development of algorithms in these and other areas. Also, we do not give a detailed review of computational methods in inelasticity nor a broad presentation of these methods, but rather only focus on our experiences in the field.

We wrote this book for self-study by engineers and students, and for use in graduate courses on computational inelasticity, with emphasis in certain areas. For example, Chapters 1 to 5 can be used in a course devoted to inelastic deformations of metals. In case large strains are also considered, Chapter 7 should also be used. For subjects with emphasis on geological materials, Chapters 1 to 4 and 6 are applicable. Also, certain sections of the book can be used in courses on general plasticity to simply illustrate how inelastic response is computed in practice.

Preparing this text required of us a large effort, and we are grateful to a number of institutions and individuals.

Miloš Kojić is thankful to the Department of Mechanical Engineering of the University of Kragujevac, the Center for Scientific Research of the Serbian Academy of Science and Art and the University of Kragujevac, and the Automobile Institute Zastava, for the support of his research and for the many granted leaves of absence. He is also grateful to Professors Radovan Slavković, Miroslav Zivković, Nenad Grujović, Branislav Popović and Ivo Vlastelica, and Dusan Begović, all of the University of Kragujevac and the PAK-software research and development group, and Srdjan Divac of Harvard University for their collaboration in research and comments on this book. And he thanks, for an extraordinary effort in preparing the figures for this book, Dr. Ivo Vlastelica and Vladimir Djordjević, and for technical support in typing the manuscript Miloš Babović, Vladimir Dimitrijević, Snezana Vulović, Dr. Nebojsa Zdravković, Dr. Nenad Filipović, and Aleksandar and Nikola Kojić.

Klaus-Jürgen Bathe is thankful to the Department of Mechanical Engineering, M.I.T, for the excellent environment made available for his teaching and scholarly writing. This book required a large amount of his time, with much strenuous effort. However, the book has now reached a level, where he would like to see it published. K.J. Bathe is grateful to his many students and colleagues who have worked with him over many years. Regarding this book, he is particularly thankful to his student Phill-Seung Lee, to Professor Francisco J. Montáns, University Castilla La-Mancha, Spain, and to Professor Jay Wang, Louisiana Tech University, Ruston, Louisiana, for their comments, and to Dr. Song Wang of ADINA R&D, Inc. for his help in the example solutions. He would like to acknowledge as well that − to improve his teaching and research efforts − his involvement in ADINA R&D has been very valuable.

Finally, we would like to thank our families for their continuous support and understanding regarding our scientific endeavors.

M. Kojić and K. J. Bathe

Table of Contents

Preface iii

1. **Introduction** 1
   1.1 The Objectives of this Book 1
   1.2 Some Remarks on Explicit and Implicit Solutions of Nonlinear Response... 2
   1.3 The Topic of this Book in an Overall Context 5

2. **An Introduction to the Incremental-Iterative Solution of Nonlinear Structural Problems** 13
   2.1 Some Solution Procedures for Nonlinear Equations 13
       2.1.1 Some Solution Procedures for One Nonlinear Equation 13
       2.1.2 Newton Method for a System of Nonlinear Equations . 18
   2.2 Nonlinear Structural Problems and Incremental-Iterative Equilibrium Equations 19
       2.2.1 A Simple Example to Exemplify Nonlinear Analysis .. 20
       2.2.2 Multiple-Degree-of-Freedom System 23
   2.3 A Linearized Form of the Principle of Virtual Work. Updated Lagrangian Formulation and Finite Element Equilibrium Equations 26
       2.3.1 General Considerations........................... 27
       2.3.2 Isoparametric Continuum Finite Elements 30

3. **Fundamental Notions of Metal Plasticity** 37
   3.1 Introduction 37
   3.2 Isotropic Plasticity 38
       3.2.1 Uniaxial Elastic-Plastic Deformation 39
       3.2.2 Yield Condition for General Three-Dimensional Deformations 44
       3.2.3 Three-Dimensional Stress-Plastic Strain Relations 48
       3.2.4 Hardening in 3–D 53
       3.2.5 The von Mises Material Model with Mixed Hardening. 63
       3.2.6 Principle of Maximum Plastic Dissipation 71
       3.2.7 Examples 74
   3.3 Orthotropic Plasticity 84

	3.3.1	Hill's Orthotropic Material Model	85
	3.3.2	Orthotropic Models for Sheet Metals................	89
	3.3.3	Examples	100

4. A General Procedure for Stress Integration and Applications in Metal Plasticity 105

4.1	Introduction	105
4.2	The Governing Parameter Method	108
	4.2.1 Formulation of the Governing Parameter Method	108
	4.2.2 Time Independent Plasticity Models	111
4.3	A Brief Review of Stress Integration Procedures	121
4.4	Stress Integration for the von Mises Material Model........	128
	4.4.1 General Three-Dimensional Conditions, Plane Strain and Axisymmetry	129
	4.4.2 Shell Conditions and Plane Stress	138
	4.4.3 Elastic-Plastic Matrix	141
	4.4.4 Accuracy Considerations	147
4.5	Examples..	151
4.6	Orthotropic Material Model - Stress Integration	188
	4.6.1 Stress Integration	188
	4.6.2 Elastic-Plastic Matrix	193
	4.6.3 Examples	194

5. Creep and Viscoplasticity 201

5.1	Introduction	201
5.2	Creep and Thermoplastic Material Models	202
	5.2.1 Uniaxial Creep	203
	5.2.2 Multiaxial Creep Model	213
	5.2.3 Thermoplasticity and Creep	216
5.3	Viscoplastic Material Models...............................	217
	5.3.1 One-Dimensional Model	217
	5.3.2 General Three-Dimensional Viscoplastic Models	220
5.4	Stress Integration for Creep and Thermoplastic Models......	223
	5.4.1 General Three-Dimensional Deformations............	223
	5.4.2 Shell (Plane Stress) Conditions....................	229
	5.4.3 Elastic-Plastic-Creep Constitutive Matrix	231
	5.4.4 Examples	232
5.5	Stress Integration for Viscoplastic Models	247
	5.5.1 General Considerations............................	247
	5.5.2 The Governing Parameter Method for Stress Integration in Viscoplasticity	249
	5.5.3 Isotropic Metal...................................	249
	5.5.4 Examples	257

6.   Plasticity of Geological Materials 263
   6.1   Introduction to the Mechanical Response of Geomaterials.... 263
         6.1.1   Basic Notions in Geomechanics..................... 264
         6.1.2   Mechanical Properties of Geological Materials........ 268
         6.1.3   Short Historical Review of Soil Plasticity 279
   6.2   Cap Models.. 280
         6.2.1   Description of Cap Models......................... 280
         6.2.2   Stress Integration Procedure 285
         6.2.3   Elastic-Plastic Matrices 290
         6.2.4   Examples 295
   6.3   Cam-Clay Model 300
         6.3.1   Formulation of the Model 300
         6.3.2   Stress Integration Procedure 305
         6.3.3   Elastic-Plastic Matrices 309
         6.3.4   Examples 311
   6.4   A General Soil Plasticity Model 313
         6.4.1   Formulation of the Model 314
         6.4.2   Stress Integration 316
         6.4.3   Elastic-Plastic Matrix............................ 317
         6.4.4   Example 318

7.   Large Strain Elastic-Plastic Analysis 321
   7.1   Introduction 321
   7.2   Basic Notions in the Kinematics of Large Strain Deforma-
         tions   323
         7.2.1   Examples 341
   7.3   Stress Integration in Isotropic Plasticity Using the Logarith-
         mic Strains 349
         7.3.1   Introduction 349
         7.3.2   Stress Integration in Large Strain Plasticity.......... 353
         7.3.3   Examples 359

Appendix A1
   **A Summary of Elastic and Thermoelastic Constitutive Re-
   lations** 367

Appendix A2
   Notation - Matrices and Tensors......................... 381

Appendix A3
   List of Main Symbols 389

References 393

Index 407

1. Introduction

In this chapter we focus on the role of the integration of inelastic constitutive relations in general nonlinear finite element analysis and present the basic objectives of the book. In order to motivate the subject of the book we also briefly refer to some example solutions of complex inelastic analyses.

1.1 The Objectives of this Book

Finite element inelastic analysis is now abundantly performed in various branches of engineering design and scientific research. A number of commercial computer programs are in widespread use and many smaller research computer programs are employed for the inelastic analysis of solids and structures.

The reason that inelastic analysis has gained such importance is that it is nowadays possible to analyze realistically very complex structures for their nonlinear response and the benefits of performing a nonlinear analysis can be large. In particular, a nonlinear analysis very frequently makes it possible to achieve a safer and more economical design. Also, a nonlinear analysis is frequently needed to understand the behavior of a structure that has been in use and service but unfortunately failed due to unusual loading conditions. In general, a nonlinear analysis may be of great value and indeed necessary to more accurately model nature (Bathe 1996; Bathe 2001a)

A general nonlinear structural analysis can include the effects of large displacements, large strains and nonlinear material conditions. When the material is responding nonlinearly, an inelastic response involving plasticity, creep, thermal effects, and so on, is usually most difficult to analyze. In this case, the inelastic conditions have to be calculated in an incremental solution and state-of-the-art computing resources and computational procedures may be required.

An effective inelastic analysis procedure provides general applicability and efficiency in modeling inelastic phenomena. General applicability is reached by using material models that can represent the inelastic behavior in general loading conditions and that are based on well-established principles of mechanics. Efficiency is reached by using stable and highly accurate algorithms to solve the nonlinear equations associated with the material models.

While the basic mechanics relations of inelastic analysis were largely formulated many years ago, the widespread use of inelastic analysis has spurred much further research in the field. This research has resulted in the development of new material models, to increase the general applicability of inelastic analysis, and in the development of increasingly more efficient finite element solution schemes. In today's practice of inelastic finite element analysis using commercial computer programs, extensive libraries of material models are available. These libraries can be used to model the inelastic response of many materials, such as metals, concrete, soil and rock structures, and synthetic materials. Also, a program user is able to code an own material model and use this model in a general finite element analysis.

With this background, *the objectives of this book* can be summarized as follows:

- To present, to a certain extent, *the fundamentals of inelasticity*, from basic experimental data and mechanical principles to the formulation of material models. These models are used to describe the inelastic material behavior.

- To derive in a consistent manner *robust implicit numerical algorithms* for the effective integration of inelastic constitutive relations within a time (load) step for *strain-driven problems*. These algorithms are applied to a number of commonly used material models.

- To illustrate the computed inelastic material response through *examples which elucidate the material behavior* in typical engineering conditions.

Hence, the objective of this book is to present basic inelastic material models and efficient computational methods for these models. The presentation helps the reader to model engineering problems for inelastic analysis, to understand typical inelastic solution schemes, and to possibly program an own material model.

1.2 Some Remarks on Explicit and Implicit Solutions of Nonlinear Response

There are in essence two approaches to solve nonlinear problems and also the inelastic material response - *the explicit and the implicit approaches*. Explicit solution algorithms can be very effective to solve high velocity phenomena, such as wave propagation problems. Implicit solution algorithms are effective for the analysis of static problems and structural dynamics problems.

Consider a *general nonlinear problem* in solids and structures. Since an incremental solution is required, we use time "t" to denote the generic time of load application. In static analysis, when a time-independent inelastic analysis is performed, "t" denotes only the load level, but when a time-dependent material response is considered (such as in creep analysis) "t" denotes the actual physical time of solution (see Bathe 1996).

In an *explicit solution scheme*, the following finite element equations are solved for each time step t

$$^t\mathbf{M}\,^t\ddot{\mathbf{U}} + {}^t\mathbf{C}\,^t\dot{\mathbf{U}} + {}^t\mathbf{F} = {}^t\mathbf{R} \qquad (1.1)$$

where $^t\mathbf{F}$ are the nodal point forces corresponding to the internal element stresses at time t and $^t\mathbf{R}$ are the externally-applied nodal point forces that include all forces externally applied to the structure, such as tractions and gravity loads, $^t\mathbf{M}$ and $^t\mathbf{C}$ are the mass and damping matrices, respectively, and $^t\dot{\mathbf{U}}$ and $^t\ddot{\mathbf{U}}$ are the nodal point velocity and acceleration vectors. Note that the left superscript denotes the time considered with all quantities being time-dependent. The number of equations considered in (1.1) is equal to the number of structural degrees of freedom, and in practice thousands of degrees of freedom are used. We consider in (1.1) the dynamic equilibrium equations of the finite element system, and indeed when using an explicit solution scheme it is necessary to include inertia and/or damping forces in the analysis.

Let the time step be denoted by Δt, and assume that the solution of (1.1) has just been obtained (for the state at time t, of course). Then the displacements, velocities, accelerations, strains, stresses and all material variables have been established for time t, and the task of the solution scheme is to calculate these variables for time $t + \Delta t$. In the explicit approach, the inertia and damping effects in (1.1) provide the means to establish the required solution. Namely, the accelerations and velocities in (1.1) are discretized in time using all known displacements at times $t, t - \Delta t,...$ and the unknown displacements at time $t + \Delta t$, and hence the displacements $^{t+\Delta t}\mathbf{U}$ can directly be calculated. This solution scheme does not involve any iteration but simply a forward time integration. For stability of the integration, the time step has to be small enough, and this allows that the stresses and internal state variables are also simply forward-integrated (without iteration) based on the given material law. The solution schemes used for this stress integration need to be effective but since the time step is small, various adequate methods can be used, if necessary with sub-cycling per time step.

Considering, on the other hand, the *implicit solution of the response*, the solution procedures are more involved and effective solution schemes are much more difficult to establish. In such an approach, the finite element equilibrium equations at time $t + \Delta t$ are used to calculate all variables corresponding to time $t + \Delta t$, that is, the displacements, velocities,..., stresses and so on. While the solution algorithms to be used are considerably more complex, effective solution schemes are also significantly more powerful for many applications. In particular, they are directly applicable to static analysis.

Assuming that the solution of the response, in a static analysis, has been obtained at time t, we then consider the solution of

$$\boxed{{}^{t+\Delta t}\mathbf{F} = {}^{t+\Delta t}\mathbf{R}}$$ (1.2)

The linearization of this equation gives in the Newton-Raphson iterative solution, see Chapter 2,

$$\boxed{{}^{t+\Delta t}\mathbf{K}^{(i-1)}\Delta\mathbf{U}^{(i)} = {}^{t+\Delta t}\mathbf{R} - {}^{t+\Delta t}\mathbf{F}^{(i-1)}}$$ (1.3)

with i the iteration counter,

$$ {}^{t+\Delta t}\mathbf{U}^{(i)} = {}^{t+\Delta t}\mathbf{U}^{(i-1)} + \Delta\mathbf{U}^{(i)} $$ (1.4)

and

$$ {}^{t+\Delta t}\mathbf{K}^{(0)} = {}^{t}\mathbf{K}; \quad {}^{t+\Delta t}\mathbf{F}^{(0)} = {}^{t}\mathbf{F}; \quad {}^{t+\Delta t}\mathbf{U}^{(0)} = {}^{t}\mathbf{U} $$ (1.5)

and the iteration is continued until convergence is achieved, measured on appropriate convergence tolerances. Of course, in a dynamic analysis, simply the inertia and damping forces corresponding to time $t + \Delta t$ would be included in the governing equations and equilibrium iterations (see Bathe 1996). Hence the use of implicit integration is directly applicable to static and dynamic analyses. In these solutions, the time step size used is only governed by accuracy considerations and the requirement that convergence in the iterations needs to be achieved. In practice, relatively large load steps can be used in many static analyses and in dynamic analysis, frequently much larger time steps than in explicit time integration can be employed.

However, in order to be able to use large load and time steps it is crucial that the algorithms employed in the calculations of the stresses and the tangent constitutive relations be effective. Namely, we must have in (1.3):

- Firstly, that the *stresses* used in the evaluation of ${}^{t+\Delta t}\mathbf{F}^{(i-1)}$ be evaluated accurately with an efficient algorithm from the given displacements ${}^{t+\Delta t}\mathbf{U}^{(i-1)}$, and
- Secondly, that the *tangent constitutive relations* used in the tangent stiffness matrix ${}^{t+\Delta t}\mathbf{K}^{(i-1)}$ be evaluated in an efficient way *consistent* with the stress integration scheme.

The *accuracy* of the overall finite element solution is of course directly governed by the accuracy of the stress calculations in each solution step. And the optimum convergence in the Newton-Raphson scheme is only obtained when the evaluation of the tangent constitutive relations in the stiffness matrix is consistent with the stress integration scheme used. From a mathematical point of view, this observation is obvious since the tangent matrix is supposed to represent the derivative of the right hand side vector in (1.3) as it is actually calculated (Bathe 1996). We discuss these statements in detail in the book.

Since the first inelastic finite element analyses have been performed, much research has been expended to reach efficient stress integration schemes and associated evaluations of the tangent constitutive relations. We focus in this book on the presentation of implicit stress integration algorithms and in particular on the *"governing parameter method"* (Kojic 1996a) described in detail in Section 4.2. This method represents a generalization of *"the effective-stress-function method"*, introduced by Kojic and Bathe and first published in 1984 (Bathe, Chaudhary, Dvorkin, and Kojic 1984). The essence of the effective-stress-function method is to use in a simple way a single governing parameter, the "effective stress", to evaluate the stresses in quite general inelastic analysis conditions, but naturally, also other variables can be employed as governing parameters (see Bathe 1996; Kojic 2002a) and the governing parameter method was developed with this premise. The algorithms presented in this book are of course closely related to other inelastic analysis procedures (see Sections 4.2 and 4.3).

The governing parameter algorithm presented in the book is attractive as a solution method because it is computationally efficient, it has good convergence and accuracy characteristics and can be used with large load and time steps in static and dynamic analyses (in implicit and of course also in explicit overall solutions). The method is general and can be used for complex yield criteria, see, e.g., Sections 4.2 and 6.4. We present the method for the inelastic analysis of metals and geological materials but the procedure can also directly be developed and used for other materials.

1.3 The Topic of this Book in an Overall Context

As already mentioned, the efficient stress integration and the evaluation of the consistent tangent constitutive relations are two most important ingredients of any inelastic solution method. However, it should be realized that for a complete, efficient finite element analysis, also the finite elements and the methods for solving the algebraic finite element equations need to be effective.

Considering the two- and three-dimensional analysis of solids, the *conditions of incompressibility* arising in inelastic analyses require that specific elements be used. Here the isoparametric displacement/pressure interpolated (u/p) elements are effective because the inf-sup condition for incompressible analysis is satisfied (see Bathe 1996, 2001b). Considering shell analyses, also specific elements need to be used to circumvent spurious shear and membrane strain conditions, and the MITC shell elements are effective (Bathe 1996; Bathe et al. 2000b; Hiller and Bathe 2003).

For the solution of the algebraic equations, *sparse solver techniques* have been developed recently that are much more efficient than the traditional solvers (Bathe 1996). These solver techniques together with efficient elements and inelastic stress calculation methods now make it possible to solve on PCs very large finite element models for their nonlinear response (Bathe 1999,

2001a; Bathe et al. 2000a). We mention briefly three analyses in the following, as illustrations of today's achievements in solving complex structural inelastic response.

Figure 1.1 shows the finite element shell model of a motor car. This model was solved using an implicit solution scheme in a crush analysis for its large deformation inelastic response, involving contact and failure due to cracking. A comparison of the solution results with experimental data is given in the figure (Bathe 1998). In addition to crush analyses, which solve for the behavior of motor cars when slowly crushed, the car industries also perform abundantly crash simulations, which solve for the behavior when the cars crash against a rigid wall.

Figure 1.2 shows the East span of the San Francisco-Oakland Bay Bridge and a slab failure of the bridge during the 1989 earthquake in the region. After this event, the State of California decided to reanalyze the major toll bridges in order to evaluate their anticipated behavior in future earthquakes. A large number of inelastic static and dynamic analyses were performed of the toll bridges using various simple to very sophisticated models (Ingham 2001). The figure shows a global model using beam elements of a part of the Bay Bridge subjected to an anticipated earthquake. Based on these studies, the State of California is designing a new East span for the Bay Bridge.

Figure 1.3 gives some analysis results obtained in a metal forming problem. Such simulations are performed abundantly in various industries. This simulation involves the analysis of the sheet metal forming process during stamping and during spring-back. Large deformations, elasto-plastic response with complex contact conditions need to be solved. In this case, the solution of the complete rather slow (static) process including spring-back was carried out using implicit integration (Kawka and Bathe 2001). The implicit solution is able to model the slow physical process much more accurately than an explicit solution, for which a pseudo-dynamic analysis needs to be carried out.

The above examples give snap-shots of some inelastic analyses. These simulations, as mentioned already above, require an overall effective finite element solution scheme implemented in a computer program, with various efficient solution ingredients. However, this book only focuses on inelastic material models and associated computational procedures. Since the mechanics of the inelastic material deformations and the algorithms discussed here are embedded in a complete finite element solution scheme, we make extensive reference to Bathe (1996), where the topic of inelastic analysis is only briefly discussed, see Bathe (1996, Section 6.4). We use the same notation as in Bathe (1996) in order to facilitate the reading of the present book for those who are familiar with this reference (see also Chapelle and Bathe 2003; Bucalem and Bathe 200x).

There are many important details in an efficient finite element stress integration scheme — the design and implementation of an effective algorithm goes much beyond the writing of general equations of mechanics. We en-

a

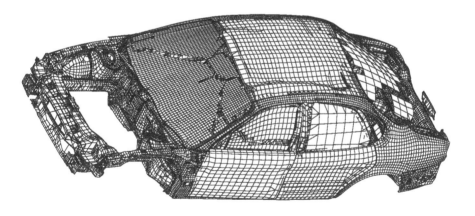

b

Fig. 1.1. Crush analysis of a motor car. **a** Photo of a typical test in the laboratory. Note that the rigid plate is crushing the car; **b** The finite element shell model of a car

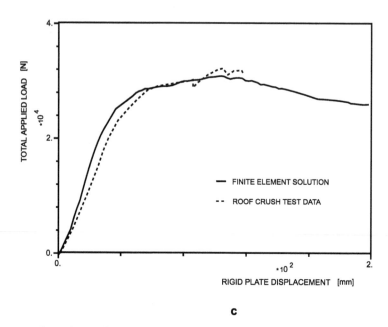

c

Fig. 1.1 (continued). c Finite element model (implicit) solution results and test data

deavor to give some details here, but in doing so, the book does not cover all topics of inelastic analysis. Indeed, the discussion given here largely focuses on the use of the classical elasto-plastic and creep material models only. We explain, to some extent, the phenomenological basis of these material models and then present the algorithms to solve for the stresses as strain-driven problems within a finite element solution, obtained with a given mesh.

Therefore, the book does not deal with advanced research topics regarding material models of inelasticity, such as non-local models, gradient plasticity theory, damage models, fracture, the thermo-dynamic foundations of inelasticity, the micro-mechanical foundations of plasticity, molecular dynamics of inelasticity, and so on. The book also does not deal with the discretization errors that arise in finite element analysis and adaptive mesh procedures to reduce these errors (for references on these advanced research topics, see for example Bathe 2001a, 2003). However, the computational procedures discussed are of course also valuable for many material models not considered explicitly in this book and for implementations of models in the advanced research topics not covered.

Fig. 1.2. Analysis of San Francisco Bay Bridge. **a** Photo of East span of Bay Bridge; **b** Photo of slab that failed during earthquake

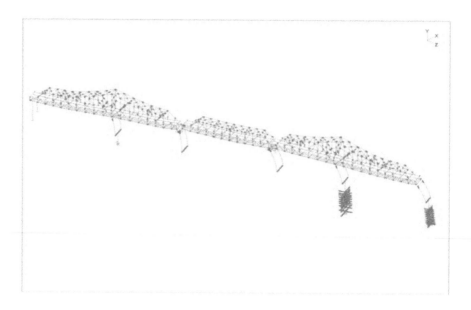

Fig. 1.2 (continued). **c** Finite element beam model of section of Bay Bridge

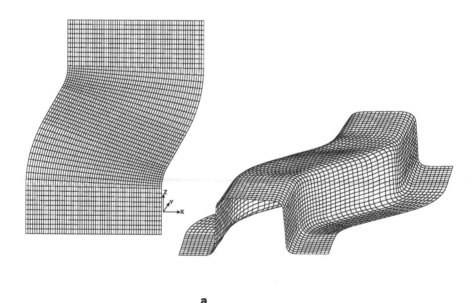

a

Fig. 1.3. Analysis of sheet metal forming of S-Rail. **a** Original flat sheet and final formed sheet

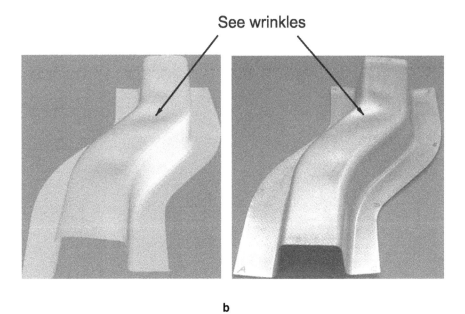

b

Fig. 1.3 (continued). b Calculated and test results regarding wrinkling

2. An Introduction to the Incremental-Iterative Solution of Nonlinear Structural Problems

This chapter gives an introduction to nonlinear structural analysis. The material selected gives some basic solution procedures and provides the fundamental relations of incremental nonlinear analysis.

We first review in Section 2.1 some commonly used solution procedures for one nonlinear equation and then present the Newton-Raphson method for a set of nonlinear equations. In Section 2.2 we introduce how we solve nonlinear structural problems through a simple example, and then show how the Newton-Raphson method is used for general finite element systems.

Section 2.3 is devoted to the linearization of the principle of virtual work equation for general nonlinear deformations of a material body and the derivation of the incremental-iterative equations. The finite element discretization of a solid continuum using the updated Lagrangian (UL) formulation is considered. The element matrices for the three-dimensional isoparametric solid finite element are derived.

The presentation given here introduces the reader to the overall solution process of nonlinear structural analysis, such that the context and details of the inelastic solution algorithms presented in the next chapters can be understood. More details of nonlinear finite element analysis using the same notation are given in Bathe (1996).

2.1 Some Solution Procedures for Nonlinear Equations

In this section we briefly present some solution procedures for one nonlinear equation and for a set of nonlinear equations.

2.1.1 Some Solution Procedures for One Nonlinear Equation

Bisection Method. A simple and robust solution procedure is the bisection method, or method of halving the interval. The procedure consists of the following steps.

Suppose that a nonlinear equation

$$\boxed{f(x) = 0} \qquad (2.1.1)$$

has only one solution in the interval $[a, b]$ and that $f(x)$ is a continuous function in this interval. Assume that $f(b) = f_b > 0$ and $f(a) = f_a < 0$, as schematically shown in Fig. 2.1.1. We start the procedure by calculating the trial solution x_0 as

$$x_0 = \frac{1}{2}(a + b) \qquad (2.1.2)$$

If $f(x_0) = f_0 > 0$ we use the new interval $[a, x_0]$ and calculate

$$x_1 = \frac{1}{2}(a + x_o) \qquad (2.1.3)$$

as shown in the figure. If f_0 were negative we would have used the interval $[x_0, b]$ as the new interval containing the zero of $f(x)$, and $x_1 = 0.5(x_0 + b)$. Denoting by x_i^+ and x_i^- the values of x with positive and negative function values of $f(x)$, respectively, and the i-th interval by $\left[x_i^-, x_i^+\right]$ we can write the relation

$$x_i^+ - x_i^- = \frac{1}{2}(x_{i-1}^+ - x_{i-1}^-) \qquad (2.1.4)$$

The bisection procedure stops when the size of the interval is less than a given tolerance ϵ, i.e., when

$$|\, x_i^+ - x_i^- \,| < \epsilon \qquad (2.1.5)$$

The error estimate is then

$$|\, x_i - x^* \,| \le \frac{1}{2^{i+1}}(b - a) \qquad (2.1.6)$$

where x^* is the solution of $f(x) = 0$.

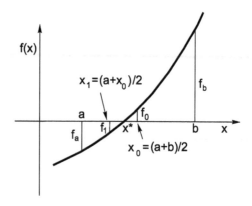

Fig. 2.1.1. Bisection procedure to solve the equation $f(x) = 0$

The bisection procedure is simple and robust, but the rate of convergence is low. The convergence rate can be increased by employing a secant approximation for a new trial value within the current interval.

Method of Successive Substitutions. The method of successive substitutions to solve the equation $f(x) = 0$ is based upon the transformation of this equation into an equivalent relation

$$x = g(x) \qquad (2.1.7)$$

Then for $f(x^*) = 0$ we have

$$x^* = g(x^*) \qquad (2.1.8)$$

Let us consider the conditions of convergence for the iterative scheme

$$x_i = g(x_{i-1}) \qquad i = 1, 2, \dots \qquad (2.1.9)$$

to the solution x^*.

Assume that the function $g(x)$ is continuous and has a continuous (hence finite) first derivative $g'(x)$ in the neighborhood of x^*. Then, using (2.1.9) and the condition (2.1.8) we can write the relation

$$x_i - x^* = g(x_{i-1}) - g(x^*) = \left(\frac{g(x_{i-1}) - g(x^*)}{x_{i-1} - x^*} \right) (x_{i-1} - x^*) \qquad (2.1.10)$$

The coefficient multiplying $(x_{i-1} - x^*)$ represents for $g(x)$ the slope of the secant between points x_{i-1} and x^*. Under the assumptions for the curve $g(x)$, there is at least one point within the interval $[x_{i-1}, x^*]$ where the tangent to the curve $g(x)$ is parallel to the secant (by the mean-value theorem). Hence we can write

$$x_i - x^* = g' \left[x^* + t \left(x_{i-1} - x^* \right) \right] \left(x_{i-1} - x^* \right) \qquad (2.1.11)$$

where $0 \leq t \leq 1$. Suppose that there is a constant c such that

$$| g'(x) | \leq c < 1 \qquad (2.1.12)$$

in the interval of iteration; then it follows from (2.1.11) that

$$| x_i - x^* | \leq c | x_{i-1} - x^* | \leq c^2 | x_{i-2} - x^* | \leq \dots \leq c^i | x_o - x^* |$$

$$(2.1.13)$$

Therefore, the iterations (2.1.9) converge to the solution x^* provided the condition (2.1.12) is satisfied.

Figure 2.1.2 shows schematically examples of convergence and divergence.

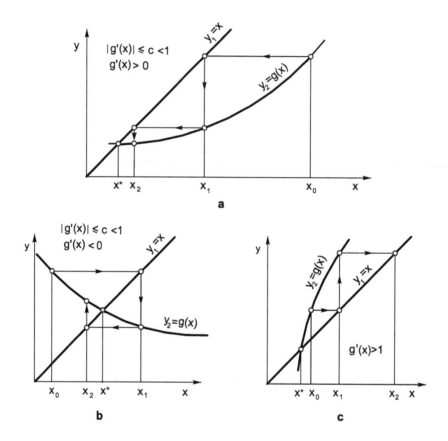

Fig. 2.1.2. Method of successive substitutions for solutions of $f(x) = x - g(x)$. **a, b** Convergence; **c** Divergence

Newton-Raphson Method. A very widely used method is the Newton-Raphson procedure (also referred to simply as the Newton method). This method is based on the calculation of tangents to the curve $y = f(x)$. Let $f(x)$ be a function such that $f(x^*) = 0$. Assume that $f(x)$ has the following properties in a neighborhood of x^*: (*i*) $f'(x)$ exists and $f'(x) \neq 0$; (*ii*) $f''(x)$ exists and is continuous; (*iii*) $f'''(x)$ exists.

Figure 2.1.3 shows schematically a function with features (*i*) to (*iii*). The following iterative relation follows from the figure:

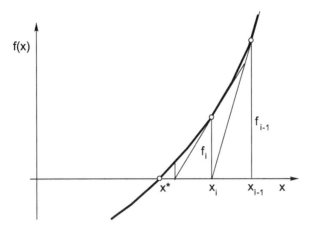

Fig. 2.1.3. Newton iterative method

$$x_i = x_{i-1} - \frac{f_{i-1}}{f'_{i-1}} \qquad (2.1.14)$$

This relation is of the type (2.1.9), hence the trial values x_i will converge to the solution x^* if the condition (2.1.12) is satisfied. Denoting by $g(x)$ the relation

$$g(x) = x - \frac{f(x)}{f'(x)} \qquad (2.1.15)$$

we obtain, from this analysis, that the condition for convergence is

$$|g'(x)| = \frac{|f(x)\,f''(x)|}{[f'(x)]^2} < 1 \qquad (2.1.16)$$

If this condition is satisfied for the interval of iteration around the root x^*, the convergence of iterations (2.1.14) is assured for the initial value of x lying within that interval. Note that $g'(x^*) = 0$, hence (2.1.16) is always satisfied for x very close to x^* for the case considered.

The Newton iteration has a quadratic rate of convergence. We prove this statement by expressing the error $x_i - x^*$ as

$$x_i - x^* = g(x_{i-1}) - g(x^*) \qquad (2.1.17)$$

Expanding $g(x_{i-1})$ as a Taylor series in terms of $(x_{i-1} - x^*)$, with the second order term as the remainder, we obtain

$$g(x_{i-1}) = g(x^*) + g'(x^*)(x_{i-1} - x^*) + \frac{1}{2}g''[x^* + t(x_{i-1} - x^*)](x_{i-1} - x^*)^2$$

(2.1.18)

where $0 \leqslant t \leqslant 1$. But $g'(x^*) = 0$ and hence

$$x_i - x^* = \frac{1}{2}g''[x^* + t(x_{i-1} - x^*)](x_{i-1} - x^*)^2$$

(2.1.19)

Based on the assumptions for the function $f(x)$ and the expression (2.1.16) for $g'(x)$, we have that $|g''(x)|$ is finite in the neighborhood of x^*. Therefore, the error is proportional to the square of the previous iteration error, i.e., the convergence rate of the Newton iteration (2.1.14) is quadratic.

Clearly, the above brief review of solution strategies shows that in selecting the solution method for a nonlinear equation we have to take into account the characteristics of the equation to be solved, the expense of each iteration and the number of iterations anticipated for the solution.

2.1.2 Newton Method for a System of Nonlinear Equations

Consider now the system of nonlinear equations

$$f_1(u_1, u_2, \ldots, u_n) = 0$$
$$f_2(u_1, u_2, \ldots, u_n) = 0$$
$$\cdots$$
$$\cdots$$
$$f_n(u_1, u_2, \ldots, u_n) = 0$$

(2.1.20a)

which we rewrite in matrix form as

$$\mathbf{f}(\mathbf{u}) = \mathbf{0}$$

(2.1.20b)

Assume that the solution of this system is $\mathbf{u}^*$ and that an approximate solution is $\mathbf{u}^{(i-1)}$. Then we can write a Taylor series expansion at $\mathbf{u} = \mathbf{u}^{(i-1)}$,

$$\mathbf{f}(\mathbf{u}^*) = \mathbf{0} = \mathbf{f}^{(i-1)} + \mathbf{K}^{(i-1)}(\mathbf{u}^* - \mathbf{u}^{(i-1)}) +$$
$$\text{second-order terms in } \left\|\mathbf{u}^* - \mathbf{u}^{(i-1)}\right\|_2$$

(2.1.21)

where $\mathbf{f}^{(i-1)}$ is $\mathbf{f}$ evaluated at $\mathbf{u}^{(i-1)}$, and

$$\mathbf{K}^{(i-1)} = \frac{\partial \mathbf{f}}{\partial \mathbf{u}}\Big|_{\mathbf{u}^{(i-1)}}$$

(2.1.22)

is the matrix of the first derivatives $\partial f_k / \partial u_j$ evaluated at $\mathbf{u} = \mathbf{u}^{(i-1)}$. The equation (2.1.21) corresponds to the i-th iteration in solving (2.1.20). Neglecting the last term in (2.1.21) as a small quantity of second order, we set

$$\mathbf{f}^{(i-1)} + \mathbf{K}^{(i-1)}(\mathbf{u}^{(i)} - \mathbf{u}^{(i-1)}) = 0 \qquad (2.1.23)$$

and the i-th approximation $\mathbf{u}^{(i)}$ is

$$\mathbf{u}^{(i)} = \mathbf{u}^{(i-1)} - \left[\mathbf{K}^{(i-1)}\right]^{-1} \mathbf{f}^{(i-1)} \qquad (2.1.24)$$

This relation represents the Newton-Raphson (or Newton) iteration for the solution of the nonlinear system of equations (2.1.20).

The above iteration scheme leads to the solution $\mathbf{u} = \mathbf{u}^*$ and the convergence is quadratic under the following conditions.

a) The matrix $\mathbf{K}^{(i-1)}$ exists and is nonsingular in the neighborhood of the solution $\mathbf{u}^*$.

b) The stiffness matrix satisfies the Lipschitz continuity condition

$$\left\|\mathbf{K}^{(i-1)} - \mathbf{K}^{(i)}\right\| \leqslant L \left\|\mathbf{u}^{(i-1)} - \mathbf{u}^{(i)}\right\|$$
$$L > 0 \qquad (2.1.25)$$

Note that these conditions are weaker than those given in Section 2.1.1, but the quadratic convergence can still be proven to hold, see Eterovic and Bathe (1991a).

Solution examples using the Newton-Raphson method are given in Bathe and Cimento (1980); Bathe (1999), (2001a) and a further discussion of the characteristics of the Newton-Raphson and other iterative methods is given in Bathe (1996).

2.2 Nonlinear Structural Problems and Incremental-Iterative Equilibrium Equations

In this section we introduce the approach for solution of nonlinear structural problems through a one degree of freedom simple example, and then generalize the basic considerations.

2.2.1 A Simple Example to Exemplify Nonlinear Analysis

Consider the simple structure (Bathe 1996) shown in Fig. 2.2.1. We suppose that the platform BD is rigid, while the bars deform due to the action of the pressure $p = p(t)$. At time t the pressure is equal to ^tp. The joints B and D can move into the y-direction only and the current equilibrium configuration is determined by the displacement ^tU. The equilibrium equation is (see Fig. 2.2.1b)

$$^tF_a \sin{}^t\beta = \frac{1}{2}{}^tR = \frac{1}{2}{}^t p \bar{L}\, b \qquad (2.2.1)$$

where b is the platform thickness (the dimension normal to the x, y plane). The axial force ^tF_a in a bar is given by

$$^tF_a = k\,{}^tU_a \qquad (2.2.2)$$

where ^tU_a is the axial displacement, and $k = EA/L$ is the axial bar stiffness; E is Young's modulus, A is the bar cross-sectional area, and L is the initial

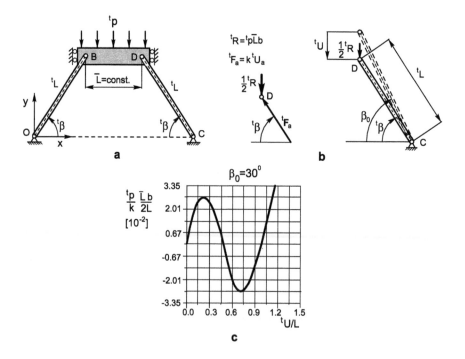

Fig. 2.2.1. Simply supported structure loaded by pressure $p(t)$. **a** Geometry and loading data; **b** Forces and displacements for the current configuration; **c** Load-displacement relation

length of the bar (that is, we have in Fig. 2.2.1b $^0L = L$). Using the geometry shown in the figure, we have

$$^tU_a = L - \left(L^2 - 2L\sin\beta_0\,^tU + \,^tU^2\right)^{1/2}$$

$$\sin\,^t\beta = \frac{L\sin\beta_0 - \,^tU}{L - \,^tU_a} \qquad\qquad (2.2.3)$$

Substituting (2.2.2) into (2.2.1) and using (2.2.3) we obtain

$$^tR - \,^tF = 0 \qquad\qquad (2.2.4)$$

where

$$^tF = 2kL\left\{\left[1 - 2\frac{^tU}{L}\sin\beta_0 + \left(\frac{^tU}{L}\right)^2\right]^{-1/2} - 1\right\}\left(\sin\beta_0 - \frac{^tU}{L}\right)$$

$$(2.2.5)$$

Note that $^tF = F(^tU)$ is the internal structural force due to the stresses within the structure and varies nonlinearly with the displacement ^tU, while the axial stiffness of the bars is constant $(k = EA/L)$. We refer to such problems as *geometrically nonlinear* structural problems. Figure 2.2.1c shows the load-displacement curve for $\beta_0 = 30°$. We see that the load first increases and then decreases to $^tp = 0$ when $^t\beta = 0$; then ^tp becomes negative, reaches the extreme value, and again $^tp = 0$ for $^tL = L$. Finally, the pressure increases causing extension of the bars.

The assumption that the axial stiffness of the bar is constant $(k = EA/L)$ for all displacements considered is hardly valid in practice. In the case of small displacement ^tU, we can use the constant axial bar stiffnesses and the force-displacement relation becomes linear

$$^tF = K\,^tU = (2\,k\sin^2\beta_0)\,^tU \qquad\qquad (2.2.6)$$

where K is the structural stiffness. The expression (2.2.6) is obtained using $^tU_a = \,^tU\sin\beta_o$ in (2.2.1), or by linearization of (2.2.5) around $^tU = 0$.

Another type of nonlinearity is due to nonlinear material behavior. If the constitutive relation is nonlinear we have

$$d^t\boldsymbol{\sigma} = \,^t\mathbf{C}\,d^t\mathbf{e} \qquad\qquad (2.2.7)$$

where $d^t\boldsymbol{\sigma}$ and $d^t\mathbf{e}$ are the stress and strain increments, and $^t\mathbf{C}$ is the constitutive matrix which depends on the stresses and strains. As an example, we use here an elastic-plastic material with the stress-strain curve for uniaxial loading shown in Fig. 2.2.2a. The stress-strain relations are

$$\sigma = E\,e \qquad \text{for } \sigma \le \sigma_{yv}$$
$$\sigma = \sigma_{yv}\,(1 - E_T/E) + E_T\,e \qquad \text{for } \sigma > \sigma_{yv} \tag{2.2.8}$$

where E_T is the tangent modulus, and σ_{yv} is the initial yield stress. Assuming small displacements ${}^tU$, the strain ${}^te$ in the problem of Fig. 2.2.1 is

$$
{}^te = \frac{\sin \beta_0}{L}\,{}^tU \tag{2.2.9}
$$

Substituting ${}^te$ into (2.2.8) and calculating the axial force as ${}^tF_a = A\,{}^t\sigma$, we express the equilibrium equation (2.2.4) as

$$
(2\frac{EA}{L}\sin^2 \beta_0)\,{}^tU = {}^tR \qquad \text{for } {}^tR \le R_y \tag{2.2.10}
$$

$$
(2\frac{E_T A}{L}\sin^2 \beta_0)\,{}^tU = {}^tR - F_0 \qquad \text{for } {}^tR > R_y \tag{2.2.11}
$$

where

$$
R_y = 2\,A\,\sigma_{yv}\sin \beta_0 \tag{2.2.12}
$$

is the force which corresponds to the start of plastic deformation, and

$$
F_0 = (1 - E_T/E)\,R_y \tag{2.2.13}
$$

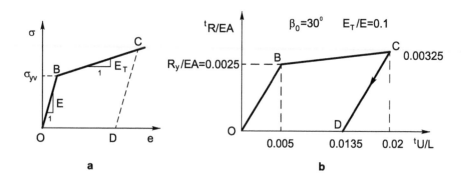

Fig. 2.2.2. Elastic-plastic deformation of the bar structure of Fig. 2.2.1.
a Elastic-plastic bilinear stress-strain relation; **b** Load-displacement relation

Figure 2.2.2b shows the load-displacement dependence for $\beta_0 = 30^o, E_T/E = 0.1$ and $R_y/EA = 0.0025$. We have an elastic deformation corresponding to the line OB, an elastic-plastic response along the line BC, and elastic unloading along the line CD (parallel to OB).

The described problem is called a *materially-nonlinear-only* (*MNO*) problem (Bathe 1996). The most general case is when the problem is materially and geometrically nonlinear.

2.2.2 Multiple-Degree-of-Freedom System

Let us now consider a general nonlinear structural analysis of a complex multi-degree of freedom (finite element) system. We assume that the system has the displacement vector $^t\mathbf{U}$ of order n. In the case of finite element analysis the system is described by n nodal point displacements. According to the principle of virtual work (see Section 2.3) we have at time t

$$\delta\,^t\mathbf{U}^T\left(\,^t\mathbf{F} - \,^t\mathbf{R}\right) = 0 \qquad (2.2.14)$$

where $\delta\,^t\mathbf{U}$ is the variation of the structural displacement vector, and $^t\mathbf{F}$ and $^t\mathbf{R}$ are, respectively, the structural internal and external forces which are work conjugate to the displacement vector $^t\mathbf{U}$. Therefore, using the fact that $\delta\,^t\mathbf{U}$ is arbitrary, we obtain the equilibrium equation

$$^t\mathbf{F} = \,^t\mathbf{R} \qquad (2.2.15)$$

This is the fundamental set of equations for any nonlinear static and dynamic analysis. The equations are nonlinear since, in general, the internal nodal force vector $^t\mathbf{F}$ and/or the external force vector $^t\mathbf{R}$ are *nonlinear functions* of the nodal displacements $^t\mathbf{U}$.

This system of nonlinear equations cannot in general directly be solved for the displacements $^t\mathbf{U}$ and we therefore proceed to solve the system response incrementally. Assume that we have calculated the soluton at time t. Then we introduce the time (load) step Δt and express the load vector in the form

$$^{t+\Delta t}\mathbf{R} = \,^t\mathbf{R} + \Delta\mathbf{R} \qquad (2.2.16)$$

where $^t\mathbf{R}$ and $^{t+\Delta t}\mathbf{R}$ correspond to the start and end of the time step, and $\Delta\mathbf{R}$ is the load increment, which we assume to be independent of the deformations. Also we write the displacement vector in the incremental form,

$$^{t+\Delta t}\mathbf{U} = \,^t\mathbf{U} + \Delta\mathbf{U} \qquad (2.2.17)$$

where the superscripts t and $t + \Delta t$ are used as in (2.2.16). Since we assume that the displacements at the start of the time step are known, our goal is to calculate the displacement increments ΔU such that the equilibrium equation at the end of the time step

$$\boxed{^{t+\Delta t}\mathbf{F} = {}^{t+\Delta t}\mathbf{R}} \qquad (2.2.18)$$

is satisfied. For this purpose we linearize the left hand side by writing the Taylor series expansion for $^{t+\Delta t}\mathbf{F}$ around $^t\mathbf{U}$ (see (2.1.21)), to obtain

$$^{t+\Delta t}\mathbf{F} \approx {}^t\mathbf{F} + \frac{\partial\,^t\mathbf{F}}{\partial\,^t\mathbf{U}}\Delta\mathbf{U}^{(1)} \qquad (2.2.19)$$

where $\Delta\mathbf{U}^{(1)}$ is the first approximation of the true displacement increment $\Delta\mathbf{U}$. We introduce the tangent structural stiffness matrix

$$^t\mathbf{K} = \frac{\partial\,^t\mathbf{F}}{\partial\,^t\mathbf{U}} \qquad (2.2.20a)$$

with components

$$^t K_{ij} = \frac{\partial\,^t F_i}{\partial\,^t U_j} \qquad (2.2.20b)$$

This stiffness matrix, as indicated, is evaluated at the displacements $^t\mathbf{U}$. Substituting (2.2.19) into (2.2.18) we obtain

$$^t\mathbf{K}\Delta\mathbf{U}^{(1)} = {}^{t+\Delta t}\mathbf{R} - {}^t\mathbf{F} \qquad (2.2.21)$$

Following the iterative procedure of Section 2.1.2 we obtain the equilibrium equation for the i-th iteration as

$$\boxed{^{t+\Delta t}\mathbf{K}^{(i-1)}\Delta\mathbf{U}^{(i)} = {}^{t+\Delta t}\mathbf{R} - {}^{t+\Delta t}\mathbf{F}^{(i-1)}} \qquad (2.2.22)$$

and then the displacements are

$$\boxed{^{t+\Delta t}\mathbf{U}^{(i)} = {}^{t+\Delta t}\mathbf{U}^{(i-1)} + \Delta\mathbf{U}^{(i)}} \qquad (2.2.23)$$

The equilibrium iterations continue until the equilibrium equation (2.2.18) is satisfied within a specified tolerance.

The iterative procedure described above represents the Newton-Raphson iterative scheme, summarized in Table 2.2.1. We have indicated three criteria to stop the iteration. Constants ϵ_F, ϵ_D and ϵ_E are force, displacement and energy tolerances, as measures of the solution accuracy (see Bathe 1996).

Table 2.2.1. Newton-Raphson iterative scheme (see Bathe 1996)

---

1. Initialization for the current time step Δt

$i = 0$

$$^{t+\Delta t}\mathbf{F}^{(0)} = {}^{t}\mathbf{F}, \quad {}^{t+\Delta t}\mathbf{K}^{(0)} = {}^{t}\mathbf{K}, \quad {}^{t+\Delta t}\mathbf{U}^{(0)} = {}^{t}\mathbf{U}$$

2. Iteration i

$i = i + 1$

$$^{t+\Delta t}\mathbf{K}^{(i-1)}\Delta\mathbf{U}^{(i)} = {}^{t+\Delta t}\mathbf{R} - {}^{t+\Delta t}\mathbf{F}^{(i-1)}$$
$$^{t+\Delta t}\mathbf{U}^{(i)} = {}^{t+\Delta t}\mathbf{U}^{(i-1)} + \Delta\mathbf{U}^{(i)}$$

3. Check for convergence

a) Force criterion
$$\left\| {}^{t+\Delta t}\mathbf{R} - {}^{t+\Delta t}\mathbf{F}^{(i)} \right\| \leq \varepsilon_F \left\| {}^{t+\Delta t}\mathbf{R} - {}^{t}\mathbf{F} \right\|$$

b) Displacement criterion
$$\left\| \Delta\mathbf{U}^{(i)} \right\| \leq \varepsilon_D \left\| {}^{t+\Delta t}\mathbf{U}^{(i)} \right\|$$

c) Energy criterion
$$\Delta\mathbf{U}^{(i)T}({}^{t+\Delta t}\mathbf{R} - {}^{t+\Delta t}\mathbf{F}^{(i)}) \leq \varepsilon_E \Delta\mathbf{U}^{(1)T}({}^{t+\Delta t}\mathbf{R} - {}^{t}\mathbf{F})$$

4. If convergence criteria are not satisfied, go to step 2; otherwise,

5. Start iterations for next time step; go to step 1

The stiffness matrix $^{t+\Delta t}\mathbf{K}^{(i-1)}$ represents the tangent stiffness matrix. The procedure used in (2.2.22) and (2.2.23) is called the *full Newton-Raphson iteration* and provides a high convergence rate (see Section 2.1 and Fig. 2.2.3). However, the calculation of $^{t+\Delta t}\mathbf{K}^{(i-1)}$ for each iteration may require a significant computational effort and modifications of the above procedure may be more effective (e.g., Bathe 1996). For example, the iterations may be performed with the stiffness matrix $^{t}\mathbf{K}$ recalculated only at certain times. Then the iterative scheme for this so-called *modified Newton-Raphson iteration* is represented by the equation

$$^{t}\mathbf{K}\Delta\mathbf{U}^{(i)} = {}^{t+\Delta t}\mathbf{R} - {}^{t+\Delta t}\mathbf{F}^{(i-1)} \tag{2.2.24}$$

The modified Newton-Raphson iteration requires a smaller number of factorizations, but the convergence rate may be low and the solution may diverge when the full Newton-Raphson method would converge.

There are other important procedures to speed up the convergence and to be able to obtain solutions when the Newton-Raphson method fails (such as line searching, arc length methods); for more details see Bathe (1996).

Iteration i	1	2	3	4	5
$\dfrac{{}^{t+\Delta t}R - {}^{t+\Delta t}F^{(i)}}{2kL}$ Full Newton	5.00×10^{-3}	7.68×10^{-4}	3.70×10^{-5}	1.05×10^{-7}	8.57×10^{-13}
Modified Newton	5.00×10^{-3}	7.68×10^{-4}	2.53×10^{-4}	9.11×10^{-5}	3.38×10^{-5}
$\dfrac{{}^{t+\Delta t}R - {}^{t+\Delta t}F^{(i)}}{2kL} \; {}_L\Delta U^{(i)}$ Full Newton	1.84×10^{-4}	6.26×10^{-6}	1.61×10^{-8}	1.31×10^{-13}	8.68×10^{-24}
Modified Newton	1.84×10^{-4}	4.34×10^{-6}	4.71×10^{-7}	6.11×10^{-8}	8.40×10^{-9}

a

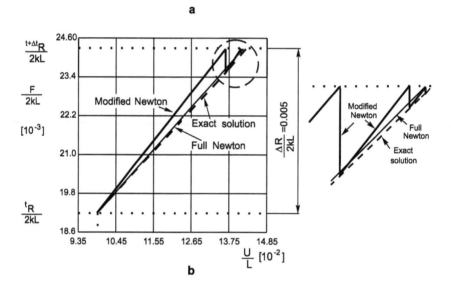

b

Fig. 2.2.3. Iterative solution of a simple structural problem (Fig. 2.2.1).
a Out-of-balance force and out-of-balance energy; **b** Full Newton and modified Newton iterative solutions

2.3 A Linearized Form of the Principle of Virtual Work. Updated Lagrangian Formulation and Finite Element Equilibrium Equations

In the previous section we briefly considered the linearization and solution of the general nonlinear system of (finite element) equilibrium equations (2.2.18). In this section we derive the linearized governing finite element equations used in Table 2.2.1 from the principle of virtual work. This derivation gives explicit forms of the finite element stiffness matrix and internal force vector, with the constitutive relations and the stresses whose determination is the subject of all subsequent chapters. We present expressions for the finite element stiffness matrix and nodal force vector for the case of 3-D solid finite elements.

Various forms of incremental equations can be used for large displacement and large strain analysis. We present here the updated Lagrangian (UL) formulation based on incremental 2nd Piola-Kirchhoff stresses and Green-

Lagrange strains. This presentation shows the essence of the approach used in large deformation analysis and is therefore valuable. However, when we discuss the inelastic large deformation analysis in Chapter 7, we imply the use of the updated Lagrangian-Hencky formulation for which the finite element matrices are more complex. On the other hand, in Chapters 3 to 6, small deformation conditions are assumed in the inelastic solutions (that is, the strains are the infinitesimal engineering strains) and in this situation, all inelastic nonlinear formulations reduce to the materially-nonlinear-only analysis (see (2.3.25) to (2.3.27) below, and Bathe 1996). Vice versa, the concepts and algorithms presented in Chapters 3 to 6 are of course also applicable in large deformation analyses.

2.3.1 General Considerations

We consider large displacements, i.e., geometrically nonlinear problems, and include nonlinear constitutive relations. The presented linearization employs the Green-Lagrange strain and 2nd Piola-Kirchhoff stress as the work conjugate strain and stress measures. We use here the last updated configuration as the reference configuration, and consequently we employ the updated Lagrangian (UL) formulation. In an analogous way the basic relations of the total Lagrangian (TL) formulation, with the initial configuration $^0\mathcal{B}$ as the reference configuration, can be derived (see Bathe 1996).

Consider the general motion of a material body shown in Fig. 2.3.1. Our objective is to form the incremental equations of motion and determine the

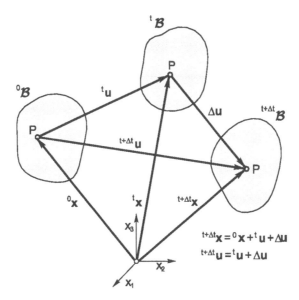

Fig. 2.3.1. Motion of material body with respect to the stationary coordinate system x_1, x_2, x_3

configurations corresponding to the load levels $\Delta t, 2\Delta t, 3\Delta t, \ldots$. We assume that the configuration of the body (system) ${}^t\mathcal{B}$ at the start of the time step is known, and we seek to determine the configuration ${}^{t+\Delta t}\mathcal{B}$. The principle of virtual work corresponding to the configuration ${}^{t+\Delta t}\mathcal{B}$ has the form (see Washizu 1975; Bathe 1996)

$$\boxed{{}^{t+\Delta t}\delta W_{int} = {}^{t+\Delta t}\delta W_{ext}} \tag{2.3.1}$$

where ${}^{t+\Delta t}\delta W_{int}$ and ${}^{t+\Delta t}\delta W_{ext}$ represent, respectively, the virtual work corresponding to the internal forces (stresses) and the external loads — of course, for the assumed virtual displacements. The external loads are known and the virtual work ${}^{t+\Delta t}\delta W_{ext}$ can be expressed in terms of the displacement variations (see (2.2.14)). The internal virtual work is (we imply summation over i,j=1,2,3, see Appendix A2 (A2.7))

$$\boxed{{}^{t+\Delta t}\delta W_{int} = \int_{{}^{t+\Delta t}V} {}^{t+\Delta t}\tau_{ij}\, \delta\,{}_{t+\Delta t}e_{ij}\, d^{t+\Delta t}V} \tag{2.3.2}$$

where the $\delta\,{}_{t+\Delta t}e_{ij}$ are virtual infinitesimal strains referred to the configuration at time $t+\Delta t$, and the ${}^{t+\Delta t}\tau_{ij}$ are the true (Cauchy) stresses at the end of the time step. Since the configuration ${}^{t+\Delta t}\mathcal{B}$ is unknown, we use the known configuration ${}^t\mathcal{B}$ for reference of stresses and incremental strains (during the equilibrium iterations using the full Newton-Raphson method we use the last calculated configuration for the i-th iteration, that is ${}^{t+\Delta t}\mathcal{B}^{(i-1)}$). Consider the Green-Lagrange strain ${}^{t+\Delta t}_{t}\epsilon$ with respect to the reference configuration ${}^t\mathcal{B}$ (see Bathe 1996)

$$ {}^{t+\Delta t}_{t}\epsilon_{ij} = \Delta\,{}_t\epsilon_{ij} \tag{2.3.3}$$

The increments $\Delta\,{}_t\epsilon_{ij}$ can be expressed as

$$\Delta\,{}_t\epsilon_{ij} = \Delta\,{}_t e_{ij} + \Delta\,{}_t\eta_{ij} \tag{2.3.4}$$

where

$$\Delta\,{}_t e_{ij} = \frac{1}{2}\left(\frac{\partial(\Delta u_i)}{\partial\,{}^t x_j} + \frac{\partial(\Delta u_j)}{\partial\,{}^t x_i} \right) \tag{2.3.5}$$

and

$$\Delta\,{}_t\eta_{ij} = \frac{1}{2}\frac{\partial(\Delta u_k)}{\partial\,{}^t x_i}\frac{\partial(\Delta u_k)}{\partial\,{}^t x_j} \tag{2.3.6}$$

are the *linear* and *nonlinear increments* of the Green-Lagrange strain. We next express the internal virtual work (2.3.2) as

$$t+\Delta t \, \delta W_{int} = \int_{t_V} {}^{t+\Delta t}_t S_{ij} \, \delta {}^{t+\Delta t}_t \epsilon_{ij} \, d^t V \qquad (2.3.7)$$

where the 2nd Piola-Kirchhoff stress ${}^{t+\Delta t}_t S_{ij}$ (a description of this stress measure is given in Bathe 1996) can be decomposed as

$$t+\Delta t \atop t S_{ij} = {}^t \tau_{ij} + \Delta_t S_{ij} \qquad (2.3.8)$$

Next, we express $\delta {}^{t+\Delta t}_t \epsilon_{ij}$ as

$$\delta {}^{t+\Delta t}_t \epsilon_{ij} = \delta(\Delta_t e_{ij}) + \delta(\Delta_t \eta_{ij}) \qquad (2.3.9)$$

and linearize the expression for the internal virtual work (2.3.7) as

$$t+\Delta t \, \delta W_{int} = \int_{t_V} \delta(\Delta_t e_{ij}) \, \Delta_t S_{ij} \, d^t V + \int_{t_V} \delta(\Delta_t e_{ij}) \, {}^t \tau_{ij} \, d^t V$$
$$+ \int_{t_V} \delta(\Delta_t \eta_{ij}) \, {}^t \tau_{ij} \, d^t V \qquad (2.3.10)$$

where we have neglected the term,

$$\int_{t_V} \delta(\Delta_t \eta_{ij}) \, \Delta_t S_{ij} \, d^t V \approx 0 \qquad (2.3.11)$$

The next approximation lies in the calculation of $\Delta_t S_{ij}$,

$$\Delta_t S_{ij} \approx {}^t C_{ijrs} \Delta_t e_{rs} \qquad (2.3.12)$$

where ${}^t C_{ijrs}$ is the stress-strain (constitutive) tensor. We use the upper left index t to emphasize that the tangent constitutive tensor corresponds to the stresses and strains at time t (instead of the lower left index, i.e., ${}_t C_{ijrs}$, which would follow from ${}_t C_{ijrs} = \partial(\Delta_t S_{ij})/\partial(\Delta_t e_{rs})$ and could be equally used, see Bathe 1996).

We next substitute (2.3.12) into (2.3.10) and employ (2.3.1). Performing the same steps within the equilibrium iterations, the equation corresponding to the i-th equilibrium iteration with the configuration $^{t+\Delta t}\mathcal{B}^{(i-1)}$ as reference has the following form

$$
\int_{^{t+\Delta t}V^{(i-1)}} \delta(\Delta_{t+\Delta t}e^{(i-1)}_{mn})\ ^{t+\Delta t}C^{(i-1)}_{mnrs}\ \Delta_{t+\Delta t}e^{(i)}_{rs}\ d^{t+\Delta t}V^{(i-1)}
$$

$$
+ \int_{^{t+\Delta t}V^{(i-1)}} \delta(\Delta_{t+\Delta t}\eta^{(i)}_{mn})\ ^{t+\Delta t}\tau^{(i-1)}_{mn}\ d^{t+\Delta t}V^{(i-1)}
$$

$$
= {}^{t+\Delta t}\delta W_{ext} - \int_{^{t+\Delta t}V^{(i-1)}} \delta(\Delta_{t+\Delta t}e^{(i-1)}_{mn})\ ^{t+\Delta t}\tau^{(i-1)}_{mn}\ d^{t+\Delta t}V^{(i-1)}
$$

$$(2.3.13)$$

Obviously, we use the current reference configuration to calculate the strains and the volume integrals, and $^{t+\Delta t}C^{(i-1)}_{mnrs}$ is the constitutive tensor corresponding to the iteration.

The iterations (2.3.13) continue until convergence tolerances are satisfied (see Table 2.2.1). Since the evaluation of the right hand side involves the calculation of the stresses $^{t+\Delta t}\tau^{(i-1)}_{mn}$, the accuracy of the converged solution depends on the accuracy of the stresses calculated by the integration of the constitutive relations within the time step. The optimal convergence rate in the iterations is achieved provided the system stiffness matrix obtained from the volume integrals on the left hand side is properly evaluated, where the accurate evaluation of the tangent constitutive relation $^{t+\Delta t}C^{(i-1)}_{mnrs}$ is important (see Sections 4.4.3 and 4.5; Example 4.5.6, in particular).

2.3.2 Isoparametric Continuum Finite Elements

The relation (2.3.13) is used to formulate the incremental finite element equilibrium equations. Considering isoparametric displacement-based continuum finite elements we have that the interpolation for displacements within a finite element (details are given in Bathe 1996) is given as

$$^t\mathbf{u} = \mathbf{H}\,^t\mathbf{U}$$

$$(2.3.14a)$$

where $^t\mathbf{u}$ are the displacements at a material point, $^t\mathbf{U}$ are the element nodal point displacements, and $\mathbf{H}$ is the interpolation matrix. The component form of this equation is

$$^t u_i = \sum_{k=1}^{N} h_k\,^t U_i^k \qquad i = 1, 2, 3$$

$$(2.3.14b)$$

where ^tu_i and $^tU_i^k$ are the components of the displacement vector at a material particle and the k-th node, h_k are the interpolation functions, and N is the number of nodes of the element. From the displacement interpolation we obtain

$$\Delta_t \hat{\mathbf{e}} = {}_t^t\mathbf{B}_L \Delta\mathbf{U} \tag{2.3.15}$$

where ${}_t^t\mathbf{B}_L$ represents the *linear strain-displacement matrix*, and $\Delta_t \hat{\mathbf{e}}$ is the *strain increment vector* (see (2.3.31) and (A1.3)). The terms of the matrix ${}_t^t\mathbf{B}_L$ contain the corresponding derivatives of the interpolation functions with respect to the coordinates ^tx_i, and $\Delta\mathbf{U}$ is the vector of the nodal displacement increments. Also

$$\delta(\Delta_t \hat{\mathbf{e}}) = {}_t^t\mathbf{B}_L \,\delta\mathbf{U} \tag{2.3.16}$$

Using (2.3.6) we have

$$\delta(\Delta_t\eta_{ij}) = \frac{1}{2}\left[\frac{\partial(\Delta u_k)}{\partial\,^tx_i}\frac{\partial}{\partial\,^tx_j}\delta(\Delta u_k)\right] + \frac{1}{2}\left[\frac{\partial(\Delta u_k)}{\partial\,^tx_j}\frac{\partial}{\partial\,^tx_i}\delta(\Delta u_k)\right] \tag{2.3.17}$$

With use of (2.3.17) we write the product $\delta(\Delta_t\eta_{ij})\,^t\tau_{ij}$ in the matrix form

$$\delta(\Delta_t\eta_{ij})\,^t\tau_{ij} = \delta\mathbf{U}^T\,{}_t^t\mathbf{B}_{NL}^T\,{}^t\tilde{\boldsymbol{\tau}}\,{}_t^t\mathbf{B}_{NL}\Delta\mathbf{U} \tag{2.3.18}$$

where ${}_t^t\mathbf{B}_{NL}$ contains the same derivatives of the interpolation functions as ${}_t^t\mathbf{B}_L$, and $^t\tilde{\boldsymbol{\tau}}$ is a 9×9 matrix containing the stress components $^t\tau_{ij}$ (see Bathe 1996). Note that the matrix $^t\tilde{\boldsymbol{\tau}}$ differs from the symmetric 3×3 matrix $\boldsymbol{\tau}$ used to represent the stress tensor, see (2.3.41) and (A1.60). The matrix ${}_t^t\mathbf{B}_{NL}$ is called the *nonlinear strain-displacement matrix*.

The external virtual work can be expressed as

$$^{t+\Delta t}\delta W_{ext} = \delta\mathbf{U}^T\,{}^{t+\Delta t}\mathbf{R} \tag{2.3.19}$$

where we assume deformation independent loading. Substituting (2.3.16), (2.3.18) and (2.3.19) into (2.3.13), and using that $\delta\mathbf{U}$ is arbitrary, we obtain, due to the linearization, the equilibrium equation of the form (2.2.21)

$$\left({}_t^t\mathbf{K}_L + {}_t^t\mathbf{K}_{NL}\right)\Delta\mathbf{U}^{(1)} = {}^{t+\Delta t}\mathbf{R} - {}^t\mathbf{F} \tag{2.3.20}$$

where

$$
{}_t^t\mathbf{K}_L = \int_{{}^tV} {}_t^t\mathbf{B}_L^T \, {}^t\mathbf{C} \, {}_t^t\mathbf{B}_L \, d^tV \tag{2.3.21}
$$

$$
{}_t^t\mathbf{K}_{NL} = \int_{{}^tV} {}_t^t\mathbf{B}_{NL}^T \, {}^t\tilde{\boldsymbol{\tau}} \, {}_t^t\mathbf{B}_{NL} \, d^tV \tag{2.3.22}
$$

$$
{}^t\mathbf{F} = \int_{{}^tV} {}_t^t\mathbf{B}_L^T \, {}^t\boldsymbol{\tau} \, d^tV \tag{2.3.23}
$$

are the *element linear stiffness matrix, geometrically nonlinear stiffness matrix*, and the *nodal force vector*, respectively. Here ${}^t\boldsymbol{\tau}$ is the stress vector defined according to (A1.1). Note that, referring to the notation in (2.3.13), we have used in (2.3.21) to (2.3.23) ${}^t\mathbf{C} = {}^{t+\Delta t}\mathbf{C}^{(0)}$ and ${}^t\tau_{mn} = {}^{t+\Delta t}\tau_{mn}^{(0)}$. The vector $\Delta\mathbf{U}^{(1)}$ is the first approximation of the true displacement increment $\Delta\mathbf{U}$. The incremental-iterative equilibrium equation (2.2.22) is now

$$
\left({}_{t+\Delta t}^{t+\Delta t}\mathbf{K}_L^{(i-1)} + {}_{t+\Delta t}^{t+\Delta t}\mathbf{K}_{NL}^{(i-1)} \right) \Delta\mathbf{U}^{(i)} = {}^{t+\Delta t}\mathbf{R} - {}^{t+\Delta t}\mathbf{F}^{(i-1)} \tag{2.3.24}
$$

The stiffness matrices and the nodal force vector are calculated using the appropriate matrices ${}_{t+\Delta t}^{t+\Delta t}\mathbf{B}_L^{(i-1)}$, ${}_{t+\Delta t}^{t+\Delta t}\mathbf{B}_{NL}^{(i-1)}$, ${}^{t+\Delta t}\tilde{\boldsymbol{\tau}}^{(i-1)}$ and ${}^{t+\Delta t}\mathbf{C}^{(i-1)}$, and the stress vector ${}^{t+\Delta t}\boldsymbol{\tau}^{(i-1)}$.

Note that in case of nonlinear problems with small displacements (*MNO* analysis) the above *UL* formulation reduces to

$$
{}^{t+\Delta t}\mathbf{K}_L^{(i-1)} \Delta\mathbf{U}^{(i)} = {}^{t+\Delta t}\mathbf{R} - {}^{t+\Delta t}\mathbf{F}^{(i-1)} \tag{2.3.25}
$$

with

$$
{}^{t+\Delta t}\mathbf{K}_L^{(i-1)} = \int_V \mathbf{B}_L^T \, {}^{t+\Delta t}\mathbf{C}^{(i-1)} \, \mathbf{B}_L \, dV \tag{2.3.26}
$$

$$
{}^{t+\Delta t}\mathbf{F}^{(i-1)} = \int_V \mathbf{B}_L^T \, {}^{t+\Delta t}\boldsymbol{\sigma}^{(i-1)} \, dV \tag{2.3.27}
$$

In these expressions we have neglected the volume change and the geometric stiffness matrix. The stresses ${}^{t+\Delta t}\boldsymbol{\sigma}^{(i-1)}$ now correspond to the initial areas.

The above equilibrium equations (2.3.20), (2.3.24) and (2.3.25) correspond to one finite element. The equilibrium equations for an assemblage of finite elements have the same form, but of course with the system stiffness matrices and nodal point force vectors obtained by the appropriate summations over all finite elements (see Bathe 1996 for details).

Three-Dimensional Solid Element. Let us present the explicit form of the above element matrices for the three-dimensional solid finite element. A detailed description of finite elements is given in Bathe (1996). This element represents an *isoparametric* finite element, since the same interpolation functions are used to interpolate the geometry (coordinates of material points) and the displacements. The position vector $^{t}\mathbf{x}$ (with the coordinates $^{t}x_1$, $^{t}x_2$, $^{t}x_3$) is expressed as

$$^{t}\mathbf{x} = \mathbf{H}\,^{t}\mathbf{X} \qquad (2.3.28)$$

where $^{t}\mathbf{X}$ contains the coordinates of all nodal points of the element, $^{t}x_i^k$,

$$^{t}\mathbf{X}^T = \left[\, ^{t}x_1^1\ ^{t}x_2^1\ ^{t}x_3^1\ ^{t}x_1^2\ ^{t}x_2^2\ ^{t}x_3^2 \ldots \ ^{t}x_1^N\ ^{t}x_2^N\ ^{t}x_3^N \right] \qquad (2.3.29)$$

Here N is the number of element nodes. The matrix $\mathbf{H}$ can be written in the form

$$\mathbf{H} = [h_1\mathbf{I}_3\ h_2\mathbf{I}_3\ \ldots h_N\mathbf{I}_3] \qquad (2.3.30)$$

where $h_1,\ldots,h_N$ are the interpolation functions, and $\mathbf{I}_3$ is the (3×3) identity matrix. Figure 2.3.2 shows an *8-node element* $(N = 8)$ with the corresponding interpolation functions.

We next define the linear *strain increment vector* (see also (A1.3)) of the Green-Lagrange strain,

$$\Delta_t\hat{\mathbf{e}}^T = [\Delta_t e_{11}\ \Delta_t e_{22}\ \Delta_t e_{33}\ \Delta_t\gamma_{12}\ \Delta_t\gamma_{23}\ \Delta_t\gamma_{31}] \qquad (2.3.31)$$

The vector of nodal displacement increments and the linear strain-displacement matrix in (2.3.15) are

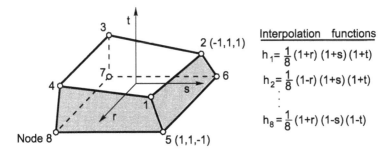

Fig. 2.3.2. Three-dimensional solid element (8 nodes)

$$\Delta\mathbf{U}^T = \begin{bmatrix} \Delta U_1^1 \ \Delta U_2^1 \ \Delta U_3^1 \ \ldots \Delta U_1^N \ \Delta U_2^N \ \Delta U_3^N \end{bmatrix} \tag{2.3.32}$$

and

$$_t^t\mathbf{B}_L = \begin{bmatrix} _t^t\mathbf{B}_L^1 \ _t^t\mathbf{B}_L^2 \ \ldots \ _t^t\mathbf{B}_L^N \end{bmatrix} \tag{2.3.33}$$

with

$$_t^t\mathbf{B}_L^k = \begin{bmatrix} _t h_{k,1} & 0 & 0 \\ 0 & _t h_{k,2} & 0 \\ 0 & 0 & _t h_{k,3} \\ _t h_{k,2} & _t h_{k,1} & 0 \\ 0 & _t h_{k,3} & _t h_{k,2} \\ _t h_{k,3} & 0 & _t h_{k,1} \end{bmatrix} \tag{2.3.34}$$

The derivatives $\partial h_k / \partial\, {}^t x_j$ are denoted by $_t h_{k,j}$. These derivatives can be expressed as

$$\frac{\partial h_k}{\partial\, {}^t x_i} = {}^t J_{ij}^{-1} \frac{\partial h_k}{\partial r_j} \tag{2.3.35}$$

where r_1, r_2 and r_3 stand for the natural coordinates r, s and t shown in Fig. 2.3.2, and the $^t J_{ij}^{-1}$ are the components of the inverse Jacobian matrix. The components of the Jacobian of the transformation $x_k(r_1, r_2, r_3)$ follow from the interpolation (2.3.28),

$$^t J_{ij} = \frac{\partial h_k}{\partial r_i} {}^t x_j^k \tag{2.3.36}$$

The matrix $_t^t\mathbf{B}_{NL}$ in (2.3.18) is

$$_t^t\mathbf{B}_{NL} = \begin{bmatrix} _t^t\tilde{\mathbf{B}}_{NL} & \tilde{\mathbf{0}} & \tilde{\mathbf{0}} \\ \tilde{\mathbf{0}} & _t^t\tilde{\mathbf{B}}_{NL} & \tilde{\mathbf{0}} \\ \tilde{\mathbf{0}} & \tilde{\mathbf{0}} & _t^t\tilde{\mathbf{B}}_{NL} \end{bmatrix} \tag{2.3.37}$$

where

$$_t^t\tilde{\mathbf{B}}_{NL} = \begin{bmatrix} _t h_{1,1} & 0 & 0 & _t h_{2,1} & \cdots & _t h_{N,1} \\ _t h_{1,2} & 0 & 0 & _t h_{2,2} & \cdots & _t h_{N,2} \\ _t h_{1,3} & 0 & 0 & _t h_{2,3} & \cdots & _t h_{N,3} \end{bmatrix} \tag{2.3.38}$$

and

$$\tilde{\mathbf{0}} = \begin{bmatrix} 0 \\ 0 \\ 0 \end{bmatrix} \tag{2.3.39}$$

The stress vector $^t\boldsymbol{\tau}$ and the stress matrix $^t\tilde{\boldsymbol{\tau}}$ in (2.3.23) and (2.3.22) are

$$^t\boldsymbol{\tau}^T = \begin{bmatrix} ^t\tau_{11} & ^t\tau_{22} & ^t\tau_{33} & ^t\tau_{12} & ^t\tau_{23} & ^t\tau_{31} \end{bmatrix} \tag{2.3.40}$$

and

$$
{}^t\tilde{\boldsymbol{\tau}} = \begin{bmatrix} {}^t\tilde{\boldsymbol{\tau}}_3 & \mathbf{0} & \mathbf{0} \\ \mathbf{0} & {}^t\tilde{\boldsymbol{\tau}}_3 & \mathbf{0} \\ \mathbf{0} & \mathbf{0} & {}^t\tilde{\boldsymbol{\tau}}_3 \end{bmatrix}
\tag{2.3.41}
$$

where

$$
{}^t\tilde{\boldsymbol{\tau}}_3 = \begin{bmatrix} {}^t\tau_{11} & {}^t\tau_{12} & {}^t\tau_{13} \\ {}^t\tau_{21} & {}^t\tau_{22} & {}^t\tau_{23} \\ {}^t\tau_{31} & {}^t\tau_{32} & {}^t\tau_{33} \end{bmatrix}
\tag{2.3.42}
$$

and $\mathbf{0}$ is the 3×3 zero-matrix

The evaluation of the element matrices and vectors (see, e.g., (2.3.21) to (2.3.23)) requires the integration over the finite element volume. The integration can be performed in analytical form only in some very special cases, hence numerical integration is necessary.

In the numerical integration, each term of the matrices and vectors given above is integrated numerically. Let $F(r, s, t)$ be a generic term of a stiffness matrix or force vector (with $F(r)$ and $F(r, s)$ for one- and two-dimensional elements). Then the integration is obtained as

$$
\int_{-1}^{+1}\int_{-1}^{+1}\int_{-1}^{+1} F(r, s, t)\, dr\, ds\, dt \approx \sum_{i,j,k} \alpha_{ijk} F(r_i, s_j, t_k)
\tag{2.3.43}
$$

where the summation is performed over all selected integration points and the α_{ijk} are integration weights. The location of the integration points, the corresponding weights and the number of integration points to be used are determined by accuracy considerations, so that the "approximately equal" sign in (2.3.43) is indeed justified; that is, the difference between the left-hand and right-hand sides of (2.3.43) is sufficiently small. The details of numerical integration schemes used in practice are described in Bathe (1996).

3. Fundamental Notions of Metal Plasticity

In this chapter we present fundamentals of the theory of plasticity with the view towards the development of a general numerical procedure for the stress calculation introduced and used in the subsequent chapters. After a brief introduction, we give in Section 3.2 the basic notions of plasticity and define the von Mises material model. Then, in Section 3.3 we present some commonly used orthotropic metal plasticity models. In each section we illustrate the theoretical concepts presented by means of various examples.

3.1 Introduction

The development and application of theories of plasticity to engineering problems started with the pioneering works of Tresca (1864); St. Venant (1870); Levy (1870); followed by seminal contributions of von Mises (1913); Prandtl (1924) and Reuss (1930). A detailed presentation of the history of strength of materials is presented in Timoshenko (1953). Today, the use of plasticity in the engineering disciplines is well established. In general, the theories of plasticity can be divided into two categories: *micromechanical theories* and *macromechanical theories*. The micromechanical theories analyze the plastic deformations on the microscopic level and seek to explain the conditions in crystals and grains of metals leading to plastic flow, e.g., Rice (1971, 1975); Asaro (1983); Aifantis (1987).

On the other hand, the macromechanical theories (also called the mathematical theories) of plasticity describe plastic deformations phenomenologically, on the macroscopic level, and establish relations among the macroscopic mechanical quantities (such as stresses, strains, etc.). These relations are based on general principles of mechanics and on experimental observations. The fundamentals of macromechanical theories of plasticity are given in many books, such as those of Hill (1950); Prager (1959); Prager and Hodge (1968); Mendelson (1968); Życzkowski (1981); Chen and Saleeb (1982); Chen and Han (1988); Lubliner (1990); Ulm and Coussy (2003). A unification of the macromechanical theories of inelastic material behavior, named disturbed state concept, is presented in Desai (2001).

Many practical problems accounting for the plastic deformations of materials have been successfully solved. Some early solution methods are based on variational theorems developed mainly in the fifth decade of the previous century, largely due to Drucker, Prager and Hill (see References). These variational solutions provide upper and lower-bound theorems for the ultimate load capacity of structures. The load carrying capacity of specimen can also be calculated using the method of characteristics, see for example Hill (1950); Prager (1955, 1956). With these classical methods, it is hard or even impossible to obtain an elastic-plastic solution which gives the total *history of deformation*, from the elastic to the ultimate load state.

In today's engineering environment, it is imperative that detailed analysis of geometrically very complicated structures be carried out. These analyses must trace out the complete response history from elastic to plastic conditions, including the progression of elasto-plasticity and large deformations, until possible collapse of the structure.

The aim of this chapter is to review fundamental notions of metal plasticity in order to establish the basis for introducing robust numerical procedures. We consider the solution of problems of elasto-plasticity modeled using classical macromechanical theories of plasticity, and assume that finite element solutions are sought.

The computational algorithms to be presented in the subsequent chapters correspond to the so-called *strain-driven* methods. Namely, we shall employ an incremental solution process (outlined in the previous chapter) in which we consider that the total strains at a material point are known at a certain time (load) step; and that the stresses corresponding to the given strains need to be calculated. This procedure is generally used in displacement-based or mixed finite element discretizations (see Chapter 2). Throughout the presentations below we follow the notation used in Bathe (1996).

3.2 Isotropic Plasticity

The classical macromechanical theories of plasticity are based on the notions of a yield surface or yield function giving the yield condition, a hardening rule (governing the change of the yield surface during deformations) and on the stress-plastic strain relations of the material.

These notions are used to formulate a material model for the calculation of the material response during plastic deformations. In the following presentation of these fundamental notions, we adopt two approaches (Sections 3.2.1 to 3.2.4).

- We start from experimental observations and give the mathematical relations to model these observations; or
- We establish mathematical relations based on a mechanical principle, and present experimental results that confirm these relations.

The fundamental relations are then employed for the definition of the von Mises material model in Section 3.2.5, in a form generally accepted for describing isotropic metal plasticity. Some relations presented in Section 3.2.5 have a general character and they are also used in the definition of other material models.

Finally, in Section 3.2.6 we briefly present the principle of maximum plastic dissipation which is of general importance for a complete presentation and understanding of the theory of plasticity.

We start with uniaxial experimental data that serve as a basis for the development of the mathematical relations to represent one-, two- and three-dimensional behavior.

3.2.1 Uniaxial Elastic-Plastic Deformation

Basic Experiments. Let us consider experimental results obtained in a uniaxial tension/compression test of a metal. According to Smith and Sidebottom (1965), if an annealed high-carbon steel specimen is subjected to tension and then compression, followed by reverse loading, the stress-strain dependence is as shown in Fig. 3.2.1 (see also McClintock and Argon 1966). The figure also shows that the same stress-strain behavior is measured if the specimen is first subjected to compression and then tension. Similar results are obtained for other metals.

Based on these experimental data we define the material models for the elastic-plastic deformation of a metal in uniaxial loading.

Uniaxial Elastic-Plastic Models. A typical stress-strain model diagram for the continuous uniaxial loading in tension or compression is shown in Fig. 3.2.2a. This diagram approximates the data obtained by direct recording during a physical experiment. Based on this diagram we introduce some important notions of plasticity.

In experimental investigations the uniaxial stress-strain curve is drawn as the relationship between the *true stress* (*Cauchy stress*, force per unit *current* area) and the *true strain* (logarithmic strain $\ln(\ell/\ell_0)$, where ℓ and ℓ_0 are the current and initial lengths of the material specimen length considered). The strain at fracture (*fracture point F* on the diagram) can be 50% and larger. In this chapter and those to follow, except in Chapter 7, we assume small strain conditions, namely that the strains are smaller than four percent[1]. In this case we can use in the development of the material model the stress σ defined as the *force per unit original area* (also called the *Kirchhoff stress*), and the *infinitesimal engineering strain*, e. The $\sigma - e$ diagram is shown in Fig. 3.2.2b.

The uniaxial stress-strain curve in Fig. 3.2.2b represents the fundamental curve in small strain and large displacement/small strain deformation conditions. The Cauchy stress-logarithmic strain curve in Fig. 3.2.2a represents the fundamental curve in large strain analyses (see Chapter 7).

---

[1] In some numerical examples given in Chapters 4 to 6 the strain values are larger than 4 percent but only to exemplify differences in values reached. In practice, all such analyses should be carried out using large strain theory.

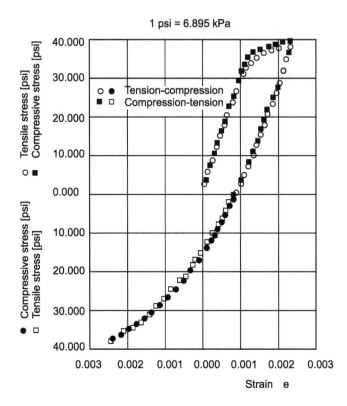

Fig. 3.2.1. Tension and compression stress-strain diagrams for annealed high-carbon steel for initial and reverse loading, according to Smith and Sidebottom (1965)

The point on the diagrams important for elastic-plastic solutions is the *elastic limit*, or *yield point* (point A in Fig. 3.2.2). Point B shown in the figure lies between the point A and the fracture point F. If the stress is below the yield stress the material returns to its undeformed configuration upon unloading. The part OA of the diagram represents the *elastic domain* of the material. In the case of loading above the yield stress, a *permanent plastic strain* e^P remains after unloading. Hence, the total strain e corresponding to the stress σ is

$$e = e^E + e^P \tag{3.2.1}$$

where

$$e^E = \frac{\sigma}{E} \tag{3.2.2}$$

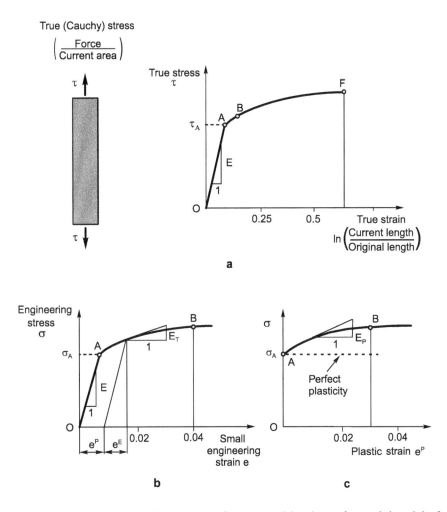

Fig. 3.2.2. Stress-strain dependence for uniaxial loading of metal (model of test results). **a** Cauchy stress-logarithmic strain diagram up to the fracture point F; **b** Stress-strain dependence for small strains; **c** Yield curve for small strains

Here E is Young's modulus which relates the elastic strain e^E and the stress σ. The plastic deformation starts at the stress corresponding to onset of yielding, that is at the *yield stress* σ_A. The part of the curve between points A and B characterizes the material behavior in the plastic domain. Figure 3.2.2b also shows the tangent to the curve between points A and B, defined as the *tangent modulus* E_T, which we assume to be ≥ 0.

From the uniaxial model stress-strain curve we can determine the fundamental dependence

$$\sigma = \sigma\left(e^{P}\right)$$

(3.2.3a)

shown in Fig. 3.2.2c. This dependence can also be written in the form

$$f_{y}\left(\sigma, e^{P}\right) = 0$$

(3.2.3b)

which represents the *yield condition* in uniaxial loading. The relation (3.2.3a) shown in Fig. 3.2.2c is called the *yield curve* of the material.

Figure 3.2.2c shows that the stress σ is increasing with the plastic deformation e^{P}. This material characteristic is known as *strain-hardening* and is an important ingredient in the description of the material behavior in the plastic regime. The instantaneous hardening at a given point on the yield curve is described by the *plastic modulus* E_P at the plastic strain level reached,

$$E_P = \frac{d\sigma}{de^{P}}$$

(3.2.4)

which can be expressed in terms of the instantaneous tangent modulus E_T and Young's modulus E, as

$$E_P = \frac{E E_T}{E - E_T}$$

(3.2.5)

This relation follows from the definition of the tangent modulus $E_T = d\sigma/de$, and from (3.2.1), (3.2.2) and (3.2.4)

$$d\sigma = E_T de = E_T \left(de^{E} + de^{P}\right)$$
$$d\sigma = E_T \left(\frac{d\sigma}{E} + \frac{d\sigma}{E_P}\right)$$

Solving for E_P we obtain (3.2.5). We note that a special case of the stress-strain relation is *the bilinear relation*, with slope E in the elastic domain and $E_T = const.$ in the plastic domain.

For some ductile materials, hardening can be neglected (that is, $\sigma = \sigma_A$, $E_P = 0$ for any e^{P}) as shown by the dashed line in Fig. 3.2.2c. Then the material is considered to be a *perfectly plastic* material.

The yield curve shown in Fig. 3.2.2c represents the basic behavior of the material. While obtained from a one-dimensional experiment, we shall see that the curve is used as a fundamental material relation to describe complex two- and three-dimensional elasto-plastic deformations.

Analytical expressions are often used for the yield curve in metal plasticity. We cite here the commonly used *Ramberg-Osgood formula*

$$e = \frac{\sigma}{E} + \alpha \frac{\sigma_R}{E} \left(\frac{\sigma}{\sigma_R} \right)^{m_R} \tag{3.2.6}$$

where σ_R is a reference stress (usually the yield stress $\sigma_{yv} = \sigma_A$ is used, see Fig. 3.2.2), and α and m_R are material constants obtained from experimental data. We will use the Ramberg-Osgood description of the yield curve in the form

$$\sigma = \sigma_{yv} + C_y \left(e^P \right)^n \tag{3.2.7}$$

or

$$e^P = A_y \left(\sigma - \sigma_{yv} \right)^{1/n} \tag{3.2.8}$$

where $A_y = C_y^{-1/n}$. The values C_y and n are material constants, which are non-negative. Note that in the case of perfect plasticity we should use $\sigma = \sigma_{yv}$ in (3.2.7), while (3.2.8) is not applicable.

Uniform loading in tension or compression was considered in the above model. Let us next consider that the material is first plastically deformed in tension and then compressed. For some materials the "new yield stress" in compression after unloading is (in magnitude) the stress at which unloading occurred. However, in general, the "new yield stress" is smaller than the yield stress reached in tension. The same behavior is observed when the material is first compressed and then subjected to tension, in which case the subsequent yield stress in tension is smaller than the yield stress reached in compression. This phenomenon is known as the *Bauschinger effect* and can be explained by changes in the metal microstructure caused by plastic deformation. This effect is important to be included in the solution of problems with cyclic loading.

For practical applications some simplified models have been developed to account for the Bauschinger effect. The two commonly used models are *isotropic hardening* and *kinematic hardening* models, see Fig. 3.2.3. In the isotropic hardening assumption, the Bauschinger effect is ignored and the same yield curve is used in tension and compression during the cyclic loading (curve $OABFF_1F_2$ in Fig. 3.2.3). In the kinematic hardening assumption the change of stress to start of yielding in the reverse loading is equal to twice the initial yield stress; for example, the yield stress at point C, σ_C is

$$\sigma_C = \sigma_B - 2\sigma_A$$

and the stress-strain path is $OABCC_1C_2$. Models that represent behavior between these two models are called *mixed hardening models* (path

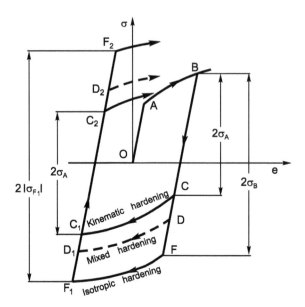

Fig. 3.2.3. Idealized uniaxial stress-strain curve for cyclic loading

$OABDD_1D_2$). We note that for all these simplified models the hardening curves have the same shape when the loading is reversed, but they start from different yield stresses.

Note that the kinematic and mixed hardening models are introducing an anisotropy into the assumed behavior of the material.

The influence of the rate of straining on the yield properties of the metal can be significant and must typically be taken into account in practical applications when the strain rate is above a certain value, characteristic for the material considered. Figure 3.2.4 shows schematically several yield curves for a mild steel which correspond to different strain rates, Mendelson (1968). It was found experimentally, for example, that for the aluminum alloy $AA7108.50$ at 340^oC the yield stress increases from 65 MPa at the strain rate of $0.1s^{-1}$ to 160 MPa at the strain rate of $2000s^{-1}$ (Djapic-Oosterkamp et al. 2000). In this book we will not explicitly include strain rate effects in the numerical algorithms, but the solution procedures can be extended to account for these effects. Namely, in the solution of problems we can determine the strain rate in the current time step and therefore identify which stress-strain curve is to be used to determine the plastic deformations.

3.2.2 Yield Condition for General Three-Dimensional Deformations

In the previous section we considered uniaxial loading and found, according to experimental data, that there exists an initial yield point A with the yield

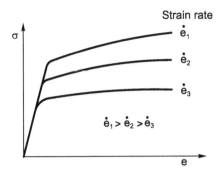

Fig. 3.2.4. Schematic uniaxial stress-strain dependence of a mild steel for different strain rates

stress of the virgin material σ_{yv}, at which plastic deformation starts. Here we extend the notion of initial yielding to multiaxial stress conditions for an initially isotropic material, when all stress components σ_{ij} may be different from zero. We want to establish a condition for *initial yielding* in the form

$$f_y\,(\sigma_{ij}) = 0 \tag{3.2.9}$$

where f_y is called the *yield function*. The equation (3.2.9) is called the *yield condition*. Hence, as long as this yield condition is not reached, we will have elastic strains only and the elastic constitutive relations in Appendix 1, see (A1.5), are applicable.

General Considerations. In order to determine an appropriate form of the yield function, we use that the material reaches initial yield in an isotropic manner; that is, the material behaves in the same way into any material direction. Since the stress components change with the coordinate system used, in order to satisfy this *isotropy condition*, the yield function can only be a function of the stress invariants, I_1, I_2 and I_3 (which are independent of the coordinate system used), i.e., we need to have

$$\boxed{f_y(I_1, I_2, I_3) = 0} \tag{3.2.10}$$

where

$$I_1 = \sigma_{ii}$$
$$I_2 = \sigma_{11}\sigma_{22} + \sigma_{22}\sigma_{33} + \sigma_{33}\sigma_{11} - (\sigma_{12}\sigma_{21} + \sigma_{23}\sigma_{32} + \sigma_{13}\sigma_{31})$$
$$I_3 = \det\sigma \tag{3.2.11}$$

We use here the usual indicial notation and summation convention (see Appendix A2, (A2.7)). This yield function is applicable to any initially isotropic material. However, we need to specify the details of the function to reflect the experimental observations.

Considering metals, it is seen experimentally that frequently the influence on yielding of the first invariant I_1, or the mean stress

$$\sigma_m = \frac{1}{3}I_1 \tag{3.2.12}$$

as well as the third invariant I_3, can be neglected. Therefore, we can reduce the yield function to depend only on I_2, or equivalently only on the second invariant of the deviatoric stresses, J_{2D}, defined as

$$J_{2D} = \frac{1}{2}S_{ij}S_{ij} \tag{3.2.13}$$

where the deviatoric stresses are

$$S_{ij} = \sigma_{ij} - \sigma_m\delta_{ij} \tag{3.2.14}$$

with δ_{ij} the Kronecker delta symbol ($\delta_{ij} = 1$ for $i = j$, $\delta_{ij} = 0$ for $i \neq j$). We write (3.2.10) as

$$f_y(J_{2D}) = 0 \tag{3.2.15}$$

This form of the yield condition is generally used to model metal plasticity, and is based on the seminal contributions of von Mises (1913). In order to introduce a specific function $f_y(J_{2D})$ we proceed as follows.

Hypothesis on Distortion Energy. A fundamental approach to determine the yield function (3.2.15) is based on the hypothesis that the metal reaches its elastic limit and begins to deform plastically when for any stress state the distortion elastic strain energy reaches a certain critical value W'_E, Mendelson (1968). The distortion elastic strain energy in the case of a general stress/strain state can be expressed in terms of the deviatoric stresses as (see (A.1.18))

$$W'_E = \frac{1}{2}S_{ij}e'^E_{ij} = \frac{1}{4G}S_{ij}S_{ij} \tag{3.2.16}$$

In the uniaxial loading condition considered earlier (tension or compression) the deviatoric stresses are at the start of yielding

$$S_{11} = \frac{2}{3}\sigma_{yv} \qquad S_{22} = S_{33} = -\frac{1}{3}\sigma_{yv} \tag{3.2.17}$$

and hence in the uniaxial case, at the start of yielding,

$$W'_E = \frac{1}{6G}\sigma_{yv}^2 \tag{3.2.18}$$

We assume that at the start of yielding, the distortion elastic strain energy has the same value for all loading conditions. Therefore, it follows from (3.2.16) and (3.2.18) that we must have

$$f_y(S_{ij}) = \frac{1}{2}S_{ij}S_{ij} - \frac{1}{3}\sigma_{yv}^2 = 0 \tag{3.2.19}$$

or, in terms of the second invariant J_{2D},

$$f_y(J_{2D}) = J_{2D} - \frac{1}{3}\sigma_{yv}^2 = 0 \tag{3.2.20}$$

This yield condition satisfies the isotropy condition and is based upon the principle of equivalence of distortion energy (i.e., (3.2.16) and (3.2.18) correspond to the same distortion energy). The form (3.2.19) of f_y is known in the literature as the *von Mises yield function*, and materials "obeying" this yield condition are called *von Mises materials*. The *von Mises yield condition (3.2.19)* is the most commonly used yield condition for metals (von Mises 1913). [2]

Experimental Verification. We quote here some experimental results which confirm that the von Mises yield condition is adequate for metals. Figure 3.2.5 shows experimental results obtained by Taylor and Quinney (1931) for three metals, subjected to tension and shear. Experiments were performed on thin-walled tubes under axial loading and torsion. It can be seen that the von Mises yield condition (the "von Mises law") closely represents the experimental results.

Other yield conditions for metals have been proposed as well, and notably the one due to Tresca. Here the hypothesis is that the yielding in the material is governed by the maximum shear stress on any plane. The practical use of this criterion is more difficult than the application of the von Mises criterion because the von Mises criterion corresponds to a simple functional form continuous in the stress variables. In general, solutions obtained using these two criteria are not far apart (for practical purposes), as we see, for example, in Fig. 3.2.5. We refer to the Tresca criterion here only because experimental

---

[2] Here we should also mention the earlier publication of Huber regarding failure of a material (Huber 1904).

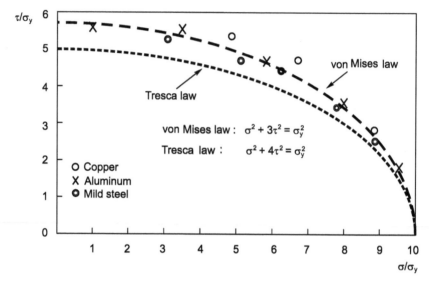

Fig. 3.2.5. Experimental results according to Taylor an Quinney (1931) from combined tension and torsion tests; σ = tension stress, τ = shear stress, σ_y = yield stress in tension

results are often interpreted using both criteria. We illustrate the application of the von Mises yield criterion in examples in Section 3.2.7.

The next important step in the formulation of the theory of plasticity is to establish the stress-plastic strain relations for general loading conditions.

3.2.3 Three-Dimensional Stress-Plastic Strain Relations

The stress-plastic strain relations that we present here rely on *experimental observations*. We start with an analysis of a simple uniaxial experiment and then consider general multiaxial loading of the material.

Let a specimen of metal be subjected to a uniaxial loading condition in small strains, as schematically shown in Fig. 3.2.6. If the stress σ exceeds the yield stress σ_{yv} (earlier also called σ_A), the material deforms plastically. To determine the permanent deformations we release the stress and measure the displacement Δ_P, shown in Fig. 3.2.6b. Measurements show that there are lateral permanent contractive strains, practically equal to one-half of the plastic strain e^P_{xx} measured in the longitudinal direction. Hence we can assume that

$$e^P_{yy} = e^P_{zz} = -\frac{1}{2}e^P_{xx} \qquad (3.2.21)$$

It follows that we can assume in a model that

$$e^P_V = e^P_{xx} + e^P_{yy} + e^P_{zz} = 0 \qquad (3.2.22)$$

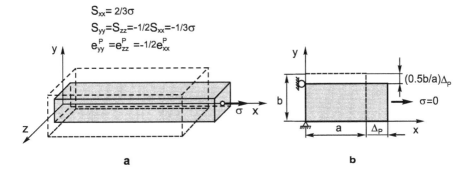

Fig. 3.2.6. Plastic deformations due to uniaxial loading of a metal. **a** Uniaxial loading; **b** Deformations after stress release

where e_V^P is the volumetric plastic strain. Hence the plastic deformation is isochoric or volume preserving. This conclusion is also reached when studying experimental results which show that yielding is not affected by the mean stress, and that the yield function has the form (3.2.15) or (3.2.20).

The deviatoric stress components that have caused these permanent plastic strains — still considering the case of uniaxial loading — are

$$S_{xx} = \frac{2}{3}\sigma$$

$$S_{yy} = S_{zz} = -\frac{1}{2}S_{xx} \qquad (3.2.23)$$

By inspection of (3.2.21) and (3.2.23) we find that the following relations can therefore be assumed:

$$\frac{e_{xx}^P}{S_{xx}} = \frac{e_{yy}^P}{S_{yy}} = \frac{e_{zz}^P}{S_{zz}} = \lambda \qquad (3.2.24)$$

where λ is a positive scalar. Hence, there exists a proportionality between the plastic strain components and the corresponding deviatoric stress components causing these strains.

This proportionality between the plastic strains and the corresponding stresses is also observed in a pure-shear experiment, schematically shown in Fig. 3.2.7, and in more complex experiments, e.g., Pugh and Robinson (1978) (see Fig. 3.2.17).

In summary, based on measurements in simple tests of metals, we can postulate the stress-plastic strain relations as

$$e_{ij}^P = \lambda S_{ij} \qquad (3.2.25)$$

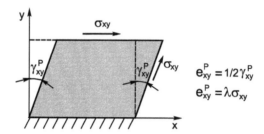

Fig. 3.2.7. Plastic deformation of metal in shear

These relations correspond to the so-called *deformation theory* of plasticity, Mendelson (1968).

However, in general, plastic deformations depend on the loading history. This means that the values of plastic strains e_{ij}^P for the current deviatoric stresses S_{ij} at a material point, depend on the manner how the stresses changed prior to reaching the current values. The dependence of plastic strains on the history of loading is illustrated in the simple example shown in Fig. 3.2.8. We suppose that the uniaxial elastic-plastic material model is defined by a bilinear stress-strain curve with isotropic hardening, and subject the material to two loading regimes:

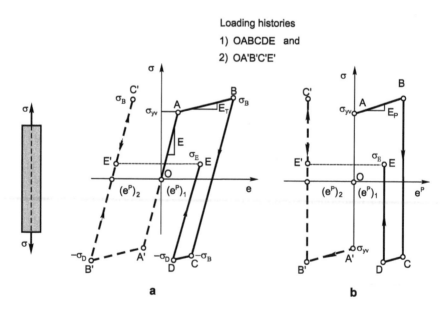

Fig. 3.2.8. Two loading histories for uniaxial loading: 1) Tension σ_B, compression σ_D, tension σ_E; 2) Compression σ_D, tension σ_B, unloading to σ_E; **a** Diagram $\sigma - e$; **b** Diagram $\sigma - e^P$

1) Tension to stress $\sigma = \sigma_B$, then reverse loading up to compressive stress $\sigma = -\sigma_D$ (with $\sigma_D > \sigma_B$), and reloading up to the tension stress $\sigma = \sigma_E$; and

2) Compression to stress $\sigma = -\sigma_D$, then reverse loading up to the tension stress $\sigma = \sigma_B$, and unloading to stress σ_E.

Here σ_B and σ_D are taken to be positive values.

Obviously, the stress state $\sigma = \sigma_E$ corresponds to different values of plastic strain e^P for the two loading conditions:

$$\left(e^P\right)_1 = \frac{1}{E_P}(2\sigma_B - \sigma_D - \sigma_{yv}) \qquad (3.2.26)$$

and

$$\left(e^P\right)_2 = -\frac{1}{E_P}(\sigma_D - \sigma_{yv}) \qquad (3.2.27)$$

for loading histories 1) and 2), respectively.

This observation about the dependence of the plastic strains on the loading history leads to a modification and generalization of the relations (3.2.25). Since, in general, the plastic strains evolve incrementally, as demonstrated in Fig. 3.2.8, we need to seek incremental plastic strain computations by a relationship which in the special case considered earlier reduces to (3.2.25). However, the relationship must also give the plastic strains when more general stress conditions, including cyclic loading, are considered

Suppose that at a given stress state in the material a small change of loading causes plastic flow. Then we will have increments of plastic strains de_{ij}^P during the change in stresses. The assumption in accordance with the proportionality relations (3.2.25) is that during a change of loading, each nonzero deviatoric stress component S_{ij} causes an increment de_{ij}^P of the corresponding plastic strain component. This assumption leads to the relation

$$de_{ij}^P = d\lambda\, S_{ij} \qquad (3.2.28a)$$

or

$$\boxed{de^P = d\lambda\, \mathbf{S}} \qquad (3.2.28b)$$

Therefore, the *increments* of the plastic strains de_{ij}^P are proportional to the current *total* deviatoric stresses S_{ij}. It is experimentally found that the relations (3.2.28) are valid in general, whereas the relations (3.2.25) are only valid when the stresses increase proportionally. The theory of plasticity based on the relations (3.2.28) is called the *incremental theory* or *flow theory* of plasticity.

In the case of proportional loading we have that the stresses σ_{ij} at a given time (load level) can be expressed as

$$\sigma_{ij} = k\,\sigma_{ij}^R$$

(3.2.29)

where k is a variable increasing in time, and the σ_{ij}^R are reference stresses corresponding to an initial elastic state. The following relations can then be written

$$\frac{S_{ij}}{S_{ks}} = \frac{S_{ij}^R}{S_{ks}^R} = r_{(ij)(ks)} \qquad \text{for all } i,j,k,s$$

and, with use of (3.2.28),

$$d\,e_{ij}^P = r_{(ij)(ks)}\,d\,e_{ks}^P \qquad \text{no sum on } i,j,k,s$$

where $r_{(ij)(ks)}$ are constants. Integrating the last equation we have

$$e_{ij}^P = r_{(ij)(ks)}\,e_{ks}^P$$

Then, using $r_{(ij)(ks)} = S_{ij}/S_{ks}$ we obtain

$$\frac{e_{ij}^P}{S_{ij}} = \frac{e_{ks}^P}{S_{ks}} \qquad \text{for all } i,j,k,s; \text{ no sum on } i,j,k,s$$

But this relation can only be satisfied if the relation (3.2.25) holds.

Let us consider further the relations (3.2.25) and (3.2.28). Namely, we might intuitively say, starting from the relations (3.2.25), that the increments of plastic strains should be proportional to the *increments* in the stresses. However, it was found *experimentally* that the relations (3.2.28) are valid for many materials (in particular metals). For example, consider the tension-torsion elasto-plastic conditions of a thin-walled tube with some material hardening used in the experimental verification of the von Mises yield condition (Fig. 3.2.5). Assume that we have a certain stress state defined by the normal stress σ and shear stress τ. If we keep the stresses constant, there will be no plastic flow. However, if, for example, we increase the torsion shear stress τ and keep the normal stress σ constant, plastic flow would continue and we would find that, in addition to the increment of shear plastic strain $d\gamma^P$, the normal components de_{xx}^P, de_{yy}^P and de_{zz}^P are different from zero as well, although the increments of normal stresses are equal to zero: $d\sigma_{xx} = d\sigma_{yy} = d\sigma_{zz} = 0$ (also $dS_{xx} = dS_{yy} = dS_{zz} = 0$). In fact, the increments de_{xx}^P, de_{yy}^P and de_{zz}^P are proportional to S_{xx}, S_{yy} and S_{zz}. Of course, in this experiment we would have $de_{xx}^P = de_{yy}^P = de_{zz}^P = 0$ if the tube were free of tension ($\sigma = 0$).

Finally, we add an observation concerning the relations (3.2.28), important for their use: when using these relations we first test for a change of stresses *causing plastic flow*, and under the condition that plastic flow occurs, the relations (3.2.28) are applicable. Detailed considerations of the conditions for yielding and plastic flow to take place are presented in Sections 3.2.4 and 3.2.6.

The relations (3.2.28) were first proposed by Prandtl (1924) for plane strain deformations and then generalized to three-dimensional conditions by Reuss (1930). They are known as the *Prandtl-Reuss equations*. The original form of these equations is given by

$$\frac{de_{11}^P}{S_{11}} = \frac{de_{22}^P}{S_{22}} = \ldots = \frac{de_{31}^P}{S_{31}} = d\lambda \tag{3.2.30}$$

that is, by (3.2.28).

Historically, the first stress-strain relations in plasticity were introduced by St. Venant (1870), who proposed that the principal directions of total strain increments coincide with the principal directions of the stresses. The general relations of the form (3.2.28), but with increments of *total* strain de_{ij} rather than de_{ij}^P, were proposed by Levy (1870) and, independently, by von Mises (1913); they are known as the Levy-von Mises equations.

Finally, we quote the results of an experimental verification of the Prandtl-Reuss equations for a more general loading condition. Taylor and Quinney (1931) subjected thin-walled tubes to combined tension and torsion non-proportional loading conditions and calculated from measurements the Lode's variables:

$$\mu_s = \frac{2S_{22} - S_{33} - S_{11}}{S_{33} - S_{11}} = \frac{2\sigma_{22} - \sigma_{33} - \sigma_{11}}{\sigma_{33} - \sigma_{11}}$$

and   (3.2.31)

$$\mu_e = \frac{2de_{22}^P - de_{33}^P - de_{11}^P}{de_{33}^P - de_{11}^P}$$

If the Prandtl-Reuss relations are valid, then $\mu_s = \mu_e$. In Fig. 3.2.9 we show experimental results for three metals. We see some deviation from the straight line given by $\mu_s = \mu_e$, but this deviation is acceptable for analyses in engineering practice.

Based upon the above discussion we can now proceed to define the hardening behavior in the model for metal plasticity under general loading conditions. The description of the hardening is necessary for modeling the material behavior in general plastic deformations.

3.2.4 Hardening in 3–D

In order to define the hardening of a metal in the process of plastic flow, after the initial yield condition has been reached, we use the hypothesis of

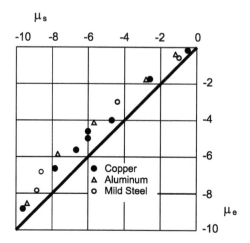

Fig. 3.2.9. Experimental results according to Taylor and Quinney(1931) from combined tension and torsion tests of thin-walled tubes

equivalence of plastic work, then present the consequences that follow and give some experimental results which confirm these consequences. We assume in this section that the same yield curve is used throughout any loading, and assume first isotropic hardening and then consider perfectly plastic material behavior as a special case of hardening.

Hypothesis of Equivalence of Plastic Work. This hypothesis states that the incremental plastic work dW_{gen}^P performed during plastic deformations in general loading conditions is equal to the plastic work dW_{uniax}^P performed in uniaxial loading conditions, i.e.,

$$dW_{gen}^P = dW_{uniax}^P = dW^P \qquad (3.2.32)$$

Using the Prandtl-Reuss equations (3.2.28) we have for the general loading conditions

$$dW^P = S_{ij}de_{ij}^P = d\lambda\, S_{ij}S_{ij} \qquad (3.2.33)$$

Note that this work dW^P corresponds to permanent distortions of the material and is irreversible. On the other hand, for the uniaxial case we have

$$dW^P = \frac{2}{3}d\lambda\, \sigma_y^2 \qquad (3.2.34)$$

where we have used (3.2.23) and the condition that the material is continuously yielding. Therefore, in uniaxial loading conditions we have continuously during the loading process that $\sigma = \sigma_y$ (see (3.2.3a) and Fig. 3.2.2). Hence, it follows from (3.2.33) and (3.2.34) that

$$S_{ij}S_{ij} = \frac{2}{3}\sigma_y^2 \qquad (3.2.35)$$

This equation can also be written in the form (3.2.19), i.e.,

$$f_y = \frac{1}{2}S_{ij}S_{ij} - \frac{1}{3}\sigma_y^2 = 0 \qquad (3.2.36)$$

The consequences of the hypothesis of equivalence of plastic work can therefore be stated as follows: Considering any state of plastic flow in general loading conditions, we can identify a yield stress σ_y on the uniaxial yield curve corresponding to the given stress state. We see that, in essence, during plastic deformation the von Mises yield criterion (3.2.36) is *continuously satisfied*, in which σ_y changes according to the uniaxial yield curve.

The hypothesis of equivalence of plastic work is in agreement with the hypothesis on the distortion energy used in Section 3.2.2. Namely, considering the distortion energy W' as a measure of the material resistance to change of shape, and supposing that the hypothesis is also valid during plastic flow, we can write the following equation

$$S_{ij}de_{ij}^E + S_{ij}de_{ij}^P = \frac{1}{6G}d(\sigma_y^2) + \frac{2}{3}d\lambda\sigma_y^2 \qquad (3.2.37)$$

where the left hand side represents dW' in general loading conditions, and the right hand side corresponds to the uniaxial loading case. The first terms on both sides correspond to the elastic deformation (already employed in Section 3.2.2), hence we obtain (3.2.35). In writing (3.2.37) we have used the condition that the increments of total strain de_{ij} are the sum of the elastic strains de_{ij}^E and plastic strains de_{ij}^P, in accordance with (3.2.1) for uniaxial loading conditions,

$$de_{ij} = de_{ij}^E + de_{ij}^P \qquad (3.2.38)$$

Also, we used that the elastic volumetric strains do not affect the distortion energy.

A geometric interpretation of the von Mises yield condition is given in Fig. 3.2.10. The yield curve $\sigma_y(e^P)$ is shown in Fig. 3.2.10a, with two points M_1 and M_2 on the curve and yield stresses $^1\sigma_y$ and $^2\sigma_y$. The yield surface $f_y = 0$, defined by (3.2.36), is represented in the principal stress space $\sigma_1, \sigma_2, \sigma_3$ by the cylinder with the "hydrostatic" axis $\sigma_1 = \sigma_2 = \sigma_3$ and the radius $R = \sqrt{2/3}\,\sigma_y$, as shown in Fig. 3.2.10b. In the deviatoric plane, with the

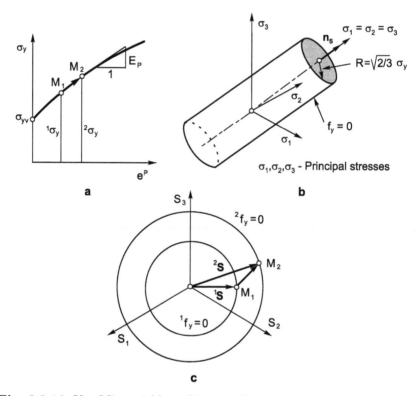

Fig. 3.2.10. Von Mises yield condition. **a** Uniaxial yield curve; **b** Yield surface in the stress space; **c** Yield surfaces in the deviatoric plane with S_1, S_2 and S_3 being the principal values

normal $\mathbf{n}_s$ to the plane and axes S_1, S_2, S_3 (principal deviatoric stresses), the yield surface is represented by a circle of radius R. Two yield surfaces $^1f_y = 0$ and $^2f_y = 0$ are shown in Fig. 3.2.10c. Note that all stress points lying in the deviatoric plane on the yield surface with the radius R correspond to one point with $\sigma_y = \sqrt{3/2}\,R$ on the yield curve. For example, all stress points in the deviatoric plane lying on the circle $^1f_y = 0$ are mapped to point M_1 on the yield curve. Hence, the change in the deviatoric stresses $(^2\mathbf{S} - ^1\mathbf{S})$ in the deviatoric plane corresponds to the arc $M_1 M_2$ on the yield curve.

Another form of the von Mises yield condition is also used in the literature. Namely, if an equivalent stress quantity, called the *effective stress* or the *von Mises stress*, $\bar{\sigma}$ is defined,

$$\bar{\sigma} = \sqrt{\frac{3}{2}S_{ij}S_{ij}} = \sqrt{\frac{3}{2}}\,\|\mathbf{S}\| \qquad (3.2.39)$$

then the yield condition (3.2.36) reduces to

$$\bar{\sigma} - \sigma_y = 0 \qquad (3.2.40)$$

In (3.2.39) $\|\mathbf{S}\|$ represents the intensity (or Euclidean norm, see Appendix A2, (A2.31)) of $\mathbf{S}$, defined as

$$\|\mathbf{S}\| = \sqrt{S_{ij}S_{ij}} \qquad (3.2.41)$$

In the above presentation we have reached the conclusion that the yield condition has the form (3.2.36), or (3.2.40), but we have not yet introduced a law for the change in the yield surface. To obtain that law or rule, we first impose the condition — based upon experimental observations of the material behavior and theoretical considerations — that for the materials considered here the plastic work W^P *must be positive*

$$dW^P > 0 \qquad (3.2.42)$$

The principle of maximum plastic dissipation introduced in Section 3.2.6 and Drucker's definition of a stable material (Drucker 1951) are based upon this fact. All results that we reach regarding the hardening behavior of a material rely on the condition (3.2.42).

The next step in the development of a hardening rule is to introduce the *increment of effective plastic strain* $d\bar{e}^P$ which corresponds to the effective stress $\bar{\sigma}$ such that the increment of work dW^P is given by

$$dW^P = \bar{\sigma}d\bar{e}^P \qquad (3.2.43)$$

Using (3.2.42) it follows that the increment of effective plastic strain must be positive. Since the plastic work dW^P, expressed by (3.2.43) for general loading conditions, is equal to dW^P for the uniaxial loading case given by (3.2.34), we obtain

$$d\lambda = \frac{3}{2}\frac{d\bar{e}^P}{\sigma_y} \qquad (3.2.44)$$

where we have used that during yielding $\sigma_y = \bar{\sigma}$.

Finally, we express in (3.2.33) S_{ij} in terms of de_{ij}^P using the Prandtl-Reuss equations (3.2.28). Substituting (3.2.28) into (3.2.33) and using (3.2.43), we obtain

$$\frac{1}{d\lambda} de_{ij}^P de_{ij}^P = \bar{\sigma} d\bar{e}^P \tag{3.2.45}$$

Hence, with (3.2.44) we obtain the expression

$$d\bar{e}^P = \sqrt{\frac{2}{3} de_{ij}^P de_{ij}^P} \tag{3.2.46}$$

for the increment of effective plastic strain in terms of the increments of plastic strain components. We note that in the case of uniaxial loading we have

$$d\bar{e}^P = |de^P| \tag{3.2.47}$$

while in two-dimensional pure shear

$$d\bar{e}^P = |d\gamma^P| / \sqrt{3} \tag{3.2.48}$$

where $d\gamma^P$ is the increment in the plastic engineering shear strain.

The yield conditions (3.2.40) and (3.2.3) show that the yield curve in simple tension represents at the same time the general yield curve

$$\sigma_y = \sigma_y \left(\bar{e}^P \right) \tag{3.2.49}$$

where

$$\bar{e}^P = \int d\bar{e}^P \tag{3.2.50}$$

is the accumulated effective plastic strain.

From the above presentation we conclude the following:

1) The *hardening* of the von Mises material is defined by the yield curve $\sigma_y \left(\bar{e}^P \right)$. The yield condition (3.2.36) can be written in the form

$$f_y = \frac{1}{2} S_{ij} S_{ij} - \frac{1}{3} \sigma_y^2 \left(\bar{e}^P \right) = 0 \tag{3.2.51}$$

and then (3.2.40) becomes

$$\bar{\sigma} - \sigma_y \left(\bar{e}^P \right) = 0 \tag{3.2.52}$$

2) The proportionality factor $d\lambda$ in the Prandtl-Reuss equations (3.2.28) is defined by (3.2.44). The factor can be calculated from the yield curve.

Since an increase in the yield stress is given as a function of the effective plastic strain, the term *strain-hardening* is often used in the literature. These results are based on two fundamental principles of plasticity:

- the equivalence of the plastic work dW^P for the one- and multi-dimensional stress conditions, and
- the physical fact that the plastic work must be positive; that is $dW^P > 0$.

The model regarding the hardening behavior is illustrated in Fig. 3.2.11. This figure shows that in the case of uniaxial tension-compression loading (Fig. 3.2.11a), the plastic strain e^P first increases and then decreases, while the effective plastic strain $\bar{e}^P$ and yield stress σ_y continuously increase during the plastic flow (Fig. 3.2.11c). In the case of general loading conditions, see Fig. 3.2.11b, the yield surface increases in size during plastic flow. Two successive stress states in the case of uniaxial tension-compression loading, and in the case of general loading, are represented by the same points M_1 and M_2 on the yield curve $\sigma_y(\bar{e}^P)$. Since, due to hardening, each new yield surface is larger in size than the previous one, it follows that the current yield surface bounds the *elastic domain* which increases during plastic flow. Namely, if during yielding we reach the yield stress ${}^1\sigma_y$, and next reach stresses such that

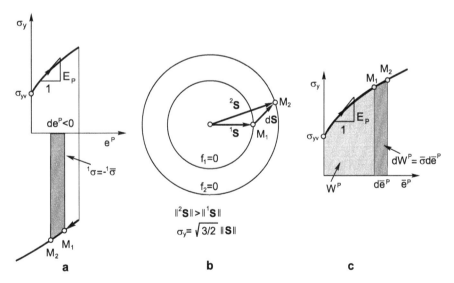

Fig. 3.2.11. Mapping of stress states on the yield curve for material with hardening. **a** Uniaxial tension-compression loading; **b** General loading conditions; **c** Yield curve $\sigma_y(\bar{e}^P)$

$$\boxed{\bar{\sigma} < {}^1\sigma_y}$$ (3.2.53)

plastic flow will not continue and we have the case of *elastic unloading*. The stress point corresponding to this case is inside the yield surface ${}^1f_y = 0$; the point in the $\sigma_y(\bar{e}^P)$ diagram lies on the line ${}^1\bar{e}^P = const.$ and below the yield curve. The case of a subsequent loading back to $\bar{\sigma} = {}^1\sigma_y$ represents *elastic loading*; and a further stress change but with $\bar{\sigma} = {}^1\sigma_y$ defines the so-called *neutral loading*, also with no plastic flow. The yield surface during neutral loading does not change, the stress point in Fig. 3.2.11b stays on the surface ${}^1f_y = 0$, with at most a change of position on the surface, while the point on the yield curve is fixed.

Note that any stress state outside the yield surface or above the yield curve is not possible.

So far we considered that the material is hardening, with $E_P > 0$. We now finally consider the model of a *perfectly plastic* material behavior, already mentioned in Section 3.2.1 (Fig. 3.2.2). This case is important in practical applications because strain-hardening may be neglected for some materials. We show in Fig. 3.2.12 two successive stress states in the case of a general loading condition, which correspond to the points M_1 and M_2 on the yield curve. We have that

$$dW^P = \sigma_{yv}\,d\bar{e}^P > 0$$ (3.2.54)

and

$$\bar{\sigma} - \sigma_{yv} = 0$$ (3.2.55)

corresponding to a yield surface of constant size. All stress states during yielding of the material are represented by only one yield surface, and the elastic domain does not change due to plastic flow. In the case of plastic flow under general loading conditions, we have that, as yielding progresses, a

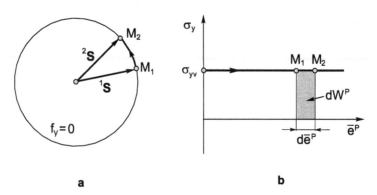

Fig. 3.2.12. Mapping of stress states on the yield curve for perfect plasticity. **a** Deviatoric plane; **b** Yield curve $\sigma_y = \sigma_{yv}$

stress point M moves but always stays on the yield surface, its image point moves along the yield line $\sigma_y = \sigma_{yv}$, and the effective plastic strain can only increase.

Let us also consider different states of deformation in a uniaxial loading condition and represent these states using: (i) the deviatoric plane, (ii) the uniaxial stress-plastic strain dependence $\sigma(e^P)$, and (iii) the yield curve $\sigma_y(\bar{e}^P)$. First, we increase the uniaxial stress σ until initial yielding is reached. The stress point moves along line OA in the deviatoric plane (Fig. 3.2.13a) and its mapping moves along the axis OA in Figs. 3.2.13b,c. If the uniaxial straining continues, the uniaxial plastic strain e^P and the effective plastic strain $\bar{e}^P$ increase along the lines AB, while the stress point in the deviatoric plane stays at the point A. Suppose that we next change the loading direction. The stress point moves along the line BC until the yield surface at point C is reached. During this elastic deformation (elastic unloading) the plastic strain e^P and effective plastic strain $\bar{e}^P$ do not change. Finally, if we continue straining in the same direction, plastic flow continues; and the stress point remains at the same position C in the deviatoric plane, with a decrease of

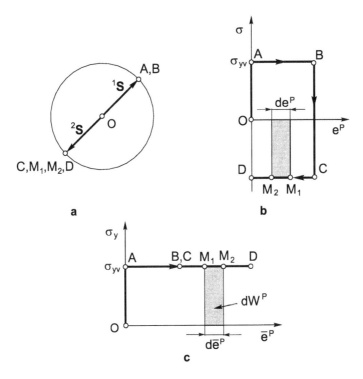

Fig. 3.2.13. Representation of stress states in the case of uniaxial loading and perfect plasticity. **a** Deviatoric plane; **b** Uniaxial stress-plastic strain dependence $\sigma(e^P)$; **c** Yield curve $\sigma_y = \sigma_{yv}$

e^P and an increase of $\bar{e}^P$. The points that correspond to plastic deformation during the reverse straining are shown in the figure. Note that during the reverse plastic flow the plastic work is also positive.

In the literature on plasticity the term *"work-hardening"* is used. A measure of the hardening is given by the plastic work W^P, geometrically represented by the area below the yield curve $\sigma_y(\bar{e}^P)$ (Fig. **3.2.11c**), since the relation $W^P[\sigma_y(\bar{e}^P)]$ is defined by the yield curve. In the case of perfect plasticity, we have $\sigma_y = \sigma_{yv} = const.$, i.e., there is no hardening, and the plastic work is simply an overall measure of plastic deformation at a material point; of course, the effective plastic strain is such a measure as well.

Experimental Verification. Many experimental investigations have been carried out in the past to verify the hypothesis (3.2.52). We quote some test results in Figs. 3.2.14 to 3.2.16. Figure 3.2.14 shows results of tests of copper tubes subjected to axial loading and internal pressure. In each test the loads are increased proportionally with a constant ratio between the principal stresses σ_1 and σ_2. The results are shown for various ratios, between 0 and 1, and are indicated by different symbols. The results shown in Fig. 3.2.15 are obtained by successive tension and shear of the material. The most severe test of the relation (3.2.52) corresponds to conditions when the loading changes sign during the test. Experimental results of such tests are shown in Fig. 3.2.16.

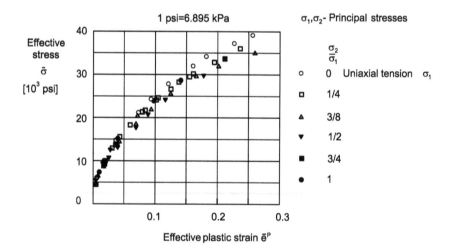

Fig. 3.2.14. Test results of copper tubes subjected to axial loading and internal pressure, according to Davis (1943)

The experimental results show that the hypothesis expressed in (3.2.52) can be used to model the material behavior in general loading conditions. The hypothesis represents physical reality closer for simpler loading conditions

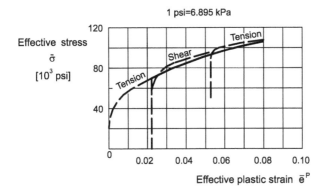

Fig. 3.2.15. Results according to Sautter, Kochendorfer and Dehlinger (1953) obtained for variable loading of material

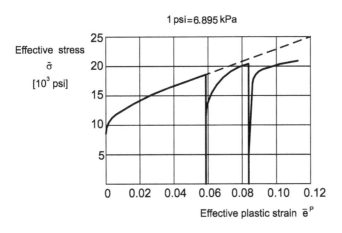

Fig. 3.2.16. Test results involving reversals of torsion of thin-walled copper tubes, according to Meyer, J. A. (1957), unpublished research at M.I.T.; cited in Crandall et al. (1972)

(like proportional loading), and gives a larger deviation from test results for complex deformation histories.

3.2.5 The von Mises Material Model with Mixed Hardening

In the previous sections we considered the basic ingredients in mathematically modeling plastic deformations. We introduced a uniaxial material model and then presented a generalization of this model to a three-dimensional model. This *generalization* is one of the *key steps* in the theory of plasticity. Although the 3-D model introduced is in agreement with experimental observations, for some metals it does not describe accurately enough the material response in the case of complex loading conditions, such as in cyclic loading when the Bauschinger effect may be pronounced.

We are now in the position to summarize a rather general model of elasto-plastic material behavior. We shall consider an initially isotropic material.

A rather general model of the von Mises type in isotropic metal plasticity is the von Mises material model with the mixed hardening assumption. This model is defined first, and then the models with kinematic and isotropic hardening, and perfect plasticity assumptions are considered as special cases.

Experimentally Determined Yield Surfaces. Experimental data obtained for steel, reported in Pugh and Robinson (1978), is shown in Fig. 3.2.17. Initial and subsequent yield surfaces were determined for the material subjected to various proportional loading conditions. It can be seen from the figure that the yield surface changes size, position and shape. The change in position is due to the Bauschinger effect.

Formulation of the von Mises Model (Mixed Hardening). In Section 3.2.2 we introduced the von Mises yield condition. The yield surface changes its size according to (3.2.51). As we stated above, this description does not take into account the Bauschinger effect introduced in Section 3.2.1. To generalize the uniaxial yield curves given in Fig. 3.2.3, we introduce the yield condition in the form (Drucker 1951; Prager 1956, 1959; Hodge 1957; Johnson and Mellor 1983)

$$f_y = \frac{1}{2}(S_{ij} - \alpha_{ij})(S_{ij} - \alpha_{ij}) - \frac{1}{3}\hat{\sigma}_y^2 = 0 \qquad (3.2.56)$$

where α_{ij} are the components of the *back stress* $\boldsymbol{\alpha}$ that define the position of the yield surface, and $\hat{\sigma}_y$ is the yield stress, as shown in Fig. 3.2.18. Note that the change of the yield surface size and position in the stress space are taken into account in a simplified form, while any change of shape is neglected. The yield condition (3.2.56) can also be written as

$$f_y = \frac{1}{2}\hat{S}_{ij}\hat{S}_{ij} - \frac{1}{3}\hat{\sigma}_y^2 = 0 \qquad (3.2.57)$$

where

$$\hat{S}_{ij} = S_{ij} - \alpha_{ij} \qquad (3.2.58)$$

are the components of the radius of the yield surface[3]. From (3.2.57) a relation analogous to (3.2.40) is obtained

---

[3] We will refer to $\hat{\mathbf{S}}$, with components $\hat{S}_{ij}$, simply as the stress radius. Note that the $\hat{S}_{ij}$ are also "deviatoric stresses", so that $\hat{S}_{ii} = 0$ (sum over i).

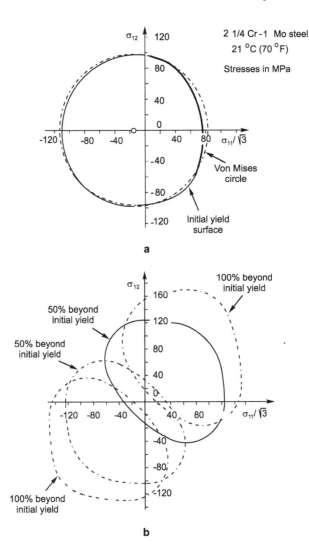

Fig. 3.2.17. Yield characteristics of annealed 2 1/4 Cr-1 Mo steel in tension-shear stress space at room temperature, according to Pugh and Robinson (1978). **a** Measured initial yield surface compared with idealized von Mises representation; **b** Yield surfaces during radial loadings

$$\hat{\sigma} = \hat{\sigma}_y \qquad\qquad (3.2.59)$$

where

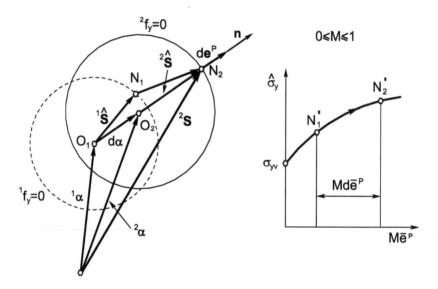

Fig. 3.2.18. Von Mises mixed hardening model represented in the deviatoric plane and by the yield curve $\hat{\sigma}_y \left(M \, \bar{e}^P \right)$. In the case of kinematic hardening ($M = 0$), the yield curve $\hat{\sigma}_y \left(M \, \bar{e}^P \right)$ is not used, see (3.2.81)

$$\hat{\sigma} = \sqrt{\frac{3}{2} \hat{S}_{ij} \hat{S}_{ij}} = \sqrt{\frac{3}{2}} \|\hat{S}\| \tag{3.2.60}$$

is the so-called *reduced effective stress*, and $\|\hat{S}\|$ is the intensity or norm of $\hat{S}$ (see (3.2.41)).

According to the assumption of mixed hardening, mentioned already in Section 3.2.1, the plastic strain increments can be divided into isotropic and kinematic parts, de^{iP} and de^{kP}, i.e.,

$$de^P = de^{iP} + de^{kP} \tag{3.2.61}$$

The simplest way to relate de^{iP} and de^{kP} is to select a constant M with

$$0 \le M \le 1 \tag{3.2.62}$$

where

$$de^{iP} = M \, de^P \tag{3.2.63}$$

and

$$de^{kP} = (1 - M) \, de^P \tag{3.2.64}$$

The constant M is called the *mixed hardening parameter* and is a material characteristic. It represents a measure of the Bauschinger effect, measured in uniaxial stress deformations. Note that $M = 1$ *and* $M = 0$ *correspond to isotropic and kinematic hardening, respectively.* It follows from this definition that the increment of plastic work dW^P is divided into isotropic and kinematic parts, dW^{iP} and dW^{kP}. These plastic work parts occur in the same ratio as do the strains de^{iP} and de^{kP}. With the assumption of mixed hardening, only the isotropic hardening part $d\bar{e}^{iP}$ of the increment $d\bar{e}^P$ of the effective plastic strain,

$$d\bar{e}^{iP} = M \, d\bar{e}^P \tag{3.2.65}$$

affects the size of the yield surface. Therefore, the yield stress $\hat{\sigma}_y$ used in (3.2.56) or (3.2.57) is a function of $\bar{e}^{iP}$,

$$\hat{\sigma}_y = \hat{\sigma}_y \left(M \, \bar{e}^P \right) \tag{3.2.66}$$

as indicated in Fig. 3.2.18.

The next step in defining the material model is to specify the *stress-plastic strain relations.* For this purpose we note that the Prandtl-Reuss equations (3.2.28) are now

$$de_{ij}^P = d\lambda \, \hat{S}_{ij} \tag{3.2.67a}$$

or

$$de^P = d\lambda \frac{\partial f_y}{\partial \boldsymbol{\sigma}} \tag{3.2.67b}$$

since

$$\frac{\partial f_y}{\partial \sigma_{ij}} = \hat{S}_{ij} \tag{3.2.68}$$

To differentiate f_y with respect to σ_{ij} we used the definition (3.2.14) of the deviatoric stresses, from which a matrix form of derivatives $\partial \mathbf{S} / \partial \boldsymbol{\sigma}$ can be obtained,

$$\left[\frac{\partial \mathbf{S}}{\partial \boldsymbol{\sigma}} \right] = \begin{bmatrix} \mathbf{A} & 0 \\ 0 & \mathbf{I_3} \end{bmatrix} \tag{3.2.69}$$

where $\mathbf{I}_3$ is the (3×3) identity matrix, and

$$\mathbf{A} = \frac{1}{3}\begin{bmatrix} 2 & -1 & -1 \\ -1 & 2 & -1 \\ -1 & -1 & 2 \end{bmatrix} \qquad (3.2.70)$$

Equation (3.2.67) represents the *flow rule*, or the *normality principle* in plasticity, considered as one of the most fundamental relations in the flow theory of plasticity. Considering (3.2.67) it also follows that the yield function $f_y[\boldsymbol{\sigma},\ \hat{\sigma}_y(\bar{e}^P)]$ is a plastic potential. In general, a function different from f_y, namely $g(\sigma_{ij})$, may be used as the plastic potential. We refer to the relation (3.2.67) as the *associated flow rule* because f_y is used. In case another function $g(\sigma_{ij}) \neq f_y(\sigma_{ij})$ is used in (3.2.67) we refer to the relation as the *nonassociated flow rule*. The relation (3.2.67) can be written in rate form as

$$\dot{\mathbf{e}}^P = \dot{\lambda}\frac{\partial f_y}{\partial \boldsymbol{\sigma}} \qquad (3.2.71)$$

where $\dot{\mathbf{e}}^P$ is the plastic strain rate tensor, and $\dot{\lambda}$ is a positive scalar.

The normality principle and the fact that the plastic work must be positive have important implications on the shape of the yield surface, in particular, the yield surface must be convex (see Section 3.2.6).

Since we have

$$d\mathbf{e}^P = d\lambda\,\hat{\mathbf{S}} \qquad (3.2.72)$$

the increment of plastic strain $d\mathbf{e}^P$, or the strain rate $\dot{\mathbf{e}}^P$, is in the direction of the stress $\hat{\mathbf{S}}$. This stress defines a unit normal to the yield surface, as geometrically shown in Fig. 3.2.18. We will use the relation (3.2.72) in the computational algorithms for stress integration.

In order to complete the formulation of the model, we need a hardening law for the back stress $\boldsymbol{\alpha}$. We use *Prager's hardening rule* in a simple form

$$d\boldsymbol{\alpha} = C\,d\mathbf{e}^{kP} = (1 - M)C\,d\mathbf{e}^P \qquad (3.2.73)$$

where C is the kinematic hardening modulus. Hence, we assume that the displacement of the yield surface is in the direction of the normal to the yield surface, as shown geometrically in Fig. 3.2.18. The modulus C is determined from the yield curve by considering uniaxial loading of the material and mixed hardening conditions. We have in uniaxial tensile conditions that the yield condition (3.2.57) reduces to

$$\frac{2}{3}\sigma_{xx} - \alpha_{xx} - \frac{2}{3}\hat{\sigma}_y = 0 \qquad (3.2.74)$$

where we have used the relations (3.2.23). Hence, incrementally,

$$\frac{2}{3} d\sigma_{xx} - d\alpha_{xx} - \frac{2}{3} d\hat{\sigma}_y = 0 \tag{3.2.75}$$

We next impose the condition that the stress σ_{xx} follows the yield curve for any value of M. Also, the yield stress $\hat{\sigma}_y$ must change according to the hardening law (3.2.66). Therefore, with use of the relation (3.2.21) from which $de_{xx}^P = de^P$, where e^P is the uniaxial plastic strain, as well as (3.2.73), we have

$$d\sigma_{xx} = E_P \, de^P$$
$$d\hat{\sigma}_y = M \hat{E}_P \, de^P \tag{3.2.76}$$
$$d\alpha_{xx} = (1 - M)C \, de^P$$

where

$$
\begin{array}{|c|}
\hline
\\
E_P = \left. \dfrac{\partial \sigma_y}{\partial e^P} \right|_{e^P} \\
\\
\hat{E}_P = \left. \dfrac{\partial \sigma_y}{\partial e^P} \right|_{Me^P} \\
\\
\hline
\end{array}
\tag{3.2.77}
$$

are the plastic moduli (the slopes on the yield curve $\sigma_y(e^P)$ recorded experimentally) corresponding, respectively, to the plastic strains e^P and Me^P. Of course, in general analyses we use the slopes on the yield stress $-$ effective plastic strain yield curve $\sigma_y(\bar{e}^P)$. Substituting (3.2.76) into (3.2.75) we obtain the equation

$$
\boxed{ C = \frac{2}{3}(E_P - M\hat{E}_P)/(1 - M) }
\tag{3.2.78}
$$

Note that in the case of a bilinear stress-strain relation (E_P is constant) and in the case of kinematic hardening, we have

$$C = \frac{2}{3} E_P \tag{3.2.79}$$

If the yield curve is represented by the Ramberg-Osgood formula (3.2.7), we have

$$C = \frac{2}{3} nC_y(1 - M^n)(e^P)^{n-1}/(1 - M) \tag{3.2.80}$$

This approach corresponds to the "splitting of plastic strain" method described in Bathe and Montans (2004). Alternatively, also the "splitting of plastic modulus" method can be used (see Bathe and Montans 2004).

In summary, the main relations that describe the von Mises material model with mixed hardening are:

- The yield condition is defined by (3.2.57).
- The stress-plastic strain relations are based on the associated flow rule and are given by (3.2.72).
- The hardening is defined according to the mixed hardening assumption, with the mixed hardening parameter M used for splitting the total increment of plastic strain de^P into the isotropic and kinematic parts, de^{iP} and de^{kP} (see (3.2.61) to (3.2.64)). The size of the yield surface depends on the isotropic part of the effective plastic strain $M\bar{e}^P$ according to (3.2.66).
- The displacement of the yield surface is governed by Prager's hardening rule (3.2.73).
- The hardening characteristics are defined by the yield curve measured in uniaxial stress conditions of the material.

As mentioned already, the von Mises model using the kinematic or isotropic hardening assumptions and the model assuming no hardening to take place (perfect plasticity) are special cases of the above mixed hardening model.

In order to model *kinematic hardening* conditions we use the hardening parameter $M = 0$. Then we assume that the yield surface has a constant size, i.e.,

$$\hat{\sigma}_y = \sigma_{yv} \qquad (3.2.81)$$

and the yield curve $\hat{\sigma}_y(M\hat{e}^P)$ in Fig. 3.2.18 is not applicable. During plastic flow the yield surface changes its position in the stress space, and hence plastic work is only performed in displacing the yield surface. The stress-plastic strain relations have the form (3.2.72), and the displacement of the yield surface is governed by Prager's rule (3.2.73). Therefore, the yield curve $\sigma_y(e^P)$ recorded experimentally in a uniaxial tension test (see Figs. 3.2.1, 3.2.2 and 3.2.10) is only used to evaluate the modulus $C(\bar{e}^P)$ according to (3.2.79). The von Mises model with kinematic hardening is adequate for a metal with a pronounced Bauschinger effect.

In case the Bauschinger effect can be neglected we assume the *isotropic hardening* material behavior. Then we use the hardening parameter $M = 1$, the yield condition has the form (3.2.51), and the stress-plastic strain relations are given by (3.2.28). Figure 3.2.11 gives a graphical representation of the yield surface change and the yield curve for this model.

Perfect plasticity conditions correspond to a material with no hardening during plastic flow. Then we take $M = 1$ and prescribe a constant yield stress $\hat{\sigma}_y = \sigma_{yv}$ as in (3.2.81), and the yield condition reduces to (3.2.55). Figures 3.2.12 and 3.2.13 show stress states and the yield curves for general and uniaxial loading conditions for a perfectly plastic material. The stress-plastic strain relations have the form (3.2.28).

Finally, we show a graphical representation of the material response using the above defined mixed hardening model, in case the material is subjected to uniaxial tension and compression loading. Figures 3.2.19a and 3.2.19b show respectively the $\sigma(e^P)$ and $\bar{\sigma}(\bar{e}^P)$ relations (see also Fig. 3.2.3). In drawing the dependence $\bar{\sigma}(\bar{e}^P)$ we have taken into account that $d\bar{e}^P = |de^P|$, see (3.2.47). It can be seen that in the loading regime AB the material response is the same for all values of M, while it becomes dependent on M when the loading changes the sign.

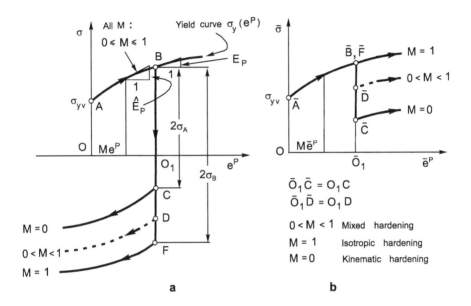

Fig. 3.2.19. Material response calculated using mixed hardening material model, in the case of uniaxial tension and compression. **a** Stress − plastic strain relation; **b** Effective stress − effective plastic strain relation

3.2.6 Principle of Maximum Plastic Dissipation

In this section we introduce the principle of maximum plastic dissipation and give some generalizations in plasticity that follow from this principle (Lubliner 1984, 1990; Eterovic and Bathe 1990, 1991b; Simo and Hughes 1998). The yield condition (3.2.56) of the von Mises model can be written in the form

$$f_y(\sigma, \bar{e}^P, \alpha) = 0 \qquad (3.2.82)$$

The effective plastic strain $\bar{e}^P$ and the back stress α represent the internal (hardening) variables of the material model. The yield condition (3.2.82) can also be written as

$$f_y(\sigma, \beta) = 0 \tag{3.2.83}$$

where β are the *internal variables*. The internal variables are represented by an array, whose terms may be scalars (such as the effective plastic strain to measure hardening) and components of tensors (such as the back stress). We note that in the case of *perfect plasticity* the yield surface does not change (see (3.2.55)) while the material deforms plastically, hence $\beta = 0$, and the yield condition (3.2.83) reduces to

$$f_y(\sigma) = 0 \tag{3.2.84}$$

We now consider isothermal elastic-plastic deformations of a stable material (a material with hardening behavior, Drucker 1951). To introduce the *principle of maximum plastic dissipation* we define the plastic dissipation as

$$\check{D}^P = \sigma \cdot \dot{e}^P + \beta \cdot \dot{\beta}^* \tag{3.2.85}$$

where $\dot{\beta}^*$ is the rate of the hardening variables work conjugate to β. Then the principle of maximum plastic dissipation states that among all *admissible states* (σ, β) the actual state corresponds to the stresses and internal variables for which $\check{D}^P$ attains its maximum. It follows from this principle that –

1) The yield surface is convex.
2) The stress-plastic strain relations have the associated character (3.2.71), i.e.,

$$\dot{e}^P = \dot{\lambda} \frac{\partial f_y}{\partial \sigma} \tag{3.2.86}$$

3) The hardening law is associated,

$$\dot{\beta}^* = \dot{\lambda} \frac{\partial f_y}{\partial \beta} \tag{3.2.87}$$

and then

$$\dot{\beta} = -\dot{\lambda} \, \mathbf{C}^P \frac{\partial f_y}{\partial \beta} \tag{3.2.88}$$

where $\mathbf{C}^P$ is the tensor of plastic moduli.

4) The loading/unloading Kuhn-Tucker conditions are satisfied,

$$f_y \leq 0 \tag{3.2.89}$$

$$\dot{\lambda} \geq 0 \tag{3.2.90}$$

$$\dot{\lambda} f_y = 0 \tag{3.2.91}$$

Figure 3.2.20a shows a convex yield surface and illustrates geometrically the associated flow rule for two stress states $^t\boldsymbol{\sigma}$ and $^{t+\Delta t}\boldsymbol{\sigma}$. The associated flow rule, or the normality principle, also follows from the microstructural rearrangements that develop during plastic flow, and the relation (3.2.86) can be derived from micromechanical considerations, Hill (1950); Rice (1971, 1975); Aifantis (1987). Of course, for a hardening material, $\mathbf{C}^P$ is a positive definite tensor.

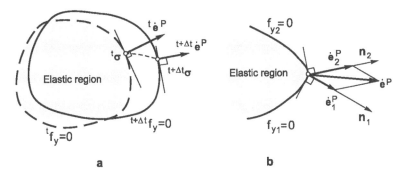

Fig. 3.2.20. Geometric interpretation of the associated plasticity. **a** Single yield surface, hardening plasticity; **b** Multi-surface yield condition

The relation (3.2.89) expresses the condition that the stress point must be inside or on the yield surface. In the case of *loading* with plastic flow or in the case of *neutral loading* (in both cases, the stress point moves in the stress space, but remains on the current yield surface) the equality sign holds. The case $f_y < 0$ corresponds to elastic unloading (or elastic conditions). Regarding (3.2.90) the condition $\dot{\lambda} > 0$ corresponds to plastic flow, while for neutral loading or *elastic unloading/elastic conditions* we have $\dot{\lambda} = 0$. The relation (3.2.91) is the *complementarity condition* and enforces the conditions that when $f_y < 0$, we must have $\dot{\lambda} = 0$ and when $f_y = 0$ we can have $\dot{\lambda} > 0$ or $\dot{\lambda} = 0$.

The principle is also valid for the perfectly plastic material, and all consequences, except (3.2.87) and (3.2.88), are applicable. The principle is in accordance with *Drucker's postulate* (Drucker 1951) of a stable material, from which the same consequences as those listed above (and the uniqueness of the solution) may be deduced, see, e.g., Mandel (1964); Hill (1968); Mendelson (1968); Chen and Han (1988).

Finally we present a generalization of the flow rule due to Koiter (1953). It is experimentally established that the elastic region for some materials is bounded by several continuous yield surfaces, say N, that mutually intersect, see Section 6.2.1. The intersecting points belong to the yield surfaces $f_{y1} = 0$, $f_{y2} = 0$, ... and $f_{yN} = 0$, and represent singular points. It is assumed that there is a contribution of each yield surface to the plastic strain rate at the common singular point, and $\dot{e}^P$ is expressed as

$$\dot{e}^P = \sum_k \dot{\lambda}_k \frac{\partial f_{yk}}{\partial \sigma} \tag{3.2.92}$$

where $\dot{\lambda}_k$ is the proportionality factor corresponding to the yield surface $f_{yk} = 0$. Figure 3.2.20b shows a geometric representation of this relation for two yield surfaces.

The above general relations will be specialized to various material models and will be used in the procedures for numerical stress integration.

3.2.7 Examples

Example 3.2.1. Compression of Metal Piece. A piece of metal is compressed in a rigid die, as schematically shown in Fig. E.3.2-1a.

  a) Assuming that the material is free to expand in the x-direction, find the pressure p_0 at the start of plastic deformation. The yield stress is σ_{yv} and Poisson's ratio is ν.
  b) Let all lateral displacements be restrained and establish how the axial strain e_{zz} depends on the applied pressure p including when there is plastic flow.

The material is of von Mises type with mixed hardening and is governed by a bilinear uniaxial stress-strain relation.

  a) *Plane Stress Deformation.* In this case we have that the normal stress σ_{xx} and normal strain e_{yy} are equal to zero. For the elastic deformation we have (see (A1.5) and (A1.24) in Appendix A1)

$$\sigma_{yy} = \nu\sigma_{zz} = -\nu p \tag{a}$$

The mean stress is

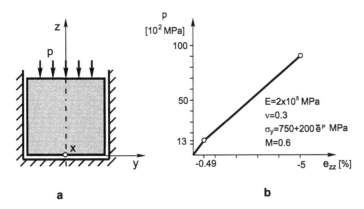

Fig. E.3.2-1. Compression of metal in a rigid die. **a** Schematic representation; **b** Dependence of pressure p on axial strain e_{zz}

$$\sigma_m = -\frac{1}{3}(1+\nu)\,p \tag{b}$$

and the deviatoric stresses are

$$S_{yy} = \frac{1}{3}(1-2\nu)p \quad S_{zz} = -\frac{1}{3}(2-\nu)p \quad S_{xx} = -S_{yy} - S_{zz} = -\sigma_m \tag{c}$$

The effective stress (3.2.39) is

$$\bar{\sigma} = \left[\frac{3}{2}\left(S_{xx}^2 + S_{yy}^2 + S_{zz}^2\right)\right]^{1/2} = (1-\nu+\nu^2)^{1/2}\,p \tag{d}$$

Yielding starts when $\bar{\sigma} = \sigma_{yv}$ (see (3.2.20)), therefore it follows from (d) that the pressure at start of yielding is

$$p_0 = \sigma_{yv}/\sqrt{1-\nu+\nu^2} \tag{e}$$

It is interesting to determine p_0 for two extreme values of ν : (i) $\nu = 0$; and (ii) $\nu = 0.5$. When $\nu = 0$ we have

$$p_0 = \sigma_{yv}$$

while for $\nu = 0.5$ (incompressible material),

$$p_0 = 2\,\sigma_{yv}/\sqrt{3}$$

b) *Restrained Compression.* The physical conditions in the lateral directions x and y are the same. Hence, we have

$$e_{xx} = e_{yy} = 0 \quad e_{xx}^E = e_{yy}^E \quad e_{xx}^P = e_{yy}^P$$

$$\sigma_{xx} = \sigma_{yy} \tag{f}$$

where e_{xx}^E, e_{yy}^E, e_{xx}^P, e_{yy}^P are the elastic and plastic strain components at any stage of deformation. Using the incompressibility condition of the plastic deformations we have

$$e_{xx}^P = e_{yy}^P = -\frac{1}{2}e_{zz}^P \tag{g}$$

From the elastic constitutive law (A1.6), with the elastic compliance matrix given in (A1.8), and with use of (f) and (g), we obtain

$$\frac{1-\nu}{E}\sigma_{xx} - \frac{\nu}{E}\sigma_{zz} - \frac{1}{2}e_{zz}^P = 0 \tag{h}$$

This equation can be written in terms of the deviatoric stresses and the mean stress. The relations between the deviatoric stresses follow from (f),

$$S_{xx} = S_{yy} = -\frac{1}{2}S_{zz} \tag{i}$$

while the mean stress can be expressed as (see (A1.12))

$$\sigma_m = \frac{E}{3(1-2\nu)}e_{zz} \tag{j}$$

Therefore (h) transforms into

$$-\frac{1+\nu}{2E}S_{zz} + \frac{1}{3}e_{zz} - \frac{1}{2}e_{zz}^P = 0 \tag{k}$$

We next employ the yield condition (3.2.57), and the hardening law (3.2.73) for the back stress $\boldsymbol{\alpha}$. We have the following relations:

$$S_{zz} = \hat{S}_{zz} + \alpha_{zz} \qquad \Delta\alpha_{zz} = \hat{C}\Delta e_{zz}^P \tag{l}$$

Since the hardening law is linear (see Fig. E.3.2-1), with a constant plastic modulus E_P, from (3.2.73) and (3.2.79) follows $\hat{C} = 2/3(1-M)E_P$. The relations (g) are satisfied at any stage of deformation, with the proportional loading conditions (3.2.29), and $E_P = const.$, therefore we have

$$\alpha_{zz} = \hat{C}e_{zz}^P \tag{m}$$

From (l), (m) and (k) we obtain

$$-\frac{1+\nu}{2E}\hat{S}_{zz} - \left(\frac{1}{2} + \frac{1+\nu}{2E}\hat{C}\right)e_{zz}^P + \frac{1}{3}e_{zz} = 0 \tag{n}$$

Also, we have

$$\hat{S}_{xx} = \hat{S}_{yy} = -\frac{1}{2}\hat{S}_{zz} \qquad (o)$$

and

$$\bar{e}^P = -e_{zz}^P \qquad (p)$$

where we have taken into account that e_{zz}^P is negative. Using that $\hat{S}_{zz}$ is negative, it follows from the yield condition (3.2.57) that

$$\hat{S}_{zz} = -\frac{2}{3}\sigma_{yv} + \frac{2}{3}MEpe_{zz}^P \qquad (q)$$

Finally, we substitute (q) into (n) and solve for e_{zz}^P,

$$e_{zz}^P = \frac{1}{3}\frac{(1+\nu)\sigma_{yv}/E + e_{zz}}{(1+\nu)Ep/(3E) + 1/2} \qquad (r)$$

Summarizing the above relations we have that for a given e_{zz} we determine e_{zz}^P from (r) and then calculate S_{zz} from (k), σ_m from (j), and $\sigma_{zz} = -p = S_{zz} + \sigma_m$. The components of the stress radius and back stress, and other components of the plastic strain and stress follow from the above given relations. It should be noted that in this example e_{zz}^P, and hence S_{zz} do not depend on the mixed hardening parameter M. The relation between e_{zz}^P and e_{zz} is linear, and then S_{zz} is linear with respect to e_{zz}. Hence e_{zz} depends linearly on $\sigma_{zz} = -p$. The linear relation $p(e_{zz})$ is shown in Fig. E.3.2-1b for the material constants given in the figure. For $e_{zz} = -0.05$, the values of stresses and plastic strains are

$$\sigma_{xx} = \sigma_{yy} = -8081.3 \qquad \sigma_{zz} = -p = -8837.3$$
$$e_{xx}^P = e_{yy}^P = -0.5e_{zz}^P \qquad e_{zz}^P = -0.03006$$

and yielding starts at

$$e_{zz}^P = 0, \ \sigma_{xx} = \sigma_{yy} = -562.5, \ \sigma_{zz} = -p = -1312.5$$
$$e_{zz} = -4.875 \times 10^{-3}.$$

Note that in the case a non-bilinear material is considered, we have a nonlinear dependence of the stresses on the strain e_{zz}, and the equations must be solved incrementally. Also note that all stresses are given in MPa.

Example 3.2.2. Torsion a Thin-Walled Tube. The thin-walled tube shown in Fig. E.3.2-2a is subjected to the torsional end moment M_t. The material is a von Mises metal. The geometric and material data are given in the same figure.

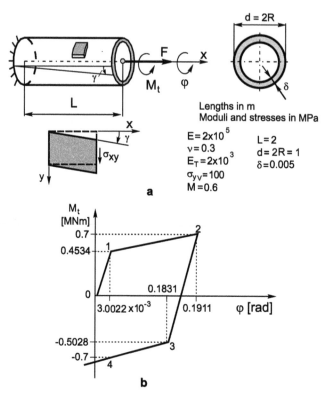

Fig. E.3.2-2. Torsion of thin-walled tube. **a** Geometric and material data; **b** Torsional moment - end rotation dependence

a) Determine the end rotation as a function of the torsional moment if the moment increases to the value $^2M_t = 0.7$ MNm and then decreases.

b) Determine the axial force necessary to cause yielding if the moment is released at the values $^2M_t = 0.7$ and $^4M_t = -0.7$ MNm, see Fig. E.3.2-2.

We first give some general relations. The only non-zero stress due to torsion is the shear stress σ_{xy} related to the torsional moment M_t as (considering that δ/R is small, hence using the linear term δ/R only)

$$\sigma_{xy} = \frac{M_t}{2\pi R^2 \delta} \tag{a}$$

The relation between the shear strain γ and the angle of rotation φ is (Fig. E.3.2-2a)

$$\varphi = \frac{L}{R}\gamma \tag{b}$$

Since the torsional moment is constant along the length, it follows that the stress state of the material is uniform.

a) Using the yield condition (3.2.56) we obtain the torsional moment at the start of yielding,

$$^1M_t = 2\pi R^2 \delta \frac{\sigma_{yv}}{\sqrt{3}} = 0.4534 \quad \text{MNm} \tag{c}$$

A further increase of the torsional moment causes plastic deformations of the material, with the elastic strain given in (A1.19),

$$e^E_{xy} = \frac{\sigma_{xy}}{2G} \tag{d}$$

while the plastic strain can be obtained from the yield condition as follows. The yield condition (3.2.56) can be expressed as

$$\sigma_{xy} - \alpha_{xy} - \frac{\hat{\sigma}_y}{\sqrt{3}} = 0 \tag{e}$$

from which

$$d\sigma_{xy} - d\alpha_{xy} - \frac{d\hat{\sigma}_y}{\sqrt{3}} = 0 \tag{f}$$

We further use the relation (3.2.66) with the bilinear stress-stain relation, and the constitutive relation (3.2.73) (with (3.2.79)), to obtain

$$d\sigma_{xy} = \frac{2}{3} E_P de^P_{xy} \tag{g}$$

Hence, the plastic strain e^P_{xy} is

$$e^P_{xy} = 1.5(\sigma_{xy} - \tau_0)/E_P \tag{h}$$

where $\tau_0 = \sigma_{yv}/\sqrt{3}$. With use of the elastic constitutive relation (d), the expression for e^P_{xy}, and the relations (a) and (b), we obtain

$$\varphi = \frac{L}{R}\left[(\frac{1}{G} + \frac{3}{E_P})\frac{M_t}{2\pi R^2 \delta} - \frac{3\tau_0}{E_P}\right] \tag{i}$$

This expression is applicable to the elastic-plastic deformations and an increasing torsional moment. In the case of elastic deformations, $M_t \leq {}^1M_t$, the terms with E_P in the denominator should be left out. The graph 0-1-2 in Fig. E.3.2-2b corresponds to the relation (i).
Some of the quantities at the point 2 on the graph are

$$\begin{array}{lll} ^2\sigma_{xy} = 89.127 & ^2\hat{\sigma}_y = 132.62 & ^2\alpha_{xy} = 12.557 \quad \text{(j)} \\ ^2e_{xy} = 2.389 \times 10^{-2} & ^2e^P_{xy} = 2.331 \times 10^{-2} & ^2\varphi = 0.1911 \end{array}$$

When the moment changes sign at point 2, yielding starts when the stress $^3\sigma_{xy}$ reaches the value (see (3.2.56) and Fig. 3.2.18)

$$^3\sigma_{xy} = -(^2\hat{\sigma}_y/\sqrt{3} - {}^2\alpha_{xy}) = -64.01 \tag{k}$$

and the moment is $^3M_t = -0.5028$. The shear strain and the rotation are

$$^3e_{xy} = 0.5\,{}^3\sigma_{xy}/G + {}^2e_{xy}^P = 2.289 \times 10^{-2} \quad {}^3\varphi = 0.1831 \tag{l}$$

The line 2-3 represents the elastic unloading. For the reverse elastic-plastic deformation, the relation (i) changes into

$$\varphi = \frac{L}{R}\left[(\frac{1}{G} + \frac{3}{E_P})\frac{M_t}{2\pi R^2\delta} - \frac{3\,{}^3\sigma_{xy}}{E_P} + 2\,{}^2e_{xy}^P\right] \tag{m}$$

and is represented by the line 3-4 in Fig. E.3.2-2b. The values corresponding to the moment ^4M_t are

$$^4e_{xy}^P = 4.662 \times 10^{-3} \quad {}^4\alpha_{xy} = 2.511 \quad {}^4\varphi = 3.266 \times 10^{-2}$$
$$^4\bar{e}^P = 4.845 \times 10^{-2} \quad {}^4\hat{\sigma}_y = \sigma_{yv} + ME_P\,{}^4\bar{e}^P = 158.72 \tag{n}$$

b) If the torsional moment is released at point 2, yielding under the axial force will start when the yield condition (3.2.56) is satisfied. From that condition we obtain the stress $^2\sigma_{xx}$ and the force ^2F,

$$^2\sigma_{xx} = \sqrt{2\hat{\sigma}_y^2 - 3\,{}^2\alpha_{xy}^2} = 130.8 \quad {}^2F = 2\pi\delta R\,{}^2\sigma_{xx} = 2.055 \text{ MN} \tag{o}$$

The stress and the axial force at start of yielding at point 4 (with the moment released), are

$$^4\sigma_{xx} = \sqrt{4\hat{\sigma}_y^2 - 3\,{}^4\alpha_{xy}^2} = 158.6 \quad {}^4F = 2\pi\delta R\,{}^4\sigma_{xx} = 2.492 \text{ MN}$$

Example 3.2.3. Pipe Structure. Two straight pipes are connected by a hinge at point B, and fixed to the wall by hinges at points A and C (Fig. E.3.2-3a). The structure is loaded by the internal pressure p.

Determine the dependence of the displacement of point B on the internal pressure p.

As it is common in the analysis of pipes, we consider that each pipe has closed ends. Therefore, the internal pressure produces the axial force

$$F_a = A_{int}\,p = \pi(R - \delta)^2\,p \tag{a}$$

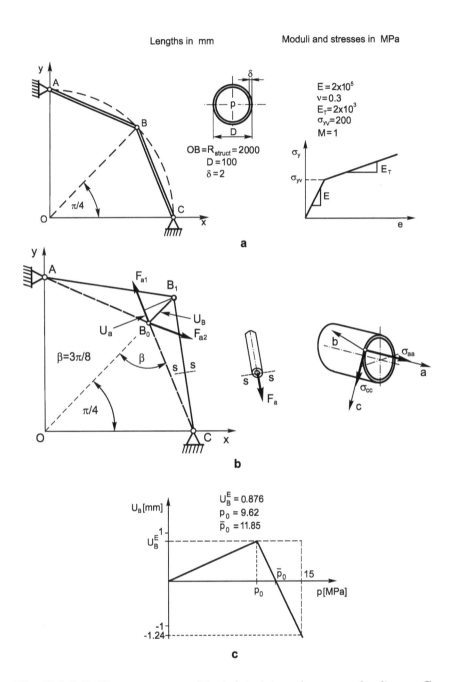

Fig. E.3.2-3. Pipe structure subjected to internal pressure loading. **a** Geometry and material data; **b** Kinematics of deformation, forces and stresses; **c** Displacement of point B as a function of internal pressure where A_{int} is the area of the internal cross-section, and $R = D/2$ is the pipe external radius.

In order to find the displacement U_B as a function of pressure p, we express U_B in terms of the axial strain e_{aa},

$$U_B = \frac{2R_{struct}\sin(\pi/8)}{\cos\beta} e_{aa} \qquad \text{(b)}$$

This relation follows from the kinematics of deformation shown in Fig. E.3.2-3b.

We next employ the fact that the pipe wall is subjected to the axial stress σ_{aa} and hoop stress σ_{cc}. Since the two pipe sections are free to expand, the axial stress σ_{aa} is due to the axial force F_a, as shown for a cross-section $s-s$ in the figure. Then σ_{aa} is

$$\sigma_{aa} = k_a\, p \qquad \text{(c)}$$

where

$$k_a = \frac{A_{int}}{A} = \frac{(R-\delta)^2}{2R_m\delta} = \frac{1}{2}(\frac{R}{\delta} - 1)\frac{R-\delta}{R-\delta/2} \approx \frac{1}{2}(\frac{R}{\delta} - \frac{3}{2} + \frac{\delta}{2R}) \qquad \text{(d)}$$

where the last expression is the approximation neglecting quadratic terms of (δ/R); $A = 2\pi R_m \delta$ is the cross-section of the pipe wall, and $R_m = R - \delta/2$ is the mean radius. The hoop stress σ_{cc} is the static equivalent to the internal pressure,

$$\sigma_{cc} = k_c\, p \qquad \text{(e)}$$

where

$$k_c = \frac{R}{\delta} - 1 \qquad \text{(f)}$$

From (c) and (e) follows that we have a case of proportional loading (see (3.2.29)).

We next relate the stresses and strains. The deviatoric stresses S_{aa}, S_{bb} and S_{cc} are :

$$S_{aa} = k_a'\, p \qquad S_{bb} = k_b'\, p \qquad S_{cc} = k_c'\, p \qquad \text{(g)}$$

where

$$k_a' = \frac{1}{3}(2k_a - k_c) \approx -\frac{1}{6}(1 - \frac{\delta}{R}) \qquad \text{(h)}$$

$$k_b' = -k_a' - k_c' \approx -\frac{1}{12}(6\frac{R}{\delta} - 7 + \frac{\delta}{R}) \qquad \text{(i)}$$

$$k_c' = \frac{1}{3}(2k_c - k_a) \approx \frac{1}{12}(6\frac{R}{\delta} - 5 - \frac{\delta}{R}) \qquad \text{(j)}$$

Using (3.2.44) and $d\bar{e}^P = (1/E_P)d\bar{\sigma}$, we obtain

$$d\lambda = \frac{3}{2}\frac{d\bar{e}^P}{\bar{\sigma}} = \frac{3}{2}\frac{\bar{k}\,dp}{E_P\bar{k}\,p} = \frac{3}{2E_P}\frac{dp}{p} \tag{k}$$

where

$$\bar{k} = \sqrt{\frac{3}{2}\left(k_a'^2 + k_b'^2 + k_c'^2\right)} \tag{l}$$

and $E_P = EE_T/(E - E_T)$ (see (3.2.5)) is the plastic modulus. Further,

$$e_{aa}^P = \int S_{aa}\,d\lambda = \frac{3k_a'}{2E_P}\int_{p_0}^{p} dp = \frac{3k_a'}{2E_P}(p - p_0)$$

$$e_{bb}^P = \frac{3k_b'}{2E_P}(p - p_0) \qquad e_{cc}^P = \frac{3k_c'}{2E_P}(p - p_0) \tag{m}$$

where p_0 corresponds to the start of yielding,

$$p_0 = \sigma_{yv}/\bar{k} \tag{n}$$

Note that $e_{aa}^P < 0$ since $k_a' < 0$. Also, the ratios between the plastic strains are

$$e_{aa}^P/e_{cc}^P \approx -\frac{1}{3}\frac{\delta}{R} \qquad e_{bb}^P/e_{cc}^P \approx -(1 - \frac{1}{3}\frac{\delta}{R}) \tag{o}$$

The plastic strain e_{cc}^P is positive while the others are negative. Also, e_{aa}^P is small compared to the other plastic strains (the ratio is of order δ/R).

We next employ the elastic constitutive relations (A1.6), and the above expressions for the plastic strains to obtain

$$e_{aa} = \frac{1}{E}(k_a - \nu k_c)\,p + e_{aa}^P \tag{p}$$

The first term represents the elastic strain e_{aa}^E which can be expressed as

$$e_{aa}^E \approx \frac{p}{E}(\frac{R}{\delta} - 1)(\frac{1}{2} - \nu - \frac{\delta}{4R}) \tag{q}$$

and we have that

$$e_{aa}^E > 0 \quad \text{for} \quad \nu < \frac{1}{2} - \frac{\delta}{4R} \tag{r}$$

Using (p) and (m) we can find the pressure $\bar{p}_0$ for which the axial strain is equal to zero,

$$\bar{p}_0 = \frac{3k_a'E}{2E_P(k_a - \nu k_c) + 3k_a'E}\,p_0 \tag{s}$$

Therefore, the strain e_{aa} increases linearly with the pressure in the elastic region, then decreases and reaches zero at the pressure value $\bar{p}_0$. A further pressure increase leads to negative values of the axial strain linearly increasing in magnitude.

The expressions for e_{bb} and e_{cc} are analogous to (p). With e_{aa} determined by (p), we can calculate the displacement U_B from (b).

Finally, we substitute the geometric and material data given in Fig. E.3.2-3a in the above expressions. The values of p_0, stresses, strains and displacement U_B corresponding to p_0, are :

$$p_0 = 9.622 \quad \sigma_{aa} = 113.11 \quad \sigma_{cc} = 230.92 \quad \text{MPa}$$
$$e_{aa} = 2.191 \times 10^{-4} \quad e_{cc} = 9.850 \times 10^{-4} \quad e_{bb} = -5.160 \times 10^{-4}$$
$$U_B = 0.876 \quad \text{mm} \tag{t}$$

The pressure $\bar{p}_0$ calculated from (s) and stresses and strains are:

$$\bar{p}_0 = 11.848 \quad \sigma_{aa} = 139.27 \quad \sigma_{cc} = 284.35$$
$$e_{aa} = 0.0 \quad e_{cc} = 2.118 \times 10^{-2} \quad e_{bb} = -1.992 \times 10^{-2}$$
$$e_{aa}^P = -2.698 \times 10^{-4} \quad e_{cc}^P = 1.997 \times 10^{-2} \quad e_{bb}^P = -1.970 \times 10^{-2}$$
$$U_B = 0.0 \tag{u}$$

Finally, the values corresponding to pressure $p = 15$ are:

$$\sigma_{aa} = 176.33 \quad \sigma_{cc} = 360.00$$
$$e_{aa} = -3.103 \times 10^{-4} \quad e_{cc} = 5.031 \times 10^{-2} \quad e_{bb} = -4.840 \times 10^{-2}$$
$$e_{aa}^P = -6.520 \times 10^{-4} \quad e_{cc}^P = 4.825 \times 10^{-2} \quad e_{bb}^P = -4.759 \times 10^{-2}$$
$$U_B = -1.241 \tag{v}$$

A graphical representation of the displacement U_B in terms of pressure p is given in Fig. E.3.2-3c.

3.3 Orthotropic Plasticity

So far we assumed that the material is initially isotropic and that during plastic flow some anisotropic response may be developed due to the Bauschinger effect. This anisotropic plasticity behavior is modeled by the von Mises material model with mixed hardening assumptions. However, we have in engineering practice metals which already prior to and at the onset of yielding, and then also thereafter, show anisotropic elastic and plastic response, a behavior which can be due to the technological process in the material production. For example, such anisotropy is usually found in metal sheets. This section is devoted to some basic material models used in anisotropic metal plasticity.

We first introduce Hill's orthotropic model (Hill 1950), an often used material model to describe orthotropic metal plasticity. We write the expressions

of Hill's yield condition and the stress-plastic strain relations in a form suitable for the stress integration algorithm of Section 4.6. Then we present some
orthotropic-plasticity models commonly employed in sheet metal plasticity.
Finally, we illustrate the application of Hill's orthotropic model in simple
examples.

3.3.1 Hill's Orthotropic Material Model

We assume that the material possesses three orthogonal (principal, or material) axes and that these axes remain orthogonal during the course of plastic deformations. To simplify the presentation, we use the principal axes of
anisotropy as the coordinate axes.

The yield condition due to Hill (1950) is

$$
\begin{aligned}
f_y = \frac{1}{2} [\, & F(\sigma_{yy} - \sigma_{zz})^2 + G(\sigma_{zz} - \sigma_{xx})^2 + H(\sigma_{xx} - \sigma_{yy})^2 \\
& + 2\,(L\sigma_{xy}^2 + I\sigma_{yz}^2 + K\sigma_{zx}^2) - 1\,] = 0
\end{aligned}
\tag{3.3.1a}
$$

where F, G, H, L, I and K are material constants, related to the assumption
that the yield stresses in the three principal directions are different. The
material constants can be determined by uniaxial loading in the x, y, and z
directions and by pure shear loading in the principal planes. If X, Y and Z
are the yield stresses in the material directions, and Y_{xy}, Y_{yz} and Y_{zx} are the
yield stresses in pure shear in the $(x, y), (y, z)$, and (x, z) planes, then the
following relations can be obtained from (3.3.1a),

$$
\begin{aligned}
F &= \frac{1}{2} \left(\frac{1}{Y^2} + \frac{1}{Z^2} - \frac{1}{X^2} \right) \\
G &= \frac{1}{2} \left(\frac{1}{Z^2} + \frac{1}{X^2} - \frac{1}{Y^2} \right) \\
H &= \frac{1}{2} \left(\frac{1}{X^2} + \frac{1}{Y^2} - \frac{1}{Z^2} \right) \\
L &= \frac{1}{2Y_{xy}^2} \qquad I = \frac{1}{2Y_{yz}^2} \qquad K = \frac{1}{2Y_{zx}^2}
\end{aligned}
\tag{3.3.2}
$$

In the case of isotropy, the condition (3.3.1a) reduces to the von Mises yield
criterion (3.2.36). We note that the mean stress does not enter the yield
condition (3.3.1a) and therefore has no effect on yielding, as in the case of
isotropic plasticity. Hence, instead of the stresses $\sigma_{xx}, \sigma_{yy}, \sigma_{zz}$ we may use
the deviatoric stresses S_{xx}, S_{yy}, S_{zz}.

The Hill yield condition can be written using matrix notation as

$$f_y = \frac{1}{2}(\boldsymbol{\sigma}^T \tilde{\mathbf{N}} \boldsymbol{\sigma} - 1) = 0 \qquad (3.3.1b)$$

where the stress vector $\boldsymbol{\sigma}$ is defined in (A1.1), and the matrix $\tilde{\mathbf{N}}$ contains coefficients multiplying the stress components.

In case the yield stresses X, Y, Z differ significantly, some of the coefficients F, G, H in (3.3.1a) may be negative, and we must ask whether the yield surface is still convex. Generally, the yield surface (3.3.1b) is convex if the (3×3) coefficient submatrix of the (6×6) matrix $\tilde{\mathbf{N}}$ corresponding to the normal stresses σ_{xx}, σ_{yy} and σ_{zz} is positive semidefinite, Barlat et al. (1991). Other specific conditions that the coefficients F, G, H must satisfy are discussed in Hill (1990).

We transform the yield condition (3.3.1) into a form analogous to (3.2.36) by using the stress deviator $\mathbf{S}$ and multiplying (3.3.1) by $1/3\sigma_y^2$. Then we obtain (Kojic 1992)

$$f_y = \frac{1}{2}\mathbf{S}^T \mathbf{N} \mathbf{S}' - \frac{1}{3}\sigma_y^2 = 0 \qquad (3.3.3)$$

where $\mathbf{N}$ is a matrix of the shape coefficients, and σ_y is the yield stress, still to be defined. We use here matrix notation as in (3.3.1b) for simpler writing (see Appendix A2), with

$$\mathbf{S}^T = [S_1 = S_{xx}, \ S_2 = S_{yy}, \ S_3 = S_{zz}, \ S_4 = S_{xy}, \ S_5 = S_{yz}, \ S_6 = S_{zx}]$$

$$(3.3.4)$$

and $\mathbf{S}'$ is equal to $\mathbf{S}$ but contains twice the shear terms. The dimensionless shape coefficients N_{ij} can be related to the coefficients F, G, ..., K by using the deviatoric stress components S_i. Hence

$$\mathbf{N} = \begin{bmatrix} N_1 + N_2 & -N_1 & -N_2 & & & \\ & N_1 + N_3 & -N_3 & & & \\ & & N_2 + N_3 & & & \\ & & & N_{xy} & & \\ \text{symmetric} & & & & N_{yz} & \\ & & & & & N_{zx} \end{bmatrix} \qquad (3.3.5)$$

where

$$N_1 = \frac{2}{3}H\sigma_y^2, \ N_2 = \frac{2}{3}G\sigma_y^2, \ N_3 = \frac{2}{3}F\sigma_y^2, \ \dots, \ N_{zx} = \frac{2}{3}K\sigma_y^2 \qquad (3.3.6)$$

The yield condition (3.3.3) (as well as (3.3.1)) corresponds to the initial yielding. A schematic representation of the yield surface (3.3.3) shows an elliptical form due to the shape coefficients N_{ij}, with the center of the ellipse at the stress origin. We illustrate the application of the above relations in Example 3.3.1.

A form of the yield stress σ_y which takes into account the yield stresses X, Y, ..., Y_{zx} is

$$\sigma_y = \left\{ \frac{1}{2} \left[\frac{1}{3} \left(X^2 + Y^2 + Z^2 \right) + Y_{xy}^2 + Y_{yz}^2 + Y_{zx}^2 \right] \right\}^{1/2} \tag{3.3.7}$$

In the case of isotropic conditions $\tau_y = \sigma_y/\sqrt{3}$, where τ_y and σ_y are the yield stresses in simple shear and tension, and

$$X = Y = Z = \sqrt{3}Y_{xy} = \sqrt{3}Y_{yz} = \sqrt{3}Y_{zx} = \sigma_y \tag{3.3.8}$$

Also,

$$N_1 = N_2 = N_3 = \frac{1}{3}N_{xy} = \frac{1}{3}N_{yz} = \frac{1}{3}N_{zx} = \frac{1}{3} \tag{3.3.9}$$

and of course the yield condition (3.3.3) reduces to (3.2.36).

If we assume that the flow rule (3.2.71) is still valid, we can derive the stress-plastic strain relations. From (3.2.71), (3.3.3) and (3.2.69), and with (3.3.5), we obtain

$$de^P = d\lambda \mathbf{NS} \tag{3.3.10}$$

Note that the vector de^P contains tensorial components, in accordance with (3.2.67).

If we define the equivalent stress $\bar{\sigma}_a$ as

$$\bar{\sigma}_a = \left(\frac{3}{2}\mathbf{S}^T\mathbf{NS}' \right)^{1/2} \tag{3.3.11}$$

the yield condition (3.3.3) gives the equation

$$\bar{\sigma}_a - \sigma_y = 0 \tag{3.3.12}$$

In analogy with the assumption of the equivalence of plastic work (3.2.32) for an isotropic material, we now introduce the equivalent plastic strain $\bar{e}_a^P$ using

$$dW^P = S_{ij}\,de^P_{ij} = d\lambda\,\mathbf{S}^T\mathbf{NS'} = \frac{2}{3}d\lambda\,\bar\sigma^2_a = \bar\sigma_a d\bar e^P_a \tag{3.3.13}$$

where we have used (3.3.10), (3.3.11) and the scalar product definition (A.2.29). From this equation

$$d\lambda = \frac{3}{2}\frac{d\bar e^P_a}{\bar\sigma_a} \tag{3.3.14}$$

which corresponds to the expression (3.2.44) for an isotropic material. Hence, using (3.3.10) we obtain

$$\mathbf{S} = \frac{1}{d\lambda}\mathbf{N}^{-1}de^P \tag{3.3.15}$$

Substituting $\mathbf{S}$ into (3.3.13) and using (3.3.14), we find that the increment of equivalent plastic strain $d\bar e^P_a$ is

$$d\bar e^P_a = \left[\frac{2}{3}\left(de^P\right)^T\mathbf{N}^{-1}d\hat e^P\right]^{1/2} \tag{3.3.16}$$

where $d\hat e^P$ is equal to de^P but with twice the shear strain terms (i.e., with engineering shear strains, see also (A1.3)).

Since the plastic volumetric strains are zero, one of the normal plastic strains can be expressed in terms of the other two. Hence in the calculation of $d\bar e^P_a$ we use a (5×5) matrix $\mathbf{N}$ which can be inverted, because the original (6×6) matrix is singular. For an isotropic material, the equivalent stress $\bar\sigma_a$ and increment of equivalent plastic strain $d\bar e^P_a$ reduce to the expressions for $\bar\sigma$ and $d\bar e^P$ given by (3.2.39) and (3.2.46), respectively.

In this model, the yield condition (3.3.1), or (3.3.3), is used to determine the initial yielding. The stress-plastic strain relations (3.3.10) are used assuming a perfectly plastic material behavior, with the shape coefficients N_{ij} and the yield stress σ_y considered constant during plastic flow. Hence during plastic flow we have $\bar\sigma_a = const.$ while $\bar e^P_a$ is increasing. The model can be employed for the elastic-plastic analysis of sheet metals, as we will see in the next section. The hardening of the material, observed experimentally, can be included, with some simplifications: for example, the shape coefficients may be considered constant, and the change of the yield stress σ_y may be taken from the uniaxial yield curve for the rolling direction. A mixed hardening model was proposed in Kojić et al. (1996b).

3.3.2 Orthotropic Models for Sheet Metals

Many structural components, such as used in vehicle or airplane bodies, are produced from sheets of metals. A metal sheet is obtained by cold or hot rolling of an isotropic bulk material with the result that the material in the sheet is orthotropic. The material axes correspond to the rolling direction, the in-plane orthogonal direction, and the direction normal to the sheet. The mechanical behavior of sheet metals is quite complex, and significant experimental and theoretical research has been directed to understand and analyze sheet metal response during elastic-plastic deformations.

In order to introduce the specific properties of sheet metals, we first present some typical experimental data. Then we describe some material models.

Experimental Observations. Important material characteristics of sheet metals, subjected to plastic deformations, are the yield stresses and yield curves in specific directions. The *yield stresses* σ_0, σ_{45} and σ_{90} are usually used as material characteristics corresponding to, respectively, the direction of rolling, the 45-degree direction, and the direction orthogonal to rolling. In addition, important material characteristics are also the plastic strain ratios (or strain rate ratios) defined as follows

$$r_0 = \frac{e_{yy}^P}{e_{zz}^P} \qquad (\sigma_{xx} = \sigma_0, \ \sigma_{yy} = \sigma_{xy} = 0)$$

$$r_{45} = \frac{e_{\bar{y}\bar{y}}^P}{e_{zz}^P} \qquad (\sigma_{\bar{x}\bar{x}} = \sigma_{45}, \ \sigma_{\bar{y}\bar{y}} = \sigma_{\bar{x}\bar{y}} = 0) \qquad (3.3.17)$$

$$r_{90} = \frac{e_{xx}^P}{e_{zz}^P} \qquad (\sigma_{yy} = \sigma_{90}, \ \sigma_{xx} = \sigma_{xy} = 0)$$

where x is the axis in the rolling direction, y is the in-plane axis orthogonal to x, and z is the axis normal to the sheet; $\bar{x}, \bar{y}$ are in-plane axes rotated 45 degrees with respect to the x, y axes. These r-ratios are referred to as the *Lankford coefficients*. They are evaluated in experiments invoking moderate or large plastic strains (see, e.g., Lege et al. 1989, Barlat et al. 1991, Makinouchi et al. 1993). Since the elastic strains are small, these values are practically equal to the total strain ratios, and the plastic strain ratios are also called the strain ratios. Note that these coefficients give in each case the ratio of strain in the direction orthogonal to pulling to the thickness strain of the sheet. In practical applications the coefficients are used as material data for a yield function employed, and are considered constant during the material deformations. Hence the strain rate ratios are also considered equal to the strain ratios (Makinouchi et al. 1993, Lee et al. 1996).

Table 3.3.1 lists experimental data for several materials. The data for the steels are taken from Lee et al. (1996), and for the aluminum alloy are from Makinouchi et al. (1993). Figure 3.3.1 shows the uniaxial stress-strain curves

Table 3.3.1. Experimental data for sheet metals. Initial yield stresses and the strain ratios

Material	σ_0	σ_{45} N/mm^2	σ_{90}	r_0	r_{45}	r_{90}
Draw quality mild steel	152	159	163	1.85	1.52	2.37
High strength steel	364	380	402	0.72	1.21	1.03
Aluminum alloy	137	134.5	135	0.71	0.58	0.70

for these materials beyond the initial yield. Note that the strains in the uniaxial tests are large with the final values corresponding to rupture of the material (see also Fig. 3.2.2).

Table 3.3.1 and Fig. 3.3.1 show that the differences in the yield stresses in the selected directions (of the same material) are of the order of several

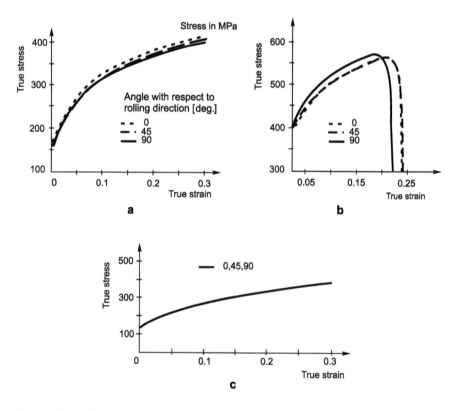

Fig. 3.3.1. Uniaxial stress-strain curves for sheet metals. **a** Draw quality mild steel; **b** High strength steel; **c** Aluminum alloy

percent (10% for the high strength steel) while the differences in the r-ratios are significantly larger. Also, Hill (1993) cites that for a brass, the yield stresses are $\sigma_0 = 126$ MPa, $\sigma_{90} = 125$ MPa, while $r_0 = 1.51$ and $r_{90} = 0.37$. The given experimental data illustrate the so-called "anomalous behavior" of sheet metal. A number of orthotropic-plastic material models have been proposed to represent these behaviors of which we describe three used in engineering practice.

The Basic Hill Model. The first step to obtain this model is to reduce Hill's yield condition (3.3.1) to plane stress conditions. Then the coefficients in the yield condition are expressed by use of the yield stresses and strain ratios specific for sheet metal.

In the case of plane stress in the x, y plane, we have $\sigma_{zz} = \sigma_{yz} = \sigma_{zx} = 0$ and the yield condition (3.3.1) becomes

$$2f_y = G\sigma_{xx}^2 + F\sigma_{yy}^2 + H\left(\sigma_{xx} - \sigma_{yy}\right)^2 + 2L\sigma_{xy}^2 - 1 = 0 \qquad (3.3.18)$$

or

$$2f_y = g\,\sigma_{xx}^2 + f\,\sigma_{yy}^2 + h\left(\sigma_{xx} - \sigma_{yy}\right)^2 + 2\ell\sigma_{xy}^2 - \sigma_0^2 = 0 \qquad (3.3.19)$$

where

$$g = G\sigma_0^2, \quad f = F\sigma_0^2, \quad h = H\sigma_0^2, \quad \ell = L\sigma_0^2 \qquad (3.3.20)$$

are dimensionless orthotropy material coefficients.

Let us consider two ways to determine the material constants $g, \dots, \ell$:

a) according to the yield stress ratios σ_0/σ_{45}, σ_0/σ_{90} and the strain ratio r_0;

b) according to the strain ratios r_0, r_{45} and r_{90}.

In each case we assume, arguably, that we can use the given r-ratios with the given initial yield stresses, see also our comment after the computed results have been presented.

In approach a) we first assume uniaxial loading to the stress σ_0. It follows from (3.3.19) that

$$g + h = 1 \qquad (3.3.21)$$

If the material is subjected to the uniaxial loading σ_{90}, we obtain

$$f + h = \left(\frac{\sigma_0}{\sigma_{90}}\right)^2 \qquad (3.3.22)$$

The stresses corresponding to uniaxial loading by the stress σ_{45} are

$$\sigma_{xx} = \sigma_{yy} = \sigma_{xy} = \frac{1}{2}\sigma_{45} \qquad (3.3.23)$$

and we obtain

$$f + g + 2\ell = 4 \left(\frac{\sigma_0}{\sigma_{45}} \right)^2 \tag{3.3.24}$$

In the system of equations (3.3.21), (3.3.22) and (3.3.24) we have four unknowns. An additional equation can be obtained by using one of the relations (3.3.17). If the definition of r_0 is employed, we obtain with the flow rule (3.2.67) and the yield condition (3.3.19):

$$\begin{aligned}
\dot{e}^P_{xx} &= \dot{\lambda}\,(g+h)\,\sigma_0 \\
\dot{e}^P_{yy} &= -\dot{\lambda}\,h\,\sigma_0 \\
\dot{e}^P_{zz} &= -\dot{e}^P_{xx} - \dot{e}^P_{yy} = -\dot{\lambda}\,g\,\sigma_0
\end{aligned} \tag{3.3.25}$$

where the last equation follows from the incompressibility condition. Hence the Lankford coefficient r_0 is

$$r_0 = \frac{\dot{e}^P_{yy}}{\dot{e}^P_{zz}} = \frac{h}{g} \tag{3.3.26}$$

The solutions for g, h, f and ℓ are

$$\begin{aligned}
g &= \frac{1}{1+r_0} \qquad h = \frac{r_0}{1+r_0} \\
f &= (\frac{\sigma_0}{\sigma_{90}})^2 - \frac{r_0}{1+r_0} \\
\ell &= \frac{1}{2} \left[4 \left(\frac{\sigma_0}{\sigma_{45}} \right)^2 - \left(\frac{\sigma_0}{\sigma_{90}} \right)^2 - \frac{1-r_0}{1+r_0} \right]
\end{aligned} \tag{3.3.27}$$

With these material constants we satisfy the yield condition under uniaxial loading in the material x and y directions, and in the 45 degree direction with respect to rolling. Also, we have the proper strain ratio r_0. However, the other strain ratios may deviate from experimental values. To illustrate this deviation we calculate r_{45} and r_{90} for the materials in Table 3.3.1. The results are shown in Table 3.3.2. It is seen that the differences between the calculated and experimental data are significant.

In the calculation of the material constants g, ..., ℓ using the approach b), we employ (3.3.21) and (3.3.26), and the expressions for r_{45} and r_{90} are

$$\begin{aligned}
r_{45} &= \frac{\ell}{f+g} - \frac{1}{2} \\
r_{90} &= \frac{h}{f}
\end{aligned} \tag{3.3.28}$$

Then, the solutions for g and h are as given in (3.3.27), while the coefficients f and ℓ are

Table 3.3.2. Calculated and experimental values of Lankford
coefficients r_{45} and r_{90}

Material	Calculated values by using material constants (3.3.27)		Experimental data	
	r_{45}	r_{90}	r_{45}	r_{90}
Draw quality mild steel	2.20	2.94	1.52	2.37
High strength steel	0.87	1.04	1.21	1.03
Aluminum alloy	0.73	0.68	0.58	0.70

$$f = \frac{r_0}{(1 + r_0)\, r_{90}}$$
$$\ell = \frac{(1 + 2r_{45})\,(r_0 + r_{90})}{2\,(1 + r_0)\, r_{90}} \qquad (3.3.29)$$

With the constants g and h given by (3.3.27), and f and ℓ given by
(3.3.29), we satisfy the yield condition in uniaxial loading σ_0. On the other
hand, the calculated yield stresses in the other directions may differ from the
experimental values. We give these differences in Table 3.3.3 for the three
materials considered. The maximum deviation from the experimental data is
smaller than in Table 3.3.2.

Table 3.3.3. Calculated and experimental values of yield stresses σ_{45}
and σ_{90} $\left[\mathrm{N/mm}^2\right]$

Material	Calculated values by using material constants g, h from (3.3.27), and f, ℓ from (3.3.29)		Experimental data	
	σ_{45}	σ_{90}	σ_{45}	σ_{90}
Draw quality mild steel	171.3	158.2	159.0	163.0
High strength steel	348.4	400.7	380.0	402.0
Aluminum alloy	142.0	136.4	134.5	135.0

In industrial applications it is important to accurately predict the strains.
Therefore, the determination of the orthotropic material coefficients $g, \ldots, \ell$
using the experimental values of the strain ratios is frequently preferable.

The yield stresses used above correspond to the initial yielding. However, the yield stresses change during plastic deformations (Fig. 3.3.1). A simplification for the description of hardening is adopted in practical applications, see, e.g., Makinouchi et al. (1993); Lee et al. (1996). Frequently, the yield curve in the rolling direction is taken as the representative hardening curve of the material for general loading conditions, and the Lankford coefficients are considered to be constant during plastic flow.

Let us note that the form (3.3.3) of Hill's yield condition can be used since it follows from (3.3.6) and (3.3.20) that

$$N_1 = \frac{2}{3}h, \quad N_2 = \frac{2}{3}g, \quad N_3 = \frac{2}{3}f, \quad N_{xy} = \frac{2}{3}\ell \tag{3.3.30}$$

We will use the form (3.3.3) of Hill's yield condition in the presentation of the numerical procedure for stress integration (Section 4.6.1).

Finally, we mention that the above described Hill's model for sheet metals is used in practice because the material constants can be easily determined (Li and Cescotto 1996). Comparing the predictions of the yield stresses and strain ratios obtained by Hill's model with experimental results, Fig. 3.3.2 shows that for the alloy considered, the deviation from the experimental results is approximately the same in character and size as obtained using another orthotropic model (Barlat's six-component model).

Hill's Sheet Metal Models. In order to achieve a better fit to experimental results, Hill proposed modified forms of the basic model described above. In Hill (1990) the proposed yield condition is

$$
\begin{aligned}
f_y &= |\sigma_{xx} + \sigma_{yy}|^m + (\sigma_b/\tau)^m \left|(\sigma_{xx} - \sigma_{yy})^2 + 4\sigma_{xy}^2\right|^{m/2} \\
&\quad + |\sigma_{xx}^2 + \sigma_{yy}^2 + 2\sigma_{xy}^2|^{m/2-1} \left[-2a\left(\sigma_{xx}^2 - \sigma_{yy}^2\right) + b\left(\sigma_{xx} - \sigma_{yy}\right)^2\right] \\
&\quad - (2\sigma_b)^m = 0
\end{aligned}
$$

$$\tag{3.3.31}$$

where a, b and m are material constants; and σ_b and τ are the in-plane equi-biaxial yield stress and shear yield stress, respectively. The conditions that the coefficients a and b must satisfy in order to ensure convexity for $m > 1$ are discussed in Hill (1990).

Another yield condition for sheet metals is proposed in Hill (1993). It is of cubic order with respect to the stress components σ_{xx} and σ_{yy}, and is given by the following expression:

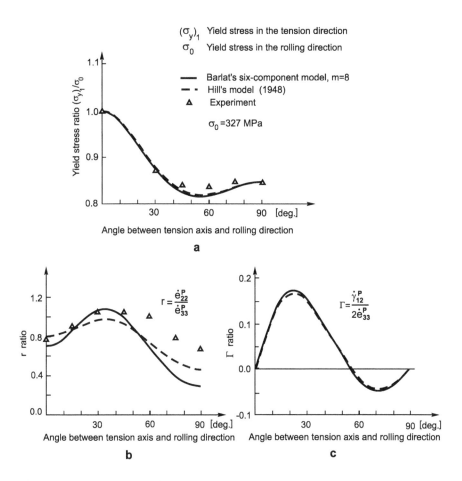

Fig. 3.3.2. Dependence of the material characteristics on the uniaxial loading direction for 2024-T3 alloy sheet. **a** The yield stress ratio $(\sigma_y)_1/\sigma_0$; **b** Normal strain rate ratio r; **c** Shear strain rate ratio Γ. Directions: (1) - tension direction, (2) - the in-plane direction orthogonal to (1), (3) - the direction normal to the sheet

$$f_y = \frac{\sigma_{xx}^2}{\sigma_0^2} - \frac{c\,\sigma_{xx}\sigma_{yy}}{\sigma_0\sigma_{90}} + \frac{\sigma_{yy}^2}{\sigma_{90}^2} + \left(p+q - \frac{p\,|\sigma_{xx}| + q\,|\sigma_{yy}|}{\sigma_b}\right)\frac{\sigma_{xx}\sigma_{yy}}{\sigma_0\sigma_{90}} - 1 = 0$$

$$(3.3.32)$$

where c, p and q are material parameters defined in terms of the yield stresses σ_0, σ_{90} and σ_b, and in terms of the strain ratios r_0 and r_{90}. The procedure to determine c, p and q in terms of σ_0, σ_{90}, σ_b, r_0 and r_{90} is based on the definitions of $\sigma_0, ..., r_{90}$ and the yield condition (3.3.32), as for the basic Hill model. The resulting expressions are:

$$c = \sigma_0 \sigma_{90} \left(\frac{1}{\sigma_0^2} + \frac{1}{\sigma_{90}^2} - \frac{1}{\sigma_b^2} \right)$$

$$p = \frac{1}{a} \left[\frac{2r_0 \left(\sigma_b - \sigma_{90} \right)}{\left(1 + r_0\right) \sigma_0^2} - \frac{2r_{90}\sigma_b}{\left(1 + r_{90}\right) \sigma_{90}^2} + \frac{c}{\sigma_0} \right] \qquad (3.3.33)$$

$$q = \frac{1}{a} \left[\frac{2r_{90} \left(\sigma_b - \sigma_0 \right)}{\left(1 + r_{90}\right) \sigma_{90}^2} - \frac{2r_0\sigma_b}{\left(1 + r_0\right) \sigma_0^2} + \frac{c}{\sigma_{90}} \right]$$

where

$$a = \frac{1}{\sigma_0} + \frac{1}{\sigma_{90}} - \frac{1}{\sigma_b}$$

The yield condition (3.3.32) is not suitable for general applications because it does not contain shear terms.

Barlat's Models. We cite here two Barlat orthotropic models that have been used in modeling sheet metal plastic deformations (Chung and Shah 1992; Nakamachi 1993). The *six-component model*, Barlat et al. (1991), is applicable to general 3-D deformations and also to plane stress (shell) conditions. The yield condition is

$$f_y = |S_1 - S_2|^m + |S_2 - S_3|^m + |S_3 - S_1|^m - 2\sigma_y^m = 0 \qquad (3.3.34)$$

where S_1, S_2 and S_3 are the principal values of a symmetric stress matrix $\tilde{\mathbf{S}}$, with the terms:

$$\tilde{S}_{xx} = \frac{1}{3} \left[c \left(\sigma_{xx} - \sigma_{yy} \right) - b \left(\sigma_{zz} - \sigma_{xx} \right) \right]$$

$$\tilde{S}_{yy} = \frac{1}{3} \left[a \left(\sigma_{yy} - \sigma_{zz} \right) - c \left(\sigma_{xx} - \sigma_{yy} \right) \right]$$

$$\tilde{S}_{zz} = \frac{1}{3} \left[b \left(\sigma_{zz} - \sigma_{xx} \right) - a \left(\sigma_{yy} - \sigma_{zz} \right) \right] \qquad (3.3.35)$$

$$\tilde{S}_{yz} = f \sigma_{yz}, \quad \tilde{S}_{zx} = g \sigma_{zx}, \quad \tilde{S}_{xy} = h \sigma_{xy}$$

and σ_y is the yield stress corresponding to the uniaxial stress in the rolling direction (taken at 0.2% plastic strain, Barlat et al. 1991). The exponent m and the material orthotropy coefficients a, b, c, f, g and h are determined by comparing the results obtained using the yield condition (3.3.34) and experimental results. Uniaxial loading of the material in several directions with respect to the rolling direction is considered. When the orthotropy coefficients are unity, the material is isotropic, and if $m = 1$, the yield condition reduces to the Tresca criterion, while $m = 2$ gives the von Mises yield condition. As the severity of the crystallographic texture increases, the exponent

m should be increased for better prediction of the material behavior. Table 3.3.4 gives values of material constants for 2024-T3 and 2008-T4 aluminum alloy sheets, while Figs. 3.3.2 and 3.3.3 show the variations of the yield stress ratio $(\sigma_y)_1/\sigma_0$ and the strain rate ratio $\dot{e}^P_{22}/\dot{e}^P_{33}$ in terms of the loading direction for 2024-T3 and 2008-T4 materials (Barlat et al. 1991). Here $(\sigma_y)_1$ is the yield stress in the loading direction (direction (1)), and $\dot{e}^P_{22}$ and $\dot{e}^P_{33}$ are the plastic strain rates in the in-plane direction orthogonal to (1) and in the direction normal to the sheet. We note that the calculated strain rate ratios correspond to the yield stresses at 0.2% plastic strain, while the experimental ratios were determined in the range of the total strain (in the pulling direction) from 0.01 to around 0.20 (Lege et al. 1989, Barlat et al. 1991).

Table 3.3.4. Material constants for 2024-T3 and 2008-T4 aluminum alloy sheets (Barlat's six component model)

Material	m	a	b	c	f	g	h
2024-T3	8	1.378	1.044	0.955	1.000	1.000	1.210
2008-T4	11	1.222	1.013	0.985	1.000	1.000	1.000

The hardening law of the material is taken to be defined by the uniaxial yield curve for the rolling direction. Figure 3.3.4 shows the yield surfaces for a 2008-T4 alloy sheet for various stress ratios σ_{xy}/σ_0. It can be seen that the yield surfaces in the chosen plane are convex.

Another Barlat's model, known as the *tri-component model*, was proposed in Barlat and Lian (1989). The yield condition is based on experimental investigations of sheet metals in plane stress conditions, and has the following form

$$f_y = a\,|K_1 + K_2|^m + a\,|K_1 - K_2|^m + (2-a)\,|2K_2|^m - 2\sigma_y^m = 0$$

$$(3.3.36)$$

where

$$K_1 = \frac{\sigma_{xx} + h\sigma_{yy}}{2}$$

$$K_2 = \left[\left(\frac{\sigma_{xx} - h\sigma_{yy}}{2}\right)^2 + p^2\sigma_{xy}^2\right]^{1/2} \qquad (3.3.37)$$

and a, m, h and p are material coefficients. As for the six-component model, if $a = h = p = 1$, the model corresponds to isotropic conditions. The yield stress σ_y is usually taken as the yield stress σ_0 in the rolling direction. The exponent m has the same character as for the six-component model, while

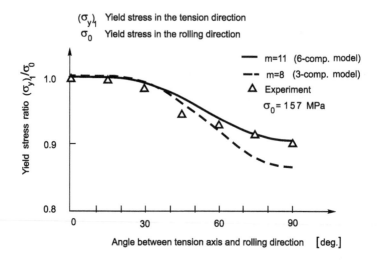

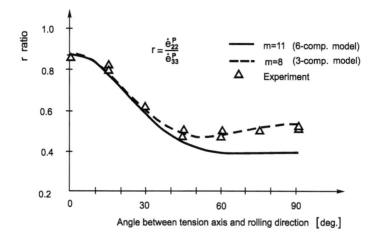

Fig. 3.3.3. Dependence of yield stress ratio $(\sigma_y)_1/\sigma_0$ and of strain rate ratio r on uniaxial loading direction, for 2008-T4 alloy sheet. Directions: (1) - tension direction, (2) - the in-plane direction orthogonal to (1), (3) - the normal direction to the sheet

the orthotropy coefficients a, h and p can be determined from the yield stresses or from the strain ratios. If yield stresses are employed, the orthotropy coefficients are as follows

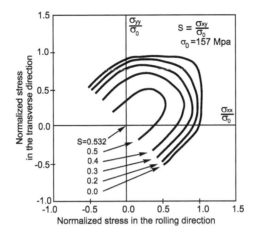

Fig. 3.3.4. Yield surfaces for 2008-T4 alloy sheet, in σ_{xx}/σ_0, σ_{yy}/σ_0 plane, for various shear stress ratios $S = \sigma_{xy}/\sigma_0$ (Barlat's six-component model)

$$a = \frac{2\left(\sigma_y/\tau_2\right)^m - 2\left(1 + \sigma_y/\sigma_{90}\right)^m}{1 + \left(\sigma_y/\sigma_{90}\right)^m - \left(1 + \sigma_y/\sigma_{90}\right)^m}$$

$$h = \frac{\sigma_y}{\sigma_{90}} \tag{3.3.38}$$

$$p = \frac{\sigma_y}{\tau_1}\left(\frac{2}{2a + 2^m c}\right)^{1/m}$$

where $c = 2 - a$; and τ_1 is the yield stress in pure shear, while τ_2 is the yield stress under shear in the loading condition $\sigma_{yy} = -\sigma_{xx} = \tau_2$, $\sigma_{xy} = 0$. When the strain ratios are used, the orthotropy coefficients a and h can be expressed in terms of the strain ratios r_0 and r_{90}

$$a = 2 - 2\left[\frac{r_0 r_{90}}{\left(1 + r_0\right)\left(1 + r_{90}\right)}\right]^{1/2}$$

$$h = \left[\frac{r_0\left(1 + r_{90}\right)}{\left(1 + r_0\right) r_{90}}\right]^{1/2} \tag{3.3.39}$$

but the coefficient p is not expressed analytically. With a and h calculated, p can be determined from a relation of strain ratios corresponding to an angle with respect to the rolling direction (e.g., 45°). Values of material coefficients for a 2008-T4 alloy sheet are given in Table 3.3.5.

Table 3.3.5. Material constants for Barlat's tri-component model

Material	m	a	h	p
2008-T4 alloy sheet	8	1.240	1.150	1.020

Figure 3.3.3 shows the prediction of a yield stress and a strain rate ratio for various directions of uniaxial loading, using the tri-component model. The orthotropy coefficients are determined by use of the strain rate ratios, so that good agreement between the calculated and experimental values of the strain rate ratios is obtained. The shape of the yield surface obtained by using the tri-component model is similar to that shown in Fig. 3.3.4 for the six-component model.

Of course, the tri-component model is simpler than the six-component model and it contains less material constants. Therefore if the response of the sheet metal is predicted to a sufficient accuracy, the tri-component model is more suitable for applications.

3.3.3 Examples

Example 3.3.1. Bending of Orthotropic Plate. An orthotropic plate is bent by a uniformly distributed force F, Fig. E.3.3-1. Determine the force which causes:

a) plastic deformations at the top and bottom surfaces;
b) yielding at the mid-plane, assuming a parabolic shear stress distribution.

Use Hill's orthotropic model, with the yield condition (3.3.3) and the shape coefficients (3.3.6). The material axes are a, b, y, and the yield stresses are:

$$X = 50 \text{ MPa} \quad Y = Z = k_y X \qquad k_y = 0.9$$
$$Y_{ab} = 30 \text{ MPa} \quad Y_{ay} = Y_{by} = k_{ab} Y_{ab} \qquad k_{ab} = 0.8$$

a) The maximum stress σ_{xx} is

$$\sigma_{xx} = \frac{FL}{I_{zz}} \frac{h}{2} = \frac{6L}{\ell h^2} F$$

and with the given data

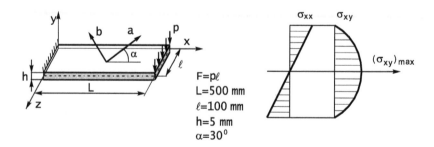

F=pℓ
L=500 mm
ℓ=100 mm
h=5 mm
α=30°

Fig. E.3.3-1. Bending of orthotropic plate. Geometric and loading data, and stress distribution

$$\sigma_{xx} = 1.2\,F \tag{a}$$

where I_{zz} is the cross-sectional moment of inertia for bending about the z axis. By applying the transformation (A1.49) we obtain the stresses σ_{aa}, σ_{bb} and σ_{ab} in the material directions

$$\sigma_{aa} = \cos^2\alpha\,\sigma_{xx} \quad \sigma_{bb} = \sin^2\alpha\,\sigma_{xx} \quad \sigma_{ab} = -\sin\alpha\cos\alpha\,\sigma_{xx} \tag{b}$$

The yield condition (3.3.3) is

$$N_{ij}\,S_iS_j + 2N_{ab}\sigma_{ab}^2 - \frac{2}{3}\sigma_y^2 = 0 \qquad \text{sum on } i,j = 1,2,3 \tag{c}$$

$$\text{no sum on a,b}$$

where i,j correspond to the a,b and y - directions, and according to (3.3.7) the yield condition is

$$\sigma_y = \left[\frac{1}{3}\left(1 + 2k_y^2\right)X^2 + \left(1 + 2k_{ab}^2\right)Y_{ab}^2\right]^{1/2}/\sqrt{2} = 46.018 \tag{d}$$

The coefficients N_{ij} can be expressed in terms of N_1, N_2 and N_3 (see (3.3.5)) defined by (3.3.6). In this example the coefficients N_1, N_2, N_3 and N_{ab} are :

$$N_1 = N_2 = \frac{1}{3}\left(\sigma_y/X\right)^2 = 0.2824$$

$$N_3 = \frac{1}{3}\left(2/k_y^2 - 1\right)\left(\sigma_y/X\right)^2 = 0.4148 \tag{e}$$

$$N_{ab} = \frac{1}{3}\left(\sigma_y/Y_{ab}\right)^2 = 0.7843$$

Then, from (b) to (e) we obtain

$$\sigma_{xx} = 50.689 \quad \text{MPa}$$

and from (a) the force F causing yielding is

$$F = 42.241 \quad \text{N}$$

b) The stress state in the midplane is pure shear, with the stress

$$\sigma_{xy} = (\sigma_{xy})_{\max} = \frac{3}{2}\frac{F}{\ell h} = 0.003F \tag{f}$$

Comparing this relation with (a), it is obvious that the yielding due to bending will be reached under a much smaller force F. However, let us proceed with the analysis below merely for demonstration purposes.

The stresses σ_{ay} and σ_{by} are

$$\sigma_{ay} = \cos \alpha \, \sigma_{xy} \qquad \sigma_{by} = -\sin \alpha \, \sigma_{xy} \tag{g}$$

The yield condition (3.3.3) is now

$$3(N_{ay} \cos^2 \alpha + N_{by} \sin^2 \alpha) \, \sigma_{xy}^2 - \sigma_y^2 = 0 \tag{h}$$

The coefficients N_{ay} and N_{by} are

$$N_{ay} = N_{by} = \frac{1}{3k_{ab}^2} (\sigma_y/Y_{ab})^2 \tag{i}$$

and then it follows from (h)

$$\sigma_{xy} = k_{ab} Y_{ab} = 24.0 \text{ MPa} \tag{j}$$

From (j) and (f) we obtain F as

$$F = k_{ab} Y_{ab}/0.003 = 8 \times 10^3 \text{ N} \tag{k}$$

Example 3.3.2. In-Plane Straining of Orthotropic Plate. The quadrilateral orthotropic plate shown in Fig. E.3.3-2 is strained in-plane so that the uniform displacements of the free edges are

$$U_x = 0.8 \text{ mm} \quad U_y = -0.4 \text{ mm}$$

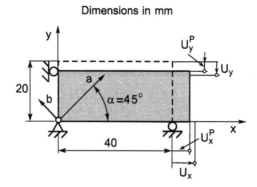

Fig. E.3.3-2. In-plane elastic-plastic deformation of orthotropic plate. Initial configuration and deformed configuration after the load release

The angle between the material axis a and the x-axis is $\alpha = 45^0$. Assuming a perfectly plastic material behavior and Hill's model, with the plasticity material constants as in Example 3.3.1, and the elastic shear modulus

$$G_{ab} = 10^5 \text{ MPa}$$

determine the plate deformation after the loading has been released.

The in-plane strains corresponding to the given displacements are

$$e_{xx} = U_x/40 = 0.02 \quad e_{yy} = U_y/20 = -0.02 \quad \gamma_{xy} = 0 \qquad \text{(a)}$$

By using the transformation of the strain components given in (A1.50) we obtain the in-plane strains for the material axes as

$$e_{aa} = e_{bb} = 0 \quad \gamma_{ab} = -0.04 \qquad \text{(b)}$$

Therefore we have pure shear of the material in the material coordinate system. Assuming elastic deformations, we obtain from the elastic constitutive law (A1.19) that (all stresses below are in MPa)

$$\sigma_{ab}^E = G_{ab}\,\gamma_{ab} = -4 \times 10^3 \qquad \text{(c)}$$

and the equivalent stress defined in (3.3.11) is

$$\bar{\sigma}_a^E = (3N_{ab})^{1/2}\,\left|\sigma_{ab}^E\right| = 6.136 \times 10^3 > \sigma_y \qquad \text{(d)}$$

where we have used the values for N_{ab} and σ_y from (e) and (d) of Example 3.3.1. Hence the material is in fact deformed plastically and we proceed to plasticity calculations.

The stresses in the material must be such that the yield condition (3.3.3) is satisfied during the plastic flow. The only non-zero stress is σ_{ab}, and we obtain from (3.3.3) and the data of Example 3.3.1 that

$$\sigma_{ab} = -Y_{ab} = -30 \qquad \text{(e)}$$

The elastic strain corresponding to this stress is

$$\gamma_{ab}^E = \sigma_{ab}/G_{ab} = -3 \times 10^{-4} \qquad \text{(f)}$$

and the plastic strain is

$$\gamma_{ab}^P = \gamma_{ab} - \gamma_{ab}^E = -0.0397 \qquad \text{(g)}$$

After the load is released the plastic strains remain in the material, in this case γ_{ab}^P. The plastic strains corresponding to the x, y coordinate system are

$$e_{xx}^P = -e_{yy}^P = 0.01985 \quad \gamma_{xy}^P = 0 \tag{h}$$

and the residual displacements of the plate edges are

$$U_x^P = -2U_y^P = 0.794 \text{ mm} \tag{i}$$

4. A General Procedure for Stress Integration and Applications in Metal Plasticity

We emphasized in Chapter 2 that a central point in any inelastic finite element analysis is the calculation of the stresses in an effective manner. The stress calculation predicts the material response described by the material model. In this chapter we present effective numerical procedures for the stress calculation in metal plasticity.

We first discuss in Section 4.1 the role of stress calculation in a finite element incremental-iterative analysis. Then, in Section 4.2 we establish a general approach for the implicit stress computation within the time (load) step, and apply the approach to a class of time independent plasticity models. We call this approach the *governing parameter method* because the stress integration problem is reduced to solving one nonlinear equation for a governing parameter. The governing parameter method is applicable to models for which the mean stress does not affect the inelastic behavior (isochoric inelastic deformation), as in metal plasticity, viscoplasticity and/or creep models, as well as to models with volumetric inelastic deformation, as in geological plasticity models. In this and in the subsequent chapters the algorithm will be presented for a number of commonly used material models.

In Section 4.3 we give a brief review of stress integration algorithms as they relate to the governing parameter method. A detailed derivation of the relations used in the governing parameter method for the stress integration of the von Mises model is presented in Section 4.4, and Section 4.5 contains a number of solved examples. Finally, in Section 4.6 we derive the computational procedure for Hill's othotropic model and give example solutions.

4.1 Introduction

In any structural analysis we need to address how to represent and model the deformations and the material behavior. For the representation of the deformations an adequate displacement field must be assumed, and the corresponding kinematic quantities, such as the strains, strain rates, deformation gradient, etc. need be calculated. This was briefly discussed in Chapter 2, while a deeper presentation, using the same notation, is given in Bathe (1996).

The material behavior is described by material mathematical models and we consider here phenomenological models based on experimental observations. In order to use a material model in an engineering analysis, it is necessary to have numerical procedures that provide an accurate calculation of the model response, according to the selected model parameters and the loading conditions. In this book we present numerical algorithms for the stress integration for *strain-driven problem formulations*; these problems arise in the displacement-based and mixed finite element formulations. We consider inelastic material deformations, and our task is to calculate the stresses and inelastic strains at the end of the time (load) step, with a known stress/strain state at the start of the time step and for given strain increments. By this calculation (stress integration) we trace the history of the material deformation at a material point, and in the whole structure, under given incremental loading conditions.

The typical calculations performed in an incremental (finite element) analysis, with nonlinear material behavior, are summarized in Table 4.1.1 (see also Bathe 1996). We consider materially-nonlinear-only (*MNO*) problems (see Chapter 2).

The variables appearing in the table are the total and inelastic strains $\mathbf{e}$ and $\mathbf{e}^{IN}$, the stresses $\boldsymbol{\sigma}$, the internal variables of the material model $\boldsymbol{\beta}$, the linear strain-displacement transformation matrix $\mathbf{B}_L$ (see Section 2.3), the integration weight due to numerical integration W, the numerical integration point associated volume ΔV, and the vector of external loads $\mathbf{R}$. As in Chapter 2, the left superscripts "t" and "$t + \Delta t$" denote the start and end of time step Δt, and the right superscript denotes the iteration number. Note that for the iteration counter $i = 1$, the quantities with the upper right index equal to zero have the values corresponding to the start of the time step, e.g., $^{t+\Delta t}\mathbf{e}^{(0)} = {}^t\mathbf{e}$, $^{t+\Delta t}\boldsymbol{\sigma}^{(0)} = {}^t\boldsymbol{\sigma}$.

Referring to the table, we see that the stress integration and calculation of the tangent constitutive relations represent key steps in an inelastic incremental analysis. These calculations must be performed in a *robust, accurate, and efficient* manner. We define an algorithm as robust if it provides solutions for stresses under any reasonable boundary and loading conditions, and for relatively large strain (or load) increments. It is very important in general engineering analysis that the algorithm does not have limitations in its range of applicability, and that it does not contain any numerical instability. Of course, the algorithm should in addition yield accurate and efficient solutions. As will be seen, our primary goal is to meet these requirements.

The solution of nonlinear structural problems is obtained by incrementing the loads (or strains) in each time step Δt. Of course, in general, there is some numerical error in the stress calculation as a consequence of the numerical approximations used in the stress evaluation. The algorithm should provide reasonable accuracy for relatively large increments in strains, and the error should rapidly diminish as the strain increments are decreased. In Section

Table 4.1.1. Equilibrium iterations showing the evaluation of the stiffness matrix and force vector by looping over the integration points (materially-nonlinear-only (*MNO*) problems)

A. Initial conditions

$$^{t+\Delta t}\mathbf{U}^{(0)} = {}^{t}\mathbf{U}; \ {}^{t+\Delta t}\mathbf{R}; \ i = 0$$

B. Equilibrium iteration loop for the complete finite element assemblage

$$i = i + 1 \qquad {}^{t+\Delta t}\mathbf{F}^{(i-1)} = \mathbf{0} \qquad {}^{t+\Delta t}\mathbf{K}_L^{(i-1)} = \mathbf{0}$$

C. Loop over all integration points of all finite elements

For each integration point ($^{t+\Delta t}\hat{\mathbf{e}}^{(i-1)}$ is the strain vector (A1.3)):

$$^{t+\Delta t}\hat{\mathbf{e}}^{(i-1)} = \mathbf{B}_L \ {}^{t+\Delta t}\mathbf{U}^{(i-1)}$$

- Stress integration

  known: $^{t}\boldsymbol{\sigma}, \ {}^{t}\mathbf{e}, \ {}^{t}\mathbf{e}^{IN}, \ {}^{t}\boldsymbol{\beta}, \ {}^{t+\Delta t}\mathbf{e}^{(i-1)}$

  evaluate: $^{t+\Delta t}\boldsymbol{\sigma}^{(i-1)}, \ {}^{t+\Delta t}\mathbf{e}^{IN(i-1)}, \ {}^{t+\Delta t}\boldsymbol{\beta}^{(i-1)}$

- Constitutive matrix

$$^{t+\Delta t}\mathbf{C}^{(i-1)} = \frac{\partial\,{}^{t+\Delta t}\boldsymbol{\sigma}^{(i-1)}}{\partial\,{}^{t+\Delta t}\hat{\mathbf{e}}^{(i-1)}}$$

- Accumulate nodal force effects

$$^{t+\Delta t}\mathbf{F}^{(i-1)} \Longleftarrow {}^{t+\Delta t}\mathbf{F}^{(i-1)} + \mathbf{B}_L \ {}^{t+\Delta t}\boldsymbol{\sigma}^{(i-1)} W \Delta V$$

- Accumulate stiffness matrix effects

$$^{t+\Delta t}\mathbf{K}_L^{(i-1)} \Longleftarrow {}^{t+\Delta t}\mathbf{K}_L^{(i-1)} +$$
$$\mathbf{B}_L^T \ {}^{t+\Delta t}\mathbf{C}^{(i-1)} \mathbf{B}_L W \Delta V$$

END OF LOOP C

Calculate increment of displacements and total current displacements

$$^{t+\Delta t}\mathbf{K}_L^{(i-1)} \Delta\mathbf{U}^{(i)} = {}^{t+\Delta t}\mathbf{R} - {}^{t+\Delta t}\mathbf{F}^{(i-1)}$$

$$^{t+\Delta t}\mathbf{U}^{(i)} = {}^{t+\Delta t}\mathbf{U}^{(i-1)} + \Delta\mathbf{U}^{(i)}$$

Convergence check: If convergence criteria are satisfied
                            perform next equilibrium iteration

END OF LOOP B

D. Increment time (load) step (go to A)

4.4.4 we give some details regarding the accuracy of the computational algorithms.

The numerical procedures for stress integration should be efficient, because these calculations are repeated at every integration point employed within the body or structure. Today's demands in engineering analysis are such that in finite element analysis millions of integration stations for stress calculation are used, and the largest part of the computational time in an inelastic analysis can correspond to the stress integration.

4.2 The Governing Parameter Method

In this section we introduce a general concept of implicit stress integration, which we call the "governing parameter method". The procedure is a generalization of the radial return method introduced for plasticity calculations by Wilkins (1964). The essence of the governing parameter method lies in that the calculation of the unknown stresses and internal material variables is reduced to the solution of a single parameter. This concept was already used in the *effective-stress-function algorithm* for thermo-elasto-plastic and creep solutions, Bathe et al. (1984); Kojic and Bathe (1987a,b); Bathe (1996). However, it can be appropriate to use another variable than the effective stress (such as the effective plastic strain, mean plastic strain, or cap position for a cap plasticity material model) and therefore we generalize the effective-stress-function method to the governing parameter method (Kojic 1996a). The procedure is applicable to all material models discussed in this book.

4.2.1 Formulation of the Governing Parameter Method

We start with the assumption that at a material point the stress/strain state at time "t" is known. An incremental analysis of the body deformation is performed, with time (load) step Δt, and we suppose that the total mechanical strains at the end of the load step are known. Therefore, the known quantities at a material point are assumed to be:

$$\boxed{{}^t\boldsymbol{\sigma},\ {}^t\mathbf{e},\ {}^t\mathbf{e}^{IN},\ {}^t\boldsymbol{\beta},\ {}^{t+\Delta t}\mathbf{e}} \tag{4.2.1a}$$

or in indicial notation

$$\boxed{{}^t\sigma_{ij},\ {}^t e_{ij},\ {}^t e_{ij}^{IN},\ {}^t\beta_{ij},\ {}^{t+\Delta t}e_{ij}} \tag{4.2.1b}$$

where ${}^t\boldsymbol{\sigma}$, ${}^t\mathbf{e}$ and ${}^t\mathbf{e}^{IN}$ are the stresses, total mechanical strains and inelastic strains at time t; ${}^t\boldsymbol{\beta}$ are internal variables at time t used to describe the history of inelastic deformation, and $^{t+\Delta t}\mathbf{e}$ are the total mechanical strains at the end of the load step[1]. The number of internal variables depends on the material model and on the type of inelastic deformation; for example, for a general 3-D elastic-plastic deformation of an initially isotropic metal with mixed hardening (described in Section 3.2.5), the internal variables are the accumulated effective plastic strain and the components of the position of the yield surface tensor. We do not include here thermal effects for ease of explanation, but these effects are included in Chapter 5.

The task of the *stress integration* is to determine the stresses $^{t+\Delta t}\boldsymbol{\sigma}$, inelastic strains $^{t+\Delta t}\mathbf{e}^{IN}$ and internal variables $^{t+\Delta t}\boldsymbol{\beta}$ at time $t + \Delta t$;[2] that is, at the end of the time step. Hence, the unknowns are

$$^{t+\Delta t}\boldsymbol{\sigma}, \quad ^{t+\Delta t}\mathbf{e}^{IN}, \quad ^{t+\Delta t}\boldsymbol{\beta} \tag{4.2.2}$$

We employ an implicit integration procedure corresponding to the Euler backward method. The basic steps in our concept of the implicit stress integration are as follows:

1) Express all unknown variables in terms of one parameter p. We assume that this step is possible for the material model considered.
2) Form a function of p whose zero provides the solution for the governing parameter, that is $^{t+\Delta t}p$.
3) Calculate the unknown variables using $^{t+\Delta t}p$.

The computational steps are summarized in Table 4.2.1.

In addition to the stress calculation, it is necessary (see Table 4.1.1) to determine the tangent constitutive relation $^{t+\Delta t}\mathbf{C}$ at the end of the time step, *consistent with the stress integration algorithm*. We call $^{t+\Delta t}\mathbf{C}$ the consistent tangent constitutive matrix, for which

$$^{t+\Delta t}C_{ijrs} = \frac{\partial \, ^{t+\Delta t}\sigma_{ij}}{\partial \, ^{t+\Delta t}e_{rs}} \tag{4.2.3}$$

With these tensor components we can construct the tangent stress-strain matrix $^{t+\Delta t}\mathbf{C}$ in Table 4.1.1. Using that the stress $^{t+\Delta t}\boldsymbol{\sigma}$ is then only a function of $^{t+\Delta t}\mathbf{e}$ and $^{t+\Delta t}p$, with the governing parameter also a function

---

[1] Note that we use here tensor components. We shall also sometimes refer to these quantitites simply as tensors although we shall mean their components; e.g., we shall sometimes refer to $^{t}\mathbf{e}$ as the strain tensor when we strictly mean its components in a chosen basis.

[2] Here we do not use, for simplicity, the iteration counter of Table 4.1.1, but really imply that the state $t + \Delta t$ corresponds to the state at iteration i.

Table 4.2.1. Computational steps in an implicit algorithm for stress integration − the governing parameter method

---

Known quantities :

$$^t\boldsymbol{\sigma}, \ ^t\mathbf{e}, \ ^t\mathbf{e}^{IN}, \ ^t\boldsymbol{\beta}, \ ^{t+\Delta t}\mathbf{e}$$

Unknown quantities :

$$^{t+\Delta t}\boldsymbol{\sigma}, \ ^{t+\Delta t}\mathbf{e}^{IN}, \ ^{t+\Delta t}\boldsymbol{\beta}$$

Step 1. Express all unknowns in terms of one unknown parameter p and known quantities

$$^{t+\Delta t}\boldsymbol{\sigma}\,(^t\boldsymbol{\sigma}, \ ^t\mathbf{e}, \ ^t\mathbf{e}^{IN}, \ ^t\boldsymbol{\beta}, \ ^{t+\Delta t}\mathbf{e}, \ p)$$

$$^{t+\Delta t}\mathbf{e}^{IN}\,(^t\boldsymbol{\sigma}, ..., p) \qquad \text{(a)}$$

$$^{t+\Delta t}\boldsymbol{\beta}\,(^t\boldsymbol{\sigma}, ..., p)$$

Step 2. Form a function $f(p)$ and solve the governing equation

$$f(p) = 0 \qquad \text{(b)}$$

Step 3. Substitute the solution $^{t+\Delta t}p$ of the governing equation in (b) to determine the unknowns in (a)

---

of the total strains, we derive the components in (4.2.3) using the chain rule for calculation of the derivatives as follows

$$^{t+\Delta t}C_{ijrs} = \frac{\partial\,^{t+\Delta t}\sigma_{ij}}{\partial\,^{t+\Delta t}e_{rs}}\bigg|_{p=const} + \frac{\partial\,^{t+\Delta t}\sigma_{ij}}{\partial\,^{t+\Delta t}p}\frac{\partial\,^{t+\Delta t}p}{\partial\,^{t+\Delta t}e_{rs}} \qquad (4.2.4)$$

where, obviously, the first term on the right hand side assumes differentiations under the condition $p = const$.

The derivatives of the governing parameter $^{t+\Delta t}p$ with respect to the strains are obtained by recognizing that the governing equation $f(p) = 0$ must be satisfied throughout the material response. Therefore $df = 0$ and hence

$$
^{t+\Delta t}\!\left(\frac{\partial f}{\partial \sigma_{rs}} \frac{\partial \sigma_{rs}}{\partial p} + \frac{\partial f}{\partial e^{IN}_{rs}} \frac{\partial e^{IN}_{rs}}{\partial p} + \frac{\partial f}{\partial \beta_{rs}} \frac{\partial \beta_{rs}}{\partial p} + \frac{\partial f}{\partial p} \right) \frac{\partial\, ^{t+\Delta t}p}{\partial\, ^{t+\Delta t}e_{ij}}
$$

$$
+ \left. \frac{\partial^{t+\Delta t} f}{\partial^{t+\Delta t} e_{ij}} \right|_{p=const} = 0
$$

$$(4.2.5)$$

From this equation we determine the derivatives $\partial\, ^{t+\Delta t}p/\partial\, ^{t+\Delta t}e_{ij}$.

We may note that the governing parameter function f is frequently equal to the yield function (see, e.g., von Mises plasticity), but can also be another function (e.g., (5.4.11), Examples 4.5.11 and 6.2.1). We exemplify the solution approach using simple uniaxial loading conditions in Example 4.5.1.

4.2.2 Time Independent Plasticity Models

Let us consider the class of time independent plasticity models for which the yield criterion can be expressed in terms of the deviatoric stresses $^{t+\Delta t}\mathbf{S}$, the mean stress $^{t+\Delta t}\sigma_m$, and internal variables $^{t+\Delta t}\beta$ and assume that the associated flow rule (3.2.71) is used.

We recall that, according to the α-method, the integral of any function $f(t)$ in an interval Δt can be approximated as

$$
\int_t^{t+\Delta t} f(t)dt = \left[(1-\alpha)\, ^t f + \alpha\, ^{t+\Delta t} f \right] \Delta t
$$

$$(4.2.6)$$

where

$$
0 \le \alpha \le 1
$$

$$(4.2.7)$$

is the integration parameter. The values $\alpha = 0$ and $\alpha = 1$ correspond to the Euler forward and Euler backward integration methods, respectively, while $\alpha = 0.5$ gives the trapezoidal rule.

In integrating the relations (3.2.71), using the Euler backward method we have the following approximation for the increment of the plastic strains $\Delta \mathbf{e}^P$ in the time step

$$
\Delta \mathbf{e}^P = \int_t^{t+\Delta t} \left(\frac{\partial f_y}{\partial \sigma} \dot{\lambda} \right) dt = \Delta\lambda \frac{\partial\, ^{t+\Delta t} f_y}{\partial\, ^{t+\Delta t}\sigma}
$$

$$(4.2.8)$$

where $\Delta\lambda$ is a positive scalar corresponding to the time step Δt. This is one of the major approximations in the algorithm. In Section 4.4.4 we will discuss consequences of this approximation with respect to the accuracy of incremental solutions.

For the class of material models considered here, the yield condition at the end of the time step can be written in the form

$$
{}^{t+\Delta t}f_y(\,{}^{t+\Delta t}\boldsymbol{\sigma}, {}^{t+\Delta t}\boldsymbol{\beta}) = \ {}^{t+\Delta t}f_y(\,{}^{t+\Delta t}\mathbf{S}, {}^{t+\Delta t}\sigma_m, {}^{t+\Delta t}\boldsymbol{\beta}) = 0 \qquad (4.2.9)
$$

and it follows from (4.2.8) that

$$
\begin{aligned}
\Delta e_{11}^P &= \frac{\Delta\lambda}{3}\,{}^{t+\Delta t}\!\left(2\frac{\partial f_y}{\partial S_{11}} - \frac{\partial f_y}{\partial S_{22}} - \frac{\partial f_y}{\partial S_{33}} + \frac{\partial f_y}{\partial\sigma_m}\right)\\[2mm]
\Delta e_{22}^P &= \frac{\Delta\lambda}{3}\,{}^{t+\Delta t}\!\left(-\frac{\partial f_y}{\partial S_{11}} + 2\frac{\partial f_y}{\partial S_{22}} - \frac{\partial f_y}{\partial S_{33}} + \frac{\partial f_y}{\partial\sigma_m}\right)\\[2mm]
\Delta e_{33}^P &= \frac{\Delta\lambda}{3}\,{}^{t+\Delta t}\!\left(-\frac{\partial f_y}{\partial S_{11}} - \frac{\partial f_y}{\partial S_{22}} + 2\frac{\partial f_y}{\partial S_{33}} + \frac{\partial f_y}{\partial\sigma_m}\right)\\[2mm]
\Delta e_{ij}^P &= \Delta\lambda\,\frac{\partial^{\,t+\Delta t}f_y}{\partial^{\,t+\Delta t}S_{ij}} \qquad i \neq j
\end{aligned}
\qquad (4.2.10)
$$

We note that in these expressions we used that the normal deviatoric stress components depend on all three normal stress components (see (3.2.69)); while the deviatoric shear stress components are equal to the actual shear stresses. In the development of the numerical procedures we will abundantly employ the expressions (4.2.10).

Figure 4.2.1 shows schematically two configurations of a generic body $\mathcal{B}$, at the start and the end of the time increment Δt, and the corresponding yield surfaces in the stress space at a material point M. Since time-independent plasticity is considered, the time step Δt represents actually a load step with the strain increment Δe. The material point M moves from point ${}^{t}M$ to point ${}^{t+\Delta t}M$, with the displacement vector $\Delta\mathbf{u}$, while the stress point (image of point ${}^{t}M$) moves in the stress space from ${}^{t}P$ to ${}^{t+\Delta t}P$, with the stress increment $\Delta\boldsymbol{\sigma}$. Now we use the so-called *return mapping* approach and formulate the computational procedure according to the governing parameter method.

The first step in the return mapping concept is to calculate the trial elastic state. We assume that only elastic deformations occurred in the time step and calculate the stresses ${}^{t+\Delta t}\boldsymbol{\sigma}^E$ according to the elastic stress-strain relationship. Hence we have (see (A1.5))

$$
{}^{t+\Delta t}\boldsymbol{\sigma}^E = \mathbf{C}^E\left({}^{t+\Delta t}\hat{\mathbf{e}} - {}^{t}\hat{\mathbf{e}}^P\right) \qquad (4.2.11)
$$

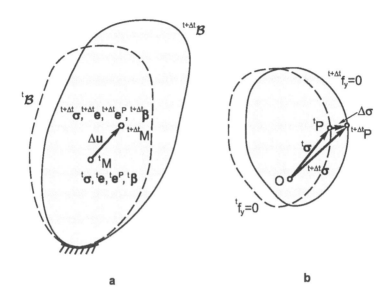

Fig. 4.2.1. Plastic deformation of body $\mathcal{B}$ during time increment Δt. **a** Physical space; **b** Stress space

where $\mathbf{C}^E$ is the elasticity *matrix*; $^{t+\Delta t}\hat{\mathbf{e}}$ is the strain *vector* and $^t\hat{\mathbf{e}}^P$ is the plastic strain *vector*, both containing the engineering shear terms (see (A1.3)). Of course, thermal and creep effects can be included, as shown in Chapter 5. Also, a nonlinear elastic constitutive law may be used to find $^{t+\Delta t}\boldsymbol{\sigma}^E$ corresponding to the strain increment $\Delta \mathbf{e}$. For simplicity of explanation, we use here $\mathbf{C}^E$ to be a constant matrix. Then we evaluate the yield function $^{t+\Delta t}f_y^E$

$$^{t+\Delta t}f_y^E = {}^{t+\Delta t}f_y\left({}^{t+\Delta t}\boldsymbol{\sigma}^E, {}^t\boldsymbol{\beta}\right) \tag{4.2.12}$$

and check for yielding in the time step. If

$$^{t+\Delta t}f_y^E \leq {}^tf_y \tag{4.2.13}$$

the complete deformation in the time step is elastic and $^{t+\Delta t}\boldsymbol{\sigma}^E$ is the solution. If

$$^{t+\Delta t}f_y^E > {}^tf_y \tag{4.2.14}$$

plastic deformation took place, and we proceed to the plasticity calculations as follows.

The stress $^{t+\Delta t}\boldsymbol{\sigma}$, with plastic flow in the time step, is

$$^{t+\Delta t}\boldsymbol{\sigma} = {}^{t+\Delta t}\boldsymbol{\sigma}^E - \mathbf{C}^E \Delta \hat{\mathbf{e}}^P \qquad (4.2.15)$$

and must satisfy the yield condition (4.2.9), where $\Delta \hat{\mathbf{e}}^P$ is the plastic strain increment *vector* (see (A1.3)). Therefore we have to correct the elastic solution $^{t+\Delta t}\boldsymbol{\sigma}^E$ which corresponds to point $^{t+\Delta t}P^E$ in Fig. 4.2.2 in order to satisfy (4.2.9); that is, we are seeking the stress point $^{t+\Delta t}P$ on the yield surface where $^{t+\Delta t}f_y = 0$. This procedure is generally called "return mapping" or "elastic predictor - plastic corrector" method. It has been shown that this method has certain advantages, especially regarding its accuracy characteristics, with respect to other algorithms that start from a stress point on the yield surface $^t f_y = 0$ and seek point $^{t+\Delta t}P$. To calculate $\Delta \mathbf{e}^P$ we employ (4.2.8), while for the hardening law we have in accordance with (3.2.88) (also with the one-index notation)

$$d\beta = -d\lambda \, \mathbf{C}^P \frac{\partial f_y}{\partial \beta} \qquad (4.2.16)$$

Using the Euler backward method, we have the increment of internal variables $\Delta\beta$ in the time step,

$$\Delta\beta = -\Delta\lambda \, \mathbf{C}^P \frac{\partial \, {}^{t+\Delta t}f_y}{\partial \, {}^{t+\Delta t}\beta} \qquad (4.2.17)$$

where, for simplicity, we have assumed $\mathbf{C}^P$ to be constant in the time step. Note that in case of perfect plasticity, there are no internal variables, and the

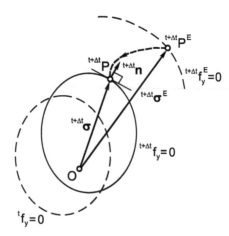

Fig. 4.2.2. A general scheme of return mapping in plasticity (stress space)

return mapping represents the closest point projection of the stress predictor $^{t+\Delta t}\boldsymbol{\sigma}^E$ to the yield surface $^{t+\Delta t}f_y = 0$.

In the governing parameter method we proceed as follows. First, the increments of plastic strain Δe_{ij}^P can be written as

$$\Delta e_{ij}^P = \left\|\Delta \mathbf{e}^P\right\| {}^{t+\Delta t}n_{ij} \tag{4.2.18}$$

where (see (A2.31))

$$\left\|\Delta \mathbf{e}^P\right\| = \left(\Delta e_{ij}^P \, \Delta e_{ij}^P\right)^{1/2} \tag{4.2.19}$$

and $^{t+\Delta t}\mathbf{n}$ is the unit normal to the yield surface $^{t+\Delta t}f_y = 0$,

$$^{t+\Delta t}\mathbf{n} = \frac{\partial \, {}^{t+\Delta t}f_y / \partial \, {}^{t+\Delta t}\boldsymbol{\sigma}}{\|\partial \, {}^{t+\Delta t}f_y / \partial \, {}^{t+\Delta t}\boldsymbol{\sigma}\|} = \frac{{}^{t+\Delta t}\mathbf{f}_{y,\sigma}}{\|\, {}^{t+\Delta t}\mathbf{f}_{y,\sigma}\|} \tag{4.2.20}$$

As in equations (4.2.11) to (4.2.17), we shall continue to use the notation $\mathbf{f}_{y,\sigma} \equiv \partial f_y / \partial \boldsymbol{\sigma}$ for simpler writing, and the one index vector notation (as $n_1 = n_{11}, ..., n_6 = n_{31}$, see (A1.1)). Comparing (4.2.8) and (4.2.18), it follows that

$$\Delta \lambda = \frac{\left\|\Delta \mathbf{e}^P\right\|}{\|\, {}^{t+\Delta t}\mathbf{f}_{y,\sigma}\|} \tag{4.2.21}$$

This relation is quite general and applicable to all plasticity models considered in this book. In case of non-associated plasticity, the relations (4.2.18), (4.2.20) and (4.2.21) have a similar form, with a plastic potential function instead of $^{t+\Delta t}f_y$.

Selecting $\|\Delta \mathbf{e}^P\|$ as the governing parameter in Table 4.2.1, we find that

$$^{t+\Delta t}\boldsymbol{\sigma} = {}^{t+\Delta t}\boldsymbol{\sigma}^E - \left\|\Delta \mathbf{e}^P\right\| \mathbf{C}^E \, {}^{t+\Delta t}\hat{\mathbf{n}} \tag{4.2.22}$$

$$^{t+\Delta t}\boldsymbol{\beta} = {}^{t}\boldsymbol{\beta} - \left\|\Delta \mathbf{e}^P\right\| \hat{\mathbf{C}}^P \, {}^{t+\Delta t}\mathbf{n}_\beta \tag{4.2.23}$$

where

$$\hat{\mathbf{C}}^P = \frac{\|\, {}^{t+\Delta t}\mathbf{f}_{y,\beta}\|}{\|\, {}^{t+\Delta t}\mathbf{f}_{y,\sigma}\|} \mathbf{C}^P \tag{4.2.24}$$

$^{t+\Delta t}\hat{\mathbf{n}}$ is equal to $^{t+\Delta t}\mathbf{n}$ but contains double the shear terms because the engineering shear strains are used in the elastic constitutive law (see (A1.5)); and $^{t+\Delta t}\mathbf{n}_\beta = {}^{t+\Delta t}\mathbf{f}_{y,\beta}/\,\|{}^{t+\Delta t}\mathbf{f}_{y,\beta}\|$, with $^{t+\Delta t}\mathbf{f}_{y,\beta} \equiv \partial^{t+\Delta t} f_y/\partial^{t+\Delta t}\beta$.

The mathematical formalism (4.2.23), with the unit normal $^{t+\Delta t}\mathbf{n}_\beta$, is applicable to a subset of internal variables. For example, in the case of a von Mises material with mixed hardening behavior, the internal variables are the back stress $\boldsymbol{\alpha}$ and the isotropic part of the effective plastic strain $M\bar{e}^P$. Then, from (3.2.57) we have that $^{t+\Delta t}\mathbf{f}_{y,\alpha} = -{}^{t+\Delta t}\mathbf{f}_{y,\sigma}$, and with $C^P = (1-M)C$ (see (3.2.73)), equation (4.2.23) gives Prager's hardening rule. Also, we have $\partial^{t+\Delta t}f_y/\partial^{t+\Delta t}(M\bar{e}^P) = -2/3M\,{}^{t+\Delta t}\hat{\sigma}_y\,{}^{t+\Delta t}\hat{E}_P$; $^{t+\Delta t}n_{M\bar{e}^P} = -1$; and from (4.2.17), (4.2.21), (3.2.59), (3.2.60), (4.2.19) and (3.2.46), we have $C^P = 1/{}^{t+\Delta t}\hat{E}_P$; hence, (4.2.23) gives $^{t+\Delta t}\bar{e}^P = {}^t\bar{e}^P + \sqrt{2/3}\,\|\Delta\mathbf{e}^P\|$.

Now substituting $^{t+\Delta t}\boldsymbol{\sigma}$ and $^{t+\Delta t}\boldsymbol{\beta}$ from (4.2.22) and (4.2.23) into the yield function (4.2.9), we can form a function $f(\|\Delta\mathbf{e}^P\|)$

$$f(\|\Delta\mathbf{e}^P\|) = {}^{t+\Delta t}f_y\left({}^{t+\Delta t}\boldsymbol{\sigma}^E - \|\Delta\mathbf{e}^P\|\,\mathbf{C}^E\,{}^{t+\Delta t}\hat{\mathbf{n}},\right.$$
$$\left.{}^t\boldsymbol{\beta} - \|\Delta\mathbf{e}^P\|\,\hat{\mathbf{C}}^P\,{}^{t+\Delta t}\mathbf{n}_\beta\right) \tag{4.2.25}$$

where $^{t+\Delta t}\hat{\mathbf{n}}$ and $^{t+\Delta t}\mathbf{n}_\beta$ correspond to $\|\Delta\mathbf{e}^P\|$. The derivative of this function is

$$f' = \frac{df}{d\left(\|\Delta\mathbf{e}^P\|\right)} = \frac{\partial^{t+\Delta t}f_y^T}{\partial^{t+\Delta t}\boldsymbol{\sigma}}\frac{\partial^{t+\Delta t}\boldsymbol{\sigma}}{\partial\left(\|\Delta\mathbf{e}^P\|\right)} + \frac{\partial^{t+\Delta t}f_y^T}{\partial^{t+\Delta t}\boldsymbol{\beta}}\frac{\partial^{t+\Delta t}\boldsymbol{\beta}}{\partial\left(\|\Delta\mathbf{e}^P\|\right)} =$$
$$-\left(\|{}^{t+\Delta t}\mathbf{f}_{y,\sigma}\|\,{}^{t+\Delta t}\mathbf{n}^T\,\hat{\mathbf{C}}^E\,{}^{t+\Delta t}\mathbf{n} + \|{}^{t+\Delta t}\mathbf{f}_{y,\beta}\|\,{}^{t+\Delta t}\mathbf{n}_\beta^T\,\hat{\mathbf{C}}^P\,{}^{t+\Delta t}\mathbf{n}_\beta\right) < 0 \tag{4.2.26}$$

where we have used matrix notation. Here $\hat{\mathbf{C}}^E$ is the usual elasticity matrix $\mathbf{C}^E$ but with twice the shear stiffness because of the definition of $^{t+\Delta t}\hat{\mathbf{n}}$. The derivative f' is less than zero because $\hat{\mathbf{C}}^E$ and $\hat{\mathbf{C}}^P$ are positive definite matrices. Hence, the governing function $f(\|\Delta\mathbf{e}^P\|)$ is a monotonically decreasing function, as for one case schematically shown in Fig. 4.2.3a. This property of the function $f(\|\Delta\mathbf{e}^P\|)$ is very important for the practical implementation of a robust computational procedure. Namely, starting with $\|\Delta\mathbf{e}^P\| = 0$ we have $f = {}^{t+\Delta t}f_y^E > 0$; then taking a large value $\|\Delta\mathbf{e}^P\|_{\text{minus}}$ to obtain $f < 0$ (point P_{minus} in Fig. 4.2.3a), we can proceed by bisection or a secant method to calculate $\|\Delta\mathbf{e}^P\|$ for which (within a tolerance) $f = 0$.

The computational steps are as follows. Select a value $\|\Delta\mathbf{e}^P\|^{(1)}$ and calculate the stresses $^{t+\Delta t}\boldsymbol{\sigma}^{(1)}$ and internal variables $^{t+\Delta t}\boldsymbol{\beta}^{(1)}$, using $^{t+\Delta t}\mathbf{n}^{(0)} = {}^{t+\Delta t}\mathbf{n}^E$ and $^{t+\Delta t}\mathbf{n}_\beta^{(0)} = {}^{t+\Delta t}\mathbf{n}_\beta^E$. Then calculate the function $f^{(1)}$ with $^{t+\Delta t}\boldsymbol{\sigma}^{(1)}$ and $^{t+\Delta t}\boldsymbol{\beta}^{(1)}$, and normals $^{t+\Delta t}\mathbf{n}^{(1)}$ and $^{t+\Delta t}\mathbf{n}_\beta^{(1)}$. Select a new

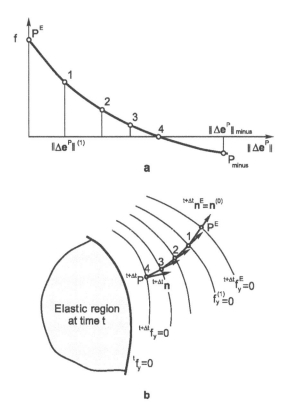

Fig. 4.2.3. Return mapping according to the governing parameter method. **a** Dependence of the governing function on increment of plastic strain $\|\Delta e^P\|$; note that the curvature of the function f could be the other way without affecting the monotonic decrease of the function. **b** Schematic representation of search for the final stress point $^{t+\Delta t}P$

value $\|\Delta e^P\|^{(2)}$ and determine $^{t+\Delta t}\sigma^{(2)}$ and $^{t+\Delta t}\beta^{(2)}$, with normals $^{t+\Delta t}n^{(1)}$ and $^{t+\Delta t}n_\beta^{(1)}$, and so on. In general, for the k-th trial value $\|\Delta e^P\|^{(k)}$ we have

$$^{t+\Delta t}\sigma^{(k)} = {}^{t+\Delta t}\sigma^{(k-1)} - \left(\|\Delta e^P\|^{(k)} - \|\Delta e^P\|^{(k-1)} \right) C^E \, {}^{t+\Delta t}\hat{n}^{(k-1)}$$

$$(4.2.27)$$

$$^{t+\Delta t}\beta^{(k)} = {}^{t+\Delta t}\beta^{(k-1)} - \left(\|\Delta e^P\|^{(k)} - \|\Delta e^P\|^{(k-1)} \right) \hat{C}^{P(k-1)} \, {}^{t+\Delta t}n_\beta^{(k-1)}$$

$$(4.2.28)$$

with $^{t+\Delta t}\boldsymbol{\sigma}^{(0)} = {}^{t+\Delta t}\boldsymbol{\sigma}^{E}$, $^{t+\Delta t}\boldsymbol{\beta}^{(0)} = {}^{t}\boldsymbol{\beta}$, $\|\Delta\mathbf{e}^{P}\|^{(0)} = 0$. The iteration continues until (4.2.9) is satisfied and the difference in two successive trials is small, i.e., until

$$\left| {}^{t+\Delta t}f_{y}^{(k)} \right| \leq \epsilon_{f}$$

$$\left| \|\Delta\mathbf{e}^{P}\|^{(k)} - \|\Delta\mathbf{e}^{P}\|^{(k-1)} \right| < \epsilon_{\Delta} \tag{4.2.29}$$

where ϵ_{f} and ϵ_{Δ} are selected numerical tolerances. When the convergence criteria are satisfied, we have fulfilled the requirements of the stress integration, which are to satisfy —

- The elastic constitutive relations (4.2.15)
- The flow rule in implicit form (4.2.8)
- The hardening law in implicit form (4.2.17)
- The yield condition (4.2.9) at the end of the time step

The above computational steps are summarized in Table 4.2.2, and depicted in Fig. 4.2.3.

Note that if the normals $^{t+\Delta t}\mathbf{n}^{(k)}$ and $^{t+\Delta t}\mathbf{n}_{\beta}^{(k)}$ are unchanged during the iterations — as it is the case for von Mises plasticity and radial loading conditions (see Section 4.4.4) — all trial stress points lie on the straight line defined by $^{t+\Delta t}\mathbf{n}^{E}$, and the computational procedure reduces to the radial return method.

To obtain insight into the stability of the method, assume that in addition to the (exact) elastically predicted stress solution $^{t+\Delta t}\boldsymbol{\sigma}^{E}$, we also consider a perturbed elastic stress state $^{t+\Delta t}\tilde{\boldsymbol{\sigma}}^{E}$ (due to computational error). Then

$$\left\| {}^{t+\Delta t}\tilde{\boldsymbol{\sigma}} - {}^{t+\Delta t}\boldsymbol{\sigma} \right\| \leq \left\| {}^{t+\Delta t}\tilde{\boldsymbol{\sigma}}^{E} - {}^{t+\Delta t}\boldsymbol{\sigma}^{E} \right\| \tag{4.2.30}$$

where $^{t+\Delta t}\tilde{\boldsymbol{\sigma}}$ is the perturbed elastic-plastic stress solution. Figure 4.2.4 gives a graphical representation of the condition (4.2.30) which shows that in plasticity calculations the stress perturbation decreases, indicating stability of the algorithm. Obviously, this result relies on the assumption that the yield surface is convex (see the principle of maximum plastic dissipation, Section 3.2.6).

Now we proceed to the determination of the consistent elastic-plastic tangent matrix. Using (4.2.4), (4.2.5), (4.2.22), and (4.2.23) we obtain

$$^{t+\Delta t}\mathbf{C}^{EP} = {}^{t+\Delta t}\mathbf{C}^{E} - \Delta\boldsymbol{\sigma}' \left[\frac{\partial(\|\Delta\mathbf{e}^{P}\|)}{\partial^{t+\Delta t}\hat{\mathbf{e}}} \right]^{T} \tag{4.2.31}$$

Table 4.2.2. Computational procedure for stress integration using the governing parameter method

---

1. **Known quantities:** ${}^{t}\boldsymbol{\sigma}$, ${}^{t}\mathbf{e}$, ${}^{t}\mathbf{e}^{P}$, ${}^{t}\boldsymbol{\beta}$, ${}^{t+\Delta t}\mathbf{e}$

2. **Calculate elastic predictor** $(k = 0)$

$${}^{t+\Delta t}\boldsymbol{\sigma}^{E} = \mathbf{C}^{E}\left({}^{t+\Delta t}\hat{\mathbf{e}} - {}^{t}\hat{\mathbf{e}}^{P}\right)$$

If ${}^{t+\Delta t}f^{E} \leq 0$ solution is elastic, ${}^{t+\Delta t}\boldsymbol{\sigma} = {}^{t+\Delta t}\boldsymbol{\sigma}^{E}$; EXIT

If ${}^{t+\Delta t}f^{E} > 0$ plastic flow occurs in time step Δt ; hence proceed
with the plasticity calculations

$${}^{t+\Delta t}\mathbf{n}^{(0)} = {}^{t+\Delta t}\mathbf{f}_{y,\sigma}^{E}/\|{}^{t+\Delta t}\mathbf{f}_{y,\sigma}^{E}\| \qquad {}^{t+\Delta t}\mathbf{n}_{\beta}^{(0)} = {}^{t+\Delta t}\mathbf{f}_{y,\beta}^{E}/\|{}^{t+\Delta t}\mathbf{f}_{y,\beta}^{E}\|$$

$${}^{t+\Delta t}\boldsymbol{\sigma}^{(0)} = {}^{t+\Delta t}\boldsymbol{\sigma}^{E},\ {}^{t+\Delta t}\boldsymbol{\beta}^{(0)} = {}^{t}\boldsymbol{\beta},\ \|\Delta\mathbf{e}^{P}\|^{(0)} = 0,\ {}^{t+\Delta t}\mathbf{e}^{P(0)} = {}^{t}\mathbf{e}^{P}$$

3. **Iterations on** $\|\Delta\mathbf{e}^{P}\|$ $(k = k + 1)$

Select $\|\Delta\mathbf{e}^{P}\|^{(k)}$

$${}^{t+\Delta t}\mathbf{e}^{P(k)} = {}^{t+\Delta t}\mathbf{e}^{P(k-1)} + (\|\Delta\mathbf{e}^{P}\|^{(k)} - \|\Delta\mathbf{e}^{P}\|^{(k-1)})\,{}^{t+\Delta t}\mathbf{n}^{(k-1)}$$

Calculate stresses and internal variables

$${}^{t+\Delta t}\boldsymbol{\sigma}^{(k)} = {}^{t+\Delta t}\boldsymbol{\sigma}^{(k-1)} - (\|\Delta\mathbf{e}^{P}\|^{(k)} - \|\Delta\mathbf{e}^{P}\|^{(k-1)})\,\mathbf{C}^{E}\,{}^{t+\Delta t}\hat{\mathbf{n}}^{(k-1)}$$

$${}^{t+\Delta t}\boldsymbol{\beta}^{(k)} = {}^{t+\Delta t}\boldsymbol{\beta}^{(k-1)} - (\|\Delta\mathbf{e}^{P}\|^{(k)} - \|\Delta\mathbf{e}^{P}\|^{(k-1)})\,\hat{\mathbf{C}}^{P(k-1)}\,{}^{t+\Delta t}\mathbf{n}_{\beta}^{(k-1)}$$

Check for convergence

$$\left|{}^{t+\Delta t}f_{y}^{(k)}\right| \leq \varepsilon_{f}; \qquad \left|\,\|\Delta\mathbf{e}^{P}\|^{(k)} - \|\Delta\mathbf{e}^{P}\|^{(k-1)}\,\right| \leq \varepsilon_{\Delta}$$

If convergence is reached go to step 4; otherwise calculate

$${}^{t+\Delta t}\mathbf{f}_{y,\sigma}^{(k)},\ {}^{t+\Delta t}\mathbf{f}_{y,\beta}^{(k)};\ {}^{t+\Delta t}\mathbf{n}^{(k)},\ {}^{t+\Delta t}\mathbf{n}_{\beta}^{(k)},\ \hat{\mathbf{C}}^{P(k)} \qquad \text{go to start of step 3}$$

4. **Consistent tangent elastic-plastic matrix**

$${}^{t+\Delta t}\mathbf{C}^{EP} = {}^{t+\Delta t}\mathbf{C}^{E} - \Delta\boldsymbol{\sigma}'\left[\frac{\partial(\|\Delta\mathbf{e}^{P}\|)}{\partial{}^{t+\Delta t}\hat{\mathbf{e}}}\right]^{T}$$

---

where $\Delta\boldsymbol{\sigma}' = \partial(\Delta\boldsymbol{\sigma})/\partial(\|\Delta\mathbf{e}^{P}\|) = -{}^{t+\Delta t}\boldsymbol{\sigma}'$, with $\Delta\boldsymbol{\sigma} = {}^{t+\Delta t}\boldsymbol{\sigma}^{E} - {}^{t+\Delta t}\boldsymbol{\sigma}$. By differentiation of (4.2.9) with respect to ${}^{t+\Delta t}\hat{\mathbf{e}}$ and solving for $\partial(\|\Delta\mathbf{e}^{P}\|)/\partial{}^{t+\Delta t}\hat{\mathbf{e}}$, we obtain

$$\boxed{\frac{\partial(\|\Delta\mathbf{e}^{P}\|)}{\partial{}^{t+\Delta t}\hat{\mathbf{e}}} = \frac{1}{a_{\sigma}}\mathbf{C}^{E}\,{}^{t+\Delta t}\mathbf{f}_{y,\sigma}} \qquad (4.2.32)$$

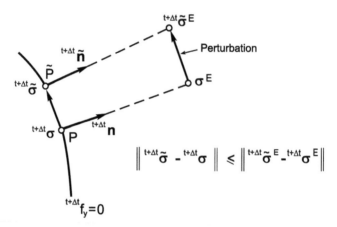

Fig. 4.2.4. Illustration of stability of the return mapping algorithm

where

$$a_\sigma = {}^{t+\Delta t}\mathbf{f}_{y,\sigma}^T \Delta\sigma' - {}^{t+\Delta t}\mathbf{f}_{y,\beta}^T \, {}^{t+\Delta t}\beta' \tag{4.2.33}$$

and ${}^{t+\Delta t}\beta' = \partial {}^{t+\Delta t}\beta / \partial(\|\Delta \mathbf{e}^P\|)$.

The application of the expressions (4.2.31) to (4.2.33) to simple uniaxial stress conditions is given in Example 4.5.1.

The computational scheme is, in general, most effective when the above expressions are evaluated analytically. This can be achieved for many commonly used material models, as presented in this book. However, as an alternative and if the analytical expressions are too complex (see, e.g., Section 6.4), a numerical evaluation is possible, in which we determine the derivatives of any variable ϕ with respect to $\|\Delta \mathbf{e}^P\|$ as

$$\phi' \approx \frac{\phi^{(k)} - \phi^{(k-1)}}{\|\Delta \mathbf{e}^P\|^{(k)} - \|\Delta \mathbf{e}^P\|^{(k-1)}} \tag{4.2.34}$$

where k is a selected iteration during the solution of the nonlinear equation (4.2.9). Note that all matrices and vectors in the expressions (4.2.31) to (4.2.33) have already been evaluated during the stress integration procedure and can now be used in the calculation of the elastic-plastic tangent matrix.

In some plasticity material models the hardening depends on both the mean plastic strain e_m^P and the equivalent deviatoric plastic strain $\bar{e}'^P$, whose increments are

$$\Delta e_m^P = \frac{1}{3}\left(\Delta e_{11}^P + \Delta e_{22}^P + \Delta e_{33}^P\right)$$

$$\Delta \bar{e}'^P = \left(\Delta e_{ij}'^P \, \Delta e_{ij}'^P\right)^{1/2} \tag{4.2.35}$$

Then

$$\left\|\Delta \mathbf{e}^P\right\| = \left[3(\Delta e_m^P)^2 + (\Delta \bar{e}'^P)^2\right]^{1/2} \tag{4.2.36}$$

and the following relations are valid:

$$\frac{\partial \left(\Delta e_m^P\right)}{\partial \left(\left\|\Delta \mathbf{e}^P\right\|\right)} = \frac{\left\|\Delta \mathbf{e}^P\right\|}{3\Delta e_m^P}$$
$$\frac{\partial \left(\Delta \bar{e}'^P\right)}{\partial \left(\left\|\Delta \mathbf{e}^P\right\|\right)} = \frac{\left\|\Delta \mathbf{e}^P\right\|}{\Delta \bar{e}'^P} \tag{4.2.37}$$

These expressions are useful in the calculation of derivatives with respect to $\left\|\Delta \mathbf{e}^P\right\|$.

4.3 A Brief Review of Stress Integration Procedures

The objective of this section is to present a brief historical review of procedures for stress integration, with a view towards the governing parameter method. More details are given for the return mapping type of algorithms, because they have been favored in modern finite element analysis procedures, and because the governing parameter method, which is developed in this book, is of that type. Research in this field has been intensive so that modeling of material behavior, even with very complex properties, has become possible in nonlinear engineering analysis. The references are grouped according to the type of method, rather than chronologically (see also Kojic 2002a).

An iterative scheme with some ingredients of the governing parameter method for solving elastic-plastic problems, called the method of successive elastic solutions, was proposed by Ilyushin (1943). Much later, this method was formulated by Mendelson (1968) for von Mises plasticity, in a form suitable for computer applications. The method consists of the following: for a load step, the increments of plastic strains Δe_{ij}^P are assumed, giving the field of the plastic strains e_{ij}^P; then, with these plastic strains, the solution for the total strains e_{ij} and stresses σ_{ij} is obtained by solving an elastic boundary-value problem (based on partial differential equations for the stresses, using the elastic constitutive relations and boundary conditions). In parallel, the effective plastic strain increment $\Delta \bar{e}^P$ is obtained from (3.2.46), the corresponding yield stress σ_y is determined with use of the yield curve, and then $\Delta \lambda$ is calculated from (3.2.44). Employing the elastic solution for the stresses and the calculated value for $\Delta \lambda$, new trial increments of plastic strains Δe_{ij}^P are determined from (3.2.28), see the block diagram in Fig. 4.3.1. The calculation cycles are repeated until convergence is reached. This approach was implemented by Mendelson (1968) for isotropic hardening to solve some simple problems within the finite difference technique.

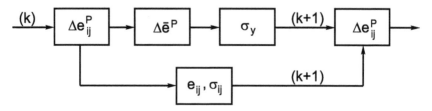

Fig. 4.3.1. Block diagram for stress integration using the method of successive elastic solutions, according to Mendelson (1968)

Another group of procedures relies on the formulation of a tangent material elastic-plastic matrix. Namely, if a yield function of the form (4.2.9) is differentiated at the start of the time step, the following equation is obtained (one-index notation, engineering shear strains):

$$d^t f_y = {}^t\mathbf{f}_{y,\sigma}^T d\boldsymbol{\sigma} - d\lambda \; {}^t\mathbf{f}_{y,\beta}^T \mathbf{C}^P \; {}^t\mathbf{f}_{y,\beta} = 0 \tag{4.3.1}$$

where (4.2.16) has been used. The increment of stress $d\boldsymbol{\sigma}$, with the use of (3.2.67), is

$$d\boldsymbol{\sigma} = \mathbf{C}^E (d\hat{\mathbf{e}} - d\lambda^t \hat{\mathbf{f}}_{y,\sigma}) \tag{4.3.2}$$

where $d\hat{\mathbf{e}}$ is the strain increment vector (see (A1.3)), and ${}^t\hat{\mathbf{f}}_{y,\sigma}$ is equal to ${}^t\mathbf{f}_{y,\sigma}$ but contains twice the shear terms. Substituting $d\boldsymbol{\sigma}$ into (4.3.1), the increment $\Delta\lambda$ in the load step, corresponding to the strain increment $\Delta\hat{\mathbf{e}}$, is obtained as

$$\Delta\lambda = \frac{{}^t\mathbf{f}_{y,\sigma}^T \mathbf{C}^E}{{}^t\mathbf{f}_{y,\sigma}^T \mathbf{C}^E \; {}^t\hat{\mathbf{f}}_{y,\sigma} + {}^t\mathbf{f}_{y,\beta}^T \mathbf{C}^P \; {}^t\mathbf{f}_{y,\beta}} \Delta\hat{\mathbf{e}} \tag{4.3.3}$$

Then, the stress increment is

$$\Delta\boldsymbol{\sigma} = {}^t\mathbf{C}^{EP} \Delta\hat{\mathbf{e}} \tag{4.3.4}$$

where the elastic-plastic stress-strain matrix is

$$\boxed{{}^t\mathbf{C}^{EP} = \mathbf{C}^E - \frac{1}{{}^tc} \; {}^t\mathbf{b} \; {}^t\hat{\mathbf{b}}^T} \tag{4.3.5}$$

Here ${}^tc$ is the denominator in (4.3.3), and the vectors ${}^t\mathbf{b}$ and ${}^t\hat{\mathbf{b}}$ are

$$ {}^t\mathbf{b} = \mathbf{C}^E \; {}^t\mathbf{f}_{y,\sigma}, \qquad {}^t\hat{\mathbf{b}} = \mathbf{C}^E \; {}^t\hat{\mathbf{f}}_{y,\sigma} \tag{4.3.6}$$

Obviously, the symmetric matrix ${}^t\mathbf{C}^{EP}$ is evaluated at the known stress/strain state at the start of the time step. We refer to this matrix as the tangent material matrix.

The above simple forward integration is particularly inefficient when perfectly plastic material conditions are considered. For this case, the tangent stiffness-radial return method and the secant stiffness method have been proposed. In the *tangent stiffness-radial return method*, introduced by Marcal (1965), we first determine the deviatoric stresses $\mathbf{S}_B$ corresponding to the stress point B on the yield surface (Fig. 4.3.2),

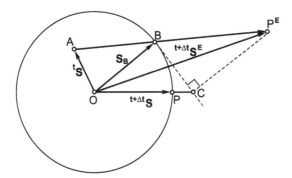

Fig. 4.3.2. Graphical representation of the tangent stiffness-radial return method (perfect plasticity)

$$\mathbf{S}_B = {}^t\mathbf{S} + 2G\,k\,\Delta\mathbf{e}' \tag{4.3.7}$$

where $\Delta\mathbf{e}'$ is the deviatoric strain increment tensor in the time step, G is the shear modulus, and k gives the fraction of $\Delta\mathbf{e}'$ that corresponds to the elastic deformation in the load step. The fraction k, $0 \le k \le 1$, can be determined by substituting $\mathbf{S}_B$ into the yield condition (3.2.36). Then, the stress ${}^{t+\Delta t}\mathbf{S}$ at the end of the load step is obtained as

$$ {}^{t+\Delta t}\mathbf{S} = s \left[{}^{t+\Delta t}\mathbf{S}^E - \frac{2G(1-k)}{R^2} (\mathbf{S}_B \cdot \Delta\mathbf{e}')\,\mathbf{S}_B \right] \tag{4.3.8}$$

where ${}^{t+\Delta t}\mathbf{S}^E$ is the elastic solution (point P^E in Fig. 4.3.2), $R = \sqrt{\frac{3}{2}}\sigma_y$ is the radius of the yield surface; $0 < s < 1$ is the coefficient determined by substituting ${}^{t+\Delta t}\mathbf{S}$ into the yield condition (3.2.36); and $\mathbf{S}_B \cdot \Delta\mathbf{e}'$ is a scalar product (see (A2.29) in Appendix A2). Figure 4.3.2 gives a geometric interpretation of the expression (4.3.8). The vector in the brackets corresponds to the stress point C, and that vector is scaled by the factor s to obtain the stress point P on the yield surface. The described computational procedure corresponds to the perfect plasticity assumption, but it can be extended to hardening models (Shreyer et al. 1979).

The *secant stiffness method* was introduced by Rice and Tracey (1973) for a perfectly plastic material. In this method the stress point B is first determined as in the tangent stiffness-radial return method. Then, the deviatoric stresses $\mathbf{S}_C$, shown in Fig. 4.3.3, are calculated as

$$\mathbf{S}_C = \frac{1}{2}\left(\mathbf{S}_B + {}^{t+\Delta t}\mathbf{S}^E\right) \tag{4.3.9}$$

Finally, the deviatoric stress ${}^{t+\Delta t}\mathbf{S}$ is

$$^{t+\Delta t}\mathbf{S} = {}^{t+\Delta t}\mathbf{S}^E - \frac{2G(1-k)}{R^2}\left(\mathbf{S}_C \cdot \Delta\mathbf{e}'\right)\mathbf{S}_C \tag{4.3.10}$$

where the fraction k has the same meaning as in (4.3.8). It can be shown by a simple geometric argument that the stress point P lies on the yield surface.

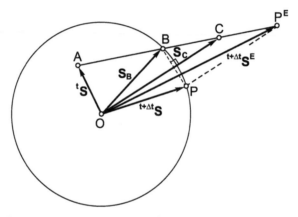

Fig. 4.3.3. Graphical representation of the secant stiffness method (perfect plasticity)

The above described explicit stress calculation schemes and variations thereof were abundantly used in the past, because they were deemed sufficiently effective for the applications considered, e.g., in Bathe et al. (1973), (1974), (1975), (1976); Krieg and Key (1976); Hinton and Owen (1980); Snyder and Bathe (1981) (for thermoplasticity and creep); Desai and Siriwardane (1984); Chen and Baladi (1985); Chen and Han (1988); Chen and Mizuno (1990). However, as the interest in nonlinear finite element analysis increased and increasingly more complex nonlinear finite element analyses were pursued, two shortcomings of these *explicit* stress calculation procedures became apparent. Firstly, the accuracy of these methods is relatively low (Krieg RD and Krieg DB 1977; Shreyer et al. 1979). Some improvements are achieved by subincrementing the total strain increment $\Delta\mathbf{e}$ and applying the stress calculation for each subincrement (Bathe et al. 1973, 1974; Bushnell 1977). This is particularly important when material hardening is included. Secondly, the use of the matrix ${}^t\mathbf{C}^{EP}$ given in (4.3.5) does not result in quadratic convergence as obtained with the algorithmic consistent tangent matrix ${}^{t+\Delta t}\mathbf{C}^{EP}$ of the form (4.2.31).

The development of *implicit* schemes for stress integration started in 1964 by Wilkins (1964). An algorithm was proposed for the perfectly plastic von Mises material model. The procedure consists of two steps: 1) calculation of the elastic solution for the time step (elastic predictor), and 2) the radial return to the yield surface (plastic corrector). This approach was implemented for various material models and generalized by a number of authors, forming a class of *return mapping* procedures. We next cite some of the references that provide information about various aspects of these algorithms. Also, we give some details for typical representative approaches of this group of procedures.

Ortiz et al. (1983) developed a return mapping algorithm for perfect and hardening viscoplasticity and plasticity. As in the algorithm published by Wilkins (1964), for a von Mises material the return mapping goes along the unit normal $\mathbf{n}^E$ to the yield surface, corresponding to the elastic predictor $^{t+\Delta t}\boldsymbol{\sigma}^E$. Then, the solution for perfectly plastic and linear hardening materials can be obtained in closed form. The procedure was generalized by Simo and Taylor (1985a, 1986) for von Mises nonlinear isotropic and kinematic hardening models, and for general 3-D and plane stress problems. A scalar equation, representing the yield condition $^{t+\alpha\Delta t}f_y = 0$ $(0 \leq \alpha \leq 1)$ was solved for $\Delta\lambda$, after which the other variables were calculated. The consistent tangent elastic-plastic matrix was derived.

An analysis of the stability and accuracy of Euler type time integration schemes when applied in this stress solution was presented by Ortiz and Popov (1985). It was shown that the algorithms are unconditionally stable when the Euler method integration parameter $\alpha \geq 1/2$, with first-order accuracy when $\alpha > 1/2$ and second order accuracy for $\alpha = 1/2$. These conclusions correspond to the general stability and accuracy characteristics of the α-method, see Collatz (1966); Bathe (1996).

Another approach of return mapping procedure is due to Simo and Ortiz (1985b). This algorithm is presented in Table 4.3.1 (see Simo and Hughes 1998). The essence of the procedure is that residuals corresponding to the nonlinear elasto-plastic flow and hardening rules (4.2.8) and (4.2.17) (see also (3.2.87) and (3.2.88)) are formed as

$$^{t+\Delta t}\mathbf{R}_\sigma = {}^t\mathbf{e}^P + \Delta\lambda \, {}^{t+\Delta t}\mathbf{f}_{y,\sigma} - {}^{t+\Delta t}\mathbf{e}^P \tag{4.3.11}$$

$$^{t+\Delta t}\mathbf{R}_\beta^* = {}^t\boldsymbol{\beta}^* + \Delta\lambda \, {}^{t+\Delta t}\mathbf{f}_{y,\beta} - {}^{t+\Delta t}\boldsymbol{\beta}^* \tag{4.3.12}$$

These residuals are calculated and the yield condition is linearized at the current k-th trial state. A linearized system of equations is solved for the increments $\Delta(\Delta\lambda)^{(k)}$, $\Delta\mathbf{e}^{P(k)}$ and $\Delta\boldsymbol{\beta}^{*(k)}$. As can be seen from Table 4.3.1

Table 4.3.1. Return mapping algorithm based on successive linearizations (Simo and Hughes 1998)

---

1. Initial quantities: $k = 0$, $^{t+\Delta t}\mathbf{e}^{P(0)} = {}^{t}\mathbf{e}^{P}$, $^{t+\Delta t}\boldsymbol{\beta}^{*(0)} = {}^{t}\boldsymbol{\beta}^{*}$, $\Delta\lambda^{(0)} = 0$

$$^{t+\Delta t}\boldsymbol{\beta}^{(0)} = {}^{t}\boldsymbol{\beta}$$

(one-index notation, tensorial shear components)

2. Calculate stresses $^{t+\Delta t}\boldsymbol{\sigma}^{(k)}$, **internal variables** $^{t+\Delta t}\boldsymbol{\beta}^{*(k)}$
and check for convergence
($\hat{\mathbf{C}}^{E(k)}$ is the elastic matrix with twice the shear terms)

$$^{t+\Delta t}\boldsymbol{\sigma}^{(k)} = \hat{\mathbf{C}}^{E(k)}\left(^{t+\Delta t}\mathbf{e} - {}^{t+\Delta t}\mathbf{e}^{P(k)}\right)$$

$$^{t+\Delta t}f_y^{(k)} = {}^{t+\Delta t}f_y\left(^{t+\Delta t}\boldsymbol{\sigma}^{(k)}, {}^{t+\Delta t}\boldsymbol{\beta}^{(k)}\right)$$

$$^{t+\Delta t}\mathbf{R}^{(k)} = \left\{ \begin{array}{c} ^{t+\Delta t}\mathbf{R}_\sigma \\ ^{t+\Delta t}\mathbf{R}_\beta^* \end{array} \right\}^{(k)} = \left\{ \begin{array}{c} ^{t}\mathbf{e}^{P} - {}^{t+\Delta t}\mathbf{e}^{P(k)} \\ ^{t}\boldsymbol{\beta}^* - {}^{t+\Delta t}\boldsymbol{\beta}^{*(k)} \end{array} \right\} + \Delta\lambda^{(k)}\left\{ \begin{array}{c} ^{t+\Delta t}\mathbf{f}_{y,\sigma} \\ ^{t+\Delta t}\mathbf{f}_{y,\beta} \end{array} \right\}^{(k)}$$

3. Calculate consistent tangent matrix $^{t+\Delta t}\mathbf{C}^{EP(k)}$

$$\left(^{t+\Delta t}\mathbf{C}^{EP(k)}\right)^{-1} = \left[\begin{array}{cc} ((\hat{\mathbf{C}}^E)^{-1} + \Delta\lambda\,{}^{t+\Delta t}\mathbf{f}_{y,\sigma\sigma}) & \Delta\lambda\,{}^{t+\Delta t}\mathbf{f}_{y,\sigma\beta} \\ \Delta\lambda\,{}^{t+\Delta t}\mathbf{f}_{y,\beta\sigma} & ((\mathbf{C}^P)^{-1} + \Delta\lambda\,{}^{t+\Delta t}\mathbf{f}_{y,\beta\beta}) \end{array} \right]^{(k)}$$

4. Compute increment of $\Delta\lambda$

$$\Delta(\Delta\lambda)^{(k)} = \frac{^{t+\Delta t}f_y^{(k)} - \left[^{t+\Delta t}\mathbf{f}_{y,\sigma}^{(k)}\ \ {}^{t+\Delta t}\mathbf{f}_{y,\beta}^{(k)}\right]^T {}^{t+\Delta t}\mathbf{C}^{EP(k)}\,{}^{t+\Delta t}\mathbf{R}^{(k)}}{\left[^{t+\Delta t}\mathbf{f}_{y,\sigma}^{(k)}\ \ {}^{t+\Delta t}\mathbf{f}_{y,\beta}^{(k)}\right]^T {}^{t+\Delta t}\mathbf{C}^{EP(k)}\left\{ \begin{array}{c} ^{t+\Delta t}\mathbf{f}_{y,\sigma} \\ ^{t+\Delta t}\mathbf{f}_{y,\beta} \end{array} \right\}^{(k)}}$$

5. Obtain increments of plastic strains and internal variables

$$\left\{ \begin{array}{c} \Delta\mathbf{e}^P \\ \Delta\boldsymbol{\beta}^* \end{array} \right\}^{(k)} = \left[\begin{array}{cc} (\hat{\mathbf{C}}^E)^{-1} & \mathbf{0} \\ \mathbf{0} & (\mathbf{C}^P)^{-1} \end{array} \right]^{(k)} \times$$

$$^{t+\Delta t}\mathbf{C}^{EP(k)}\,{}^{t+\Delta t}\left[\mathbf{R} + \Delta(\Delta\lambda)\left\{ \begin{array}{c} \mathbf{f}_{y,\sigma} \\ \mathbf{f}_{y,\beta} \end{array} \right\}\right]^{(k)}$$

6. Update variables

$$^{t+\Delta t}\mathbf{e}^{P(k+1)} = {}^{t+\Delta t}\mathbf{e}^{P(k)} + \Delta\mathbf{e}^{P(k)}, \quad {}^{t+\Delta t}\boldsymbol{\beta}^{*(k+1)} = {}^{t+\Delta t}\boldsymbol{\beta}^{*(k)} + \Delta\boldsymbol{\beta}^{*(k)}$$

$$\Delta\lambda^{(k+1)} = \Delta\lambda^{(k)} + \Delta(\Delta\lambda)^{(k)}$$

$$^{t+\Delta t}\boldsymbol{\beta}^{(k+1)} = {}^{t+\Delta t}\boldsymbol{\beta}^{(k)} - \mathbf{C}^{P(k)}\Delta\boldsymbol{\beta}^{*(k)}$$

$$k = k+1 \quad \text{go to step 2}$$

---

the computation of increments $\Delta(\Delta\lambda)^{(k)}$, $\Delta\mathbf{e}^{P(k)}$ and $\Delta\boldsymbol{\beta}^{*(k)}$ requires iterations on a system of equations of order (6 + number of internal variables) with a significant amount of numerical operations, and with possible convergence difficulties. In order to increase the computational efficiency of the algorithm, Simo and Ortiz (1985b) also proposed the so-called *cutting plane procedure*. In this method a linearization of the yield condition gives a relation for $\Delta(\Delta\lambda)^{(k+1)}$,

$$
\begin{aligned}
&{}^{t+\Delta t}f_y^{(k+1)} = {}^{t+\Delta t}f_y^{(k)} - \\
&\Delta(\Delta\lambda)^{(k+1)} \left[{}^{t+\Delta t} \left(\mathbf{f}_{y,\sigma}^T \mathbf{C}^E \, \hat{\mathbf{f}}_{y,\sigma} + \mathbf{f}_{y,\beta}^T \mathbf{C}^P \, \mathbf{f}_{y,\beta} \right) \right]^{(k)} = 0
\end{aligned}
\tag{4.3.13}
$$

This method is in some cases quite analogous to the governing parameter method (see also the text at the end of Section 4.2.1).

A summary of the return mapping algorithms based on successive linearizations, with implementations in plasticity and viscoplasticity, is given by Simo and Hughes (1998) (see also Crisfield 1991).

Independent of these developments, Kojic and Bathe formulated the *"effective-stress-function"* algorithm (Bathe et al. 1984; Kojic and Bathe 1987a,b) and showed that implicit stress integration of the α-type (4.2.6) for metal thermoplasticity and creep can be reduced to finding the zero of the effective-stress-function. The algorithm was applied to various physical conditions (three-dimensional, two-dimensional plane stress, shell, beam and pipe conditions), for von Mises general isotropic and kinematic hardening and common creep laws. However, clearly, the effective stress is not an efficient measure to use (and may not even be appropriate) for certain material models. For more generality and applications with various material models, the algorithm was extended to the governing parameter method (GPM), see Kojic (1996a), the references therein, and the references in the subsequent sections.

As shown in Table 4.2.1, in the governing parameter method the stress integration reduces to solving one nonlinear equation with respect to the governing parameter p. Of course, a number of authors calculated the stresses for various inelastic material models through solving one nonlinear equation, see, e.g., Bathe et al. (1984); Simo and Taylor (1985a); Anand (1985); Kojic and Bathe (1987a,b); Lush et al. (1989); Weber and Anand (1990a); Borja and Lee (1990); Borja (1991), and references therein, and the procedures used in these references are closely related to the governing parameter method.

The governing parameter method can of course also simply be considered to be a specific approach to solve the system of nonlinear equations obtained in the implicit stress integration. As presented in Sections 4.1 and 4.2, the governing relations of the stress integration in a time (load) step form a set of nonlinear equations. Some of these equations represent tensorial type relations for the stresses and strains. Other equations express constraints that must be satisfied, as a yield condition or creep law. The unknowns in the system are the components of tensors (stresses, inelastic strains, internal variables) and scalars, such as the effective plastic strain and the effective stress. According to the governing parameter method, we select one scalar as the governing (main) parameter p in the system and iterate with appropriate linearizations, such as those given by (4.2.27) and (4.2.28), to solve for the parameter. The iteration continues until the governing equation $f(p) = 0$ is satisfied within a selected numerical tolerance. The governing parameter may be one of the scalars in the system, or another scalar to which the other variables are related. For example, we select $\|\Delta \mathbf{e}^P\|$ for a general plasticity model (Section 4.2) because the measure of plastic flow in a time step is defined by the plastic work which is directly related to $\|\Delta \mathbf{e}^P\|$.

If, on the other hand, as pursued in other approaches, the complete governing *system* of equations is linearized with no distinction among the variables, then p is just one of these variables. The solution leads to an iterative scheme for the *system* of equations, which may not be as effective. Let us consider the following thoughts.

First, consider that we change the sequence of calculations in Fig. 4.3.1, such that the iteration scheme becomes: select $\Delta \bar{e}^P$, find $\bar{\sigma}$ from the yield curve, calculate $\Delta \lambda$ from (4.2.21), then the stresses σ_{ij} and the increments Δe_{ij}^P; iterate until the yield condition $f_y = 0$ is satisfied. The driving term $\Delta \bar{e}^P$ can be appropriately chosen during the iterations, leading to a simple and stable iterative scheme, which in fact corresponds to the governing parameter method (see Fig. 4.2.3).

Second, consider that we select $\Delta \lambda$ (or better $\|\Delta \mathbf{e}^P\|$ related to $\Delta \lambda$ by (4.2.21)) in the successive linearization procedure in Table 4.3.1, and then instead of $\Delta(\Delta \lambda)^{(i)}$, we use $\Delta \lambda^{(i)}$ as the driving term. The other variables follow, naturally, according to Table 4.2.2. By application of the governing parameter method, the cutting plane iterative scheme (4.3.13), for which the rate of convergence and the return mapping path depend on the gradients $^{t+\Delta t}\mathbf{f}_{y,\sigma}$ and $^{t+\Delta t}\mathbf{f}_{y,\beta}$, can be replaced by the simple and robust iterations of Table 4.2.2.

We illustrate the use of the governing parameter method in the text to follow.

4.4 Stress Integration for the von Mises Material Model

In this section we present the governing parameter method for metal plasticity. We give in detail the stress calculation procedure and the derivation of the consistent tangent constitutive matrix. General three-dimensional elastic-plastic deformations and the case of shell analysis conditions are considered for the case of mixed hardening (Kojic 1993, Bathe and Montans 2004). These two conditions directly give also the numerical procedures used for plane strain, axisymmetric and plane stress problems. Many of the details given here are also used in the discussion of subsequent material models.

4.4.1 General Three-Dimensional Conditions, Plane Strain and Axisymmetry

We consider a material point at which the stress-strain state at the start of the time step is known, see Fig. 4.4.1. The inelastic strains are the plastic strains e^P, and the internal variables which define the position and size of the yield surface are the back stress α and the intensity of the yield surface radius $\left\|\hat{S}\right\|$. Therefore, in accordance with (4.2.1) we consider that the known quantities are

$$\boxed{{}^t\sigma,\ {}^t e,\ {}^t e^P,\ {}^t\alpha,\ {}^t\hat{\sigma}_y,\ {}^{t+\Delta t}e}$$

(4.4.1)

where the yield stress ${}^t\hat{\sigma}_y$ defines the size of the yield surface,

$$\boxed{{}^t\hat{\sigma}_y = \sqrt{\frac{3}{2}}\left\|\ {}^t\hat{S}\right\|}$$

(4.4.2)

as shown in Fig. 4.4.1. The unknown quantities corresponding to the end of the time (load) step are (see (4.2.2)):

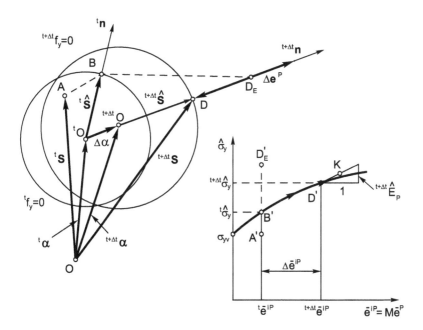

Fig. 4.4.1. Stress states at time t and time $t + \Delta t$ (mixed hardening)

$$
{}^{t+\Delta t}\boldsymbol{\sigma}, \quad {}^{t+\Delta t}\mathbf{e}^P, \quad {}^{t+\Delta t}\boldsymbol{\alpha}, \quad {}^{t+\Delta t}\hat{\sigma}_y \tag{4.4.3}
$$

We use the notation ${}^{t}\hat{\sigma}_y$ and ${}^{t+\Delta t}\hat{\sigma}_y$ to point out that the yield curve is defined as $\hat{\sigma}_y(\bar{e}^{iP})$, where $\bar{e}^{iP}$ is the isotropic part of the effective plastic strain (see (3.2.66) and Fig. 3.2.18).

Let us start with the elastic constitutive relation (see (A1.5)),

$$
{}^{t+\Delta t}\boldsymbol{\sigma} = \mathbf{C}^E \, {}^{t+\Delta t}\hat{\mathbf{e}}^E \tag{4.4.4}
$$

where $\mathbf{C}^E$ is the elastic constitutive matrix, and ${}^{t+\Delta t}\hat{\mathbf{e}}^E$ is the elastic strain vector (see (A1.3)) given by

$$
{}^{t+\Delta t}\hat{\mathbf{e}}^E = {}^{t+\Delta t}\hat{\mathbf{e}} - {}^{t+\Delta t}\hat{\mathbf{e}}^P \tag{4.4.5}
$$

Here ${}^{t+\Delta t}\hat{\mathbf{e}}$ and ${}^{t+\Delta t}\hat{\mathbf{e}}^P$ are the total strain and plastic strain vectors. Since the plastic deformation is of the deviatoric type (see definition of the von Mises yield condition in Section 3.2.5), we first consider the calculation of the mean stress (A1.14),

$$
{}^{t+\Delta t}\sigma_m = K \, {}^{t+\Delta t}e_V^E \tag{4.4.6}
$$

where K is the bulk modulus and ${}^{t+\Delta t}e_V^E$ is the volumetric elastic strain

$$
{}^{t+\Delta t}e_V^E = {}^{t+\Delta t}e_{11} + {}^{t+\Delta t}e_{22} + {}^{t+\Delta t}e_{33} \tag{4.4.7}
$$

Note that

$$
{}^{t+\Delta t}e_V^P = {}^{t+\Delta t}e_{11}^P + {}^{t+\Delta t}e_{22}^P + {}^{t+\Delta t}e_{33}^P = 0 \tag{4.4.8}
$$

It follows that, for given strains ${}^{t+\Delta t}\mathbf{e}$, the mean stress at the end of the step *does not depend* on the plastic strains. This result is very important for the development of a stress integration procedure. The result suggests that in isotropic plasticity we only need to seek a numerical solution for the deviatoric stresses ${}^{t+\Delta t}\mathbf{S}$, since the mean stress ${}^{t+\Delta t}\sigma_m$ is already determined by (4.4.6). The decoupling of the calculation of the mean stress from the calculation of the deviatoric stresses is always effective when the elastic or inelastic mean stress-strain and deviatoric stress-strain relations are defined without coupling terms.

The deviatoric stress is related to the deviatoric strain tensor by (see (A1.19))

$$^{t+\Delta t}\mathbf{S} = 2G\ ^{t+\Delta t}\mathbf{e}'^{E} = 2G\left(\ ^{t+\Delta t}\mathbf{e}' - \ ^{t+\Delta t}\mathbf{e}^{P}\right) \tag{4.4.9}$$

where $^{t+\Delta t}\mathbf{e}'$ is the deviatoric strain tensor with components

$$^{t+\Delta t}e'_{ij} = \ ^{t+\Delta t}e_{ij} - \ ^{t+\Delta t}e_m \qquad i = j$$
$$^{t+\Delta t}e'_{ij} = \frac{1}{2}\ ^{t+\Delta t}\gamma_{ij} \qquad\qquad i \neq j \tag{4.4.10}$$

and $^{t+\Delta t}\gamma_{ij}$ are the engineering shear strains. Also, $^{t+\Delta t}\mathbf{e}^{P}$ is the plastic strain tensor (shear terms equal to half of the engineering plastic strains). Equation (4.4.9) can be written as

$$^{t+\Delta t}\mathbf{S} = \ ^{t+\Delta t}\mathbf{S}^{E} - 2G\Delta\mathbf{e}^{P} \tag{4.4.11}$$

where

$$^{t+\Delta t}\mathbf{S}^{E} = 2G\ ^{t+\Delta t}\mathbf{e}'' \tag{4.4.12}$$

is the elastic solution, with no plastic deformation in the current step; hence

$$^{t+\Delta t}\mathbf{e}'' = \ ^{t+\Delta t}\mathbf{e}' - \ ^{t}\mathbf{e}^{P} \tag{4.4.13}$$

In writing (4.4.11) we have used the relation

$$^{t+\Delta t}\mathbf{e}^{P} = \ ^{t}\mathbf{e}^{P} + \Delta\mathbf{e}^{P} \tag{4.4.14}$$

where $\Delta\mathbf{e}^{P}$ is the increment of plastic strain tensor.

We next check whether plastic flow took place in the current time step. The elastic stress radius $^{t+\Delta t}\hat{\mathbf{S}}^{E}$ is

$$^{t+\Delta t}\hat{\mathbf{S}}^{E} = \ ^{t+\Delta t}\mathbf{S}^{E} - \ ^{t}\boldsymbol{\alpha} \tag{4.4.15}$$

which corresponds to the vector from ^{t}O to D_E in Fig. 4.4.1. If the stress point D_E is inside or on the yield surface $^{t}f_y = 0$, i.e., if

$$\left\|\ ^{t+\Delta t}\hat{\mathbf{S}}^{E}\right\| \leq \sqrt{\frac{2}{3}}\ ^{t}\hat{\sigma}_y \tag{4.4.16}$$

then there was no plastic deformation and $^{t+\Delta t}\mathbf{S}^E$ is the solution for the time step.

In the case of plastic flow in the time step, we continue as follows. The yield condition (3.2.56) at the end of the time step gives

$$
^{t+\Delta t}f_y = \frac{1}{2}\left(^{t+\Delta t}\mathbf{S} - {}^{t+\Delta t}\boldsymbol{\alpha}\right)\cdot\left(^{t+\Delta t}\mathbf{S} - {}^{t+\Delta t}\boldsymbol{\alpha}\right) - \frac{1}{3}\,^{t+\Delta t}\hat{\sigma}_y^2 = 0
$$

$$(4.4.17)$$

To calculate the increment of plastic strain $\Delta\mathbf{e}^P$ we employ the flow rule at the end of the step, i.e., we adopt the Euler backward integration scheme with the integration parameter $\alpha = 1$ in (4.2.6). From (4.2.10) and (4.4.17) we obtain

$$
\Delta\mathbf{e}^P = \Delta\lambda\,^{t+\Delta t}\hat{\mathbf{S}}
$$

$$(4.4.18)$$

where we have used the condition $\partial f_y/\partial\sigma_m = 0$ and (3.2.58). The quantities of the last two equations are shown in Fig. 4.4.1. It is important to note that the increment of plastic strain $\Delta\mathbf{e}^P$ is in the direction of the stress radius $^{t+\Delta t}\hat{\mathbf{S}}$ at the end of the time step.

Substituting $\Delta\mathbf{e}^P$ from (4.4.18) into (4.4.11) we obtain

$$
^{t+\Delta t}\mathbf{S} = {}^{t+\Delta t}\mathbf{S}^E - 2G\,\Delta\lambda\,^{t+\Delta t}\hat{\mathbf{S}}
$$

$$(4.4.19)$$

Next we employ the constitutive relations (3.2.73) for the back stress,

$$
\Delta\boldsymbol{\alpha} = \hat{C}\Delta\mathbf{e}^P
$$

$$(4.4.20)$$

where

$$
\hat{C} = (1 - M)\,\bar{C} = \frac{2}{3}(\bar{E}_P - M\hat{\bar{E}}_P)
$$

$$(4.4.21)$$

Here we have used (3.2.78), with $\bar{E}_P$ and $\hat{\bar{E}}_P$ as weighted plastic moduli for the time step Δt. The plastic moduli have the values between $^t E_P$ and $^{t+\Delta t}E_P$ corresponding to the accumulated effective plastic strains $^t\bar{e}^P$ and $^{t+\Delta t}\bar{e}^P$; and between $^t\hat{\bar{E}}_P$ and $^{t+\Delta t}\hat{\bar{E}}_P$ corresponding to $M\,^t\bar{e}^P$ and $M\,^{t+\Delta t}\bar{e}^P$. Using the geometric relations shown in Fig. 4.4.1 and using (4.4.20) we obtain

$$
^{t+\Delta t}\mathbf{S} = {}^t\boldsymbol{\alpha} + \left(1 + \Delta\lambda\hat{C}\right)\,^{t+\Delta t}\hat{\mathbf{S}}
$$

$$(4.4.22)$$

Next we solve for $t+\Delta t\hat{\mathbf{S}}$ using (4.4.19) and (4.4.22)

$$t+\Delta t\hat{\mathbf{S}} = \frac{t+\Delta t\hat{\mathbf{S}}^E}{1 + \left(2G + \hat{C}\right)\Delta\lambda} \tag{4.4.23}$$

where $t+\Delta t\hat{\mathbf{S}}^E$ is the elastic solution, see (4.4.15).

Taking the scalar product on both sides of (4.4.18) (see (A2.29) in Appendix A2), we obtain

$$\Delta\lambda = \frac{3}{2}\frac{\Delta\bar{e}^P}{t+\Delta t\hat{\sigma}} \tag{4.4.24}$$

where (see (3.2.46))

$$\Delta\bar{e}^P = \left(\frac{2}{3}\Delta\mathbf{e}^P \cdot \Delta\mathbf{e}^P\right)^{1/2} = (\frac{2}{3})^{1/2}\|\Delta\mathbf{e}^P\| \tag{4.4.25}$$

is the increment of the effective plastic strain, and

$$t+\Delta t\hat{\sigma} = \left(\frac{3}{2}\, t+\Delta t\hat{\mathbf{S}}\, .^{t+\Delta t}\hat{\mathbf{S}}\right)^{1/2} \tag{4.4.26}$$

defines the size of the yield surface. Note that (4.4.24) is a special case of the general expression (4.2.21) for $\Delta\lambda$. From the definition of the reduced effective stress (3.2.60), the yield condition (4.4.17), and (4.4.26), we find that

$$t+\Delta t\hat{\sigma} - {}^{t+\Delta t}\hat{\sigma}_y = 0 \tag{4.4.27}$$

is a form of the yield condition. On the other hand, hardening is defined by the yield curve (3.2.66), hence

$$t+\Delta t\hat{\sigma}_y = \hat{\sigma}_y\left[M\left({}^t\bar{e}^P + \Delta\bar{e}^P\right)\right] \tag{4.4.28}$$

Using equations (4.4.18) to (4.4.21), (4.4.23), (4.4.24) and (4.4.28), we find that the unknowns $t+\Delta t\mathbf{e}^P$, $t+\Delta t\mathbf{S}$ and $t+\Delta t\boldsymbol{\alpha}$, are functions of one parameter only, that is $\Delta\bar{e}^P$. Namely, for a given $\Delta\bar{e}^P$, we find $\hat{\sigma}_y$ from the yield curve, $\Delta\lambda$ from (4.4.24), then determine $\hat{C}$ from (4.4.21), $\Delta\boldsymbol{\alpha}$ from (4.4.20), $t+\Delta t\hat{\mathbf{S}}$ from (4.4.23), $\Delta\mathbf{e}^P$ from (4.4.18), and $t+\Delta t\mathbf{S}$ either from (4.4.19) or

(4.4.22). Therefore, we have, in the sense of Table 4.2.1, $\Delta \bar{e}^P$ as the governing parameter.

Finally, according to step 2 of Table 4.2.1, we need the governing equation (b). We obtain this equation by taking the scalar product on both sides of (4.4.23), and using (4.4.27) and (4.4.28). The governing equation is

$$f(\Delta \bar{e}^P) = \frac{^{t+\Delta t}\hat{\sigma}^E}{^{t+\Delta t}\hat{\sigma}_y + \frac{3}{2}\left(2G + \hat{C}\right)\Delta \bar{e}^P} - 1 = 0 \qquad (4.4.29)$$

where $^{t+\Delta t}\hat{\sigma}^E$ denotes the value of $\hat{\sigma}$ for $\Delta \bar{e}^P = 0$. This equation is like the yield condition (4.4.27). In deriving this equation we have assumed $\hat{\sigma}_y > 0$ since the yield stress cannot be equal to zero.

The application of the above general relations to the uniaxial stress conditions is given in Example 4.5.1.

The problem of stress integration is reduced to the solution of the nonlinear equation (4.4.29) with respect to the governing parameter $\Delta \bar{e}^P$ as presented in Section 4.2. Since $\Delta \bar{e}^P$ and $^{t+\Delta t}\hat{\sigma}_y$ are related through the yield curve (4.4.28), the stress $^{t+\Delta t}\hat{\sigma}$ may also be taken as the governing parameter, see Kojic and Bathe (1987a,b), and Bathe (1996), where the function $f(\Delta \bar{e}^P)$ is defined as the "effective-stress-function" $f(^{t+\Delta t}\hat{\sigma})$. Note that for the case of perfect plasticity conditions in the time step we have

$$^{t+\Delta t}\hat{\sigma}_y = {}^t\hat{\sigma}_y$$
$$\bar{E}^P = 0, \qquad \hat{C} = 0 \qquad (4.4.30)$$

and

$$\Delta \bar{e}^P = \frac{^{t+\Delta t}\hat{\sigma}^E - {}^t\hat{\sigma}_y}{3G} \qquad (4.4.31)$$

Figure 4.4.2 shows a schematic representation of the solution for $\Delta \bar{e}^P$ in the case of perfect plasticity.

In the general case, the solution of (4.4.29) can be found numerically by, for example, a Newton-Raphson iteration or a bisection procedure. Considering the characteristics of the governing parameter function, as we did for the general case in (4.2.25), we find that the first derivative f' of the function (4.4.29) is

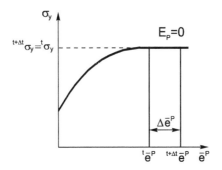

Fig. 4.4.2. Increment of the effective plastic strain in time step for perfect plasticity

$$
f' = \frac{df}{d(\Delta\bar{e}^P)} = -\frac{{}^{t+\Delta t}\hat{\sigma}^E}{\left[{}^{t+\Delta t}\hat{\sigma}_y + \frac{3}{2}\left(2G+\hat{C}\right)\Delta\bar{e}^P\right]^2}
$$
$$
\times \left[\frac{3}{2}\left(2G+\hat{C}\right) + M\;{}^{t+\Delta t}\hat{E}_P + \left(\frac{\partial\bar{E}_P}{\partial\bar{e}^P} - M\frac{\partial\hat{\bar{E}}_P}{\partial\bar{e}^P}\right)\Delta\bar{e}^P\right]
$$

(4.4.32)

where we have used (4.4.21) for $d\hat{C}/d\bar{e}^P$. If the weighted plastic modulus is taken as $\bar{E}_P = (1-\alpha)\,{}^t E_P + \alpha\,{}^{t+\Delta t} E_P$, where $0 \le \alpha \le 1$ is the parameter, then $\partial\bar{E}_P/\partial\bar{e}^P = \alpha\;\partial{}^{t+\Delta t}E_P/\partial\bar{e}^P$ and $\partial\hat{\bar{E}}_P/\partial\bar{e}^P = \alpha\;\partial{}^{t+\Delta t}\hat{E}_P/\partial\bar{e}^P$. In the case of isotropic hardening ($M=1$), the derivative f' is negative for any $\Delta\bar{e}^P$. The value $(\Delta\bar{e}^P)_0$ can be determined for which the derivative f' is zero in the case of mixed hardening,

$$
(\Delta\bar{e}^P)_0 = \frac{3G + \bar{E}_P + M({}^{t+\Delta t}E_P - \hat{\bar{E}}_P)}{\left|\partial\bar{E}_P/\partial\bar{e}^P - M\partial\hat{\bar{E}}_P/\partial\bar{e}^P\right|}
$$

(4.4.33)

where $(\partial\bar{E}_P/\partial\bar{e}^P - M\partial\hat{\bar{E}}_P/\partial\bar{e}^P) < 0$. For $\Delta\bar{e}^P > (\Delta\bar{e}^P)_0$, the derivative f' becomes positive. The value $(\Delta\bar{e}^P)_0$ is much larger than 1, which means that f' is negative for the physically meaningful range. Since $E_P > 0$ the governing function $f(\Delta\bar{e}^P)$ is a monotonic function, as schematically shown in Fig. 4.4.3. This conclusion is in accordance with the observations made regarding (4.2.26), derived earlier for general plastic deformations (see also Fig. 4.2.3).

Table 4.4.1 summarizes the computational steps for the stress calculation. We note that the computational procedure is a special case of the procedure given in Table 4.2.2 for plasticity in general.

The above described procedure is directly applicable to the case of *isotropic hardening*. Then, using $M = 1$ we have

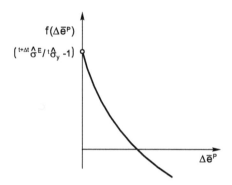

Fig. 4.4.3. Monotonic character of the governing function for mixed hardening plasticity of isotropic metal

$$
\begin{aligned}
&\alpha = 0 \\
&\hat{\sigma} = \bar{\sigma} = \sigma_y(\bar{e}^P) \\
&\hat{\mathbf{S}} = \mathbf{S} \\
&f(\Delta\bar{e}^P) = \left[{}^{t+\Delta t}\bar{\sigma}^E / \left({}^{t+\Delta t}\sigma_y + 3G\Delta\bar{e}^P\right)\right] - 1 = 0 \Big|_{M=1}
\end{aligned}
\qquad (4.4.34)
$$

If $E_P = 0$ we use the analytical solution (4.4.31) which also follows from the above relations.

In the case of *kinematic hardening*, we use $M = 0$ and also that $\hat{\sigma} = \sigma_{yv}$ (see Fig. 4.4.1). The solution for $\Delta\bar{e}^P$ follows from (4.4.29), and therefore we have

$$
\begin{aligned}
&{}^{t+\Delta t}\hat{\sigma} = \sigma_{yv} \\[2mm]
&\Delta\bar{e}^P = \left({}^{t+\Delta t}\hat{\sigma}^E - \sigma_{yv}\right) \Big/ (3G + \bar{E}_P) \Big|_{M=0}
\end{aligned}
\qquad (4.4.35)
$$

Important characteristics of the above algorithm are that the solution can be obtained for relatively large increments of strain (i.e., for relatively large increments of the effective plastic strain $\Delta\bar{e}^P$), and since the governing function $f(\Delta\bar{e}^P)$ is monotonic, the solution can be calculated very efficiently (by, e.g., Newton iteration).

With respect to the *accuracy* of the solution, the elastic constitutive relations and the yield condition are satisfied exactly at the end of the step. The stress point D in the deviatoric space (Fig. 4.4.1) is on the yield surface with the size determined by the value ${}^{t+\Delta t}\hat{\sigma}_y$ (point D' lies on the yield curve). The approximation of the solution corresponds to the use of the Euler backward method in calculating the increment of plastic strain (4.4.18), and to the use of (4.4.20) which contains the weighted moduli $\bar{E}_P$ and $\hat{E}_P$. The accuracy also depends on the degree of nonradiality of the loading within the

Table 4.4.1. Computational steps for stress integration, mixed hardening plasticity of isotropic metal

	Equation
1. Determine elastic solution	
$^{t+\Delta t}\mathbf{S}^E = 2G \, ^{t+\Delta t}\mathbf{e}''$	(4.4.12)
$^{t+\Delta t}\hat{\mathbf{S}}^E = \, ^{t+\Delta t}\mathbf{S}^E - \, ^t\boldsymbol{\alpha}$	(4.4.15)
2. Check for yielding	
$\left\| \, ^{t+\Delta t}\hat{\mathbf{S}}^E \right\| \leq \sqrt{\frac{2}{3}} \, ^t\hat{\sigma}_y$	(4.4.16)
If deformation is elastic, go to step 5.	
3. Plastic deformation in step Δt	
Find the zero of the governing function	
$f(\Delta\bar{e}^P) = 0$	(4.4.29)
4. Determine plasticity-dependent variables:	
· value $\Delta\lambda$	(4.4.24)
· stress radius $^{t+\Delta t}\hat{\mathbf{S}}$	(4.4.23)
· deviatoric stress $^{t+\Delta t}\mathbf{S}$	(4.4.19)
· increment of plastic strain $\Delta\mathbf{e}^P$	(4.4.18)
· plastic strain $^{t+\Delta t}\mathbf{e}^P = \, ^t\mathbf{e}^P + \Delta\mathbf{e}^P$	(4.4.14)
· increment of back stress $\Delta\boldsymbol{\alpha} = \hat{C}\Delta\mathbf{e}^P$	(4.4.20)
· back stress $^{t+\Delta t}\boldsymbol{\alpha} = \, ^t\boldsymbol{\alpha} + \Delta\boldsymbol{\alpha}$	
· effective plastic strain $^{t+\Delta t}\bar{e}^P = \, ^t\bar{e}^P + \Delta\bar{e}^P$	
5. Determine stresses	
$^{t+\Delta t}\sigma_m = K \, ^{t+\Delta t}e_V$	(4.4.6)
and	
$^{t+\Delta t}\sigma_{ij} = \, ^{t+\Delta t}S_{ij} + \, ^{t+\Delta t}\sigma_m\delta_{ij}$	(3.2.14)

time step, i.e., on the difference between the normals $^t\mathbf{n}$ and $^{t+\Delta t}\mathbf{n}$ in Fig. 4.4.1. In the case of radial loading within a time (load) interval $t_1 - t_2$ (the normal $\mathbf{n}$ to the yield surface does not change in this time interval) with isotropic hardening and any yield curve, or if E_P is constant within the interval in any mixed hardening, the solution at time t_2 is the same for any number of steps used in the interval $t_1 - t_2$ (see Kojic and Bathe 1987a). We give some additional details about the accuracy of the solution in Section 4.4.4

Finally, let us note that the described computational procedure for three-dimensional elastic-plastic analysis is directly applicable to two classes of two-dimensional problems (in the x, y plane): *plane strain and axisymmetric* problems. Namely, we can use all relations for three-dimensional deformations, with the following shear stresses and strains equal to zero (see Appendix A1, Fig. A1.1):

$$
\begin{aligned}
S_{xz} &= S_{yz} = 0 \\
e_{xz} &= e_{yz} = 0 \\
e_{xz}^P &= e_{yz}^P = 0
\end{aligned}
\tag{4.4.36}
$$

The condition

$$
e_{zz} = 0
\tag{4.4.37}
$$

for plane strain does not simplify the calculation of the stresses because the deviatoric components are used in the plasticity calculations, and e'_{zz}, e_{zz}^P and S_{zz} are in general different from zero.

4.4.2 Shell Conditions and Plane Stress

Many engineering structures need to be analyzed assuming a shell mathematical model, Bathe (1996); Chapelle and Bathe (1998), (2000), (2003). Considering a material point (numerical integration point) within the shell model, we now need to proceed with the stress calculation as in the previous section but using the assumption that the stress "normal" to the shell midsuface is zero. For simplicity of explanation, let us assume that we are still using the orthonormal set of axes (x, y, z), with the x and y axes in the midsurface and z normal to the shell midsurface, see Fig. 4.4.4. Then, besides the fundamental conditions used for general 3-D deformations, also the condition

$$
\sigma_{zz} = 0
\tag{4.4.38}
$$

must be satisfied.

The strain components calculated from the displacements at the end of a time step are

$$
{}^{t+\Delta t}e_{xx}, \quad {}^{t+\Delta t}e_{yy}, \quad {}^{t+\Delta t}e_{xy}, \quad {}^{t+\Delta t}e_{yz}, \quad {}^{t+\Delta t}e_{zx}
\tag{4.4.39}
$$

The total strain ${}^{t+\Delta t}e_{zz}$ can be calculated after the stresses and plastic strains have been determined. The elastic strain ${}^{t+\Delta t}e_{zz}^E$ is

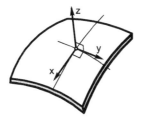

Stresses in a membrane   Sresses at a point of a shell

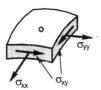   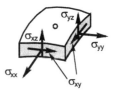

Fig. 4.4.4. Coordinate system and stresses for a membrane and shell, z-axis is normal to the membrane/shell midsurface

$$
{}^{t+\Delta t}e^{E}_{zz} = -\frac{\nu}{1-\nu}\left({}^{t+\Delta t}e^{E}_{xx} + {}^{t+\Delta t}e^{E}_{yy}\right) \tag{4.4.40}
$$

which is used to evaluate the constitutive elastic matrix $\mathbf{C}^{E}$ for the shell conditions (A1.25). Hence the mean stress ${}^{t+\Delta t}\sigma_{m}$ is

$$
{}^{t+\Delta t}\sigma_{m} = \bar{C}_{m}\left({}^{t+\Delta t}e^{E}_{xx} + {}^{t+\Delta t}e^{E}_{yy}\right) \tag{4.4.41}
$$

where

$$
\bar{C}_{m} = \frac{E}{3(1-\nu)} \tag{4.4.42}
$$

Now we can express the deviatoric stress components in the form

$$
\begin{aligned}
{}^{t+\Delta t}S_{xx} &= \bar{C}'^{E}_{11}\,{}^{t+\Delta t}e^{E}_{xx} + \bar{C}'^{E}_{12}\,{}^{t+\Delta t}e^{E}_{yy} \\
{}^{t+\Delta t}S_{yy} &= \bar{C}'^{E}_{21}\,{}^{t+\Delta t}e^{E}_{xx} + \bar{C}'^{E}_{22}\,{}^{t+\Delta t}e^{E}_{yy} \\
{}^{t+\Delta t}S_{zz} &= -\,{}^{t+\Delta t}S_{xx} - {}^{t+\Delta t}S_{yy} \\
{}^{t+\Delta t}S_{ij} &= 2G\,{}^{t+\Delta t}e^{E}_{ij} \qquad i \neq j
\end{aligned} \tag{4.4.43}
$$

The 2×2 elastic matrix $\bar{\mathbf{C}}'^{E}_{2}$ with components $\bar{C}'^{E}_{ij}$, $i,j = 1,2$, is (we use here the overbar to distinguish from general 3-D conditions, see (4.4.4) and

(4.4.9))

$$\bar{C}_2^{\prime E} = \frac{E}{3\left(1-\nu^2\right)} \begin{bmatrix} 2-\nu & -(1-2\nu) \\ -(1-2\nu) & 2-\nu \end{bmatrix} \tag{4.4.44}$$

The simpler form for the deviatoric stresses (4.4.9) is not applicable here (except for the shear components) and we use the elastic constitutive relations (4.4.43).

The stress-plastic strain relations (4.4.18) remain unchanged, as well as the constitutive relation (4.4.20) for the back stress. Therefore, the equation (4.4.22) is applicable here. The first two equations (4.4.43) can be written as

$$\begin{aligned} {}^{t+\Delta t}S_{xx} &= {}^{t+\Delta t}S_{xx}^E - \Delta\lambda \left(\bar{C}_{11}^{\prime E} \, {}^{t+\Delta t}\hat{S}_{xx} + \bar{C}_{12}^{\prime E} \, {}^{t+\Delta t}\hat{S}_{yy} \right) \\ {}^{t+\Delta t}S_{yy} &= {}^{t+\Delta t}S_{yy}^E - \Delta\lambda \left(\bar{C}_{21}^{\prime E} \, {}^{t+\Delta t}\hat{S}_{xx} + \bar{C}_{22}^{\prime E} \, {}^{t+\Delta t}\hat{S}_{yy} \right) \end{aligned} \tag{4.4.45}$$

where

$$\begin{aligned} {}^{t+\Delta t}S_{xx}^E &= \bar{C}_{11}^{\prime E} \left({}^{t+\Delta t}e_{xx} - {}^{t}e_{xx}^P \right) + \bar{C}_{12}^{\prime E} \left({}^{t+\Delta t}e_{yy} - {}^{t}e_{yy}^P \right) \\ {}^{t+\Delta t}S_{yy}^E &= \bar{C}_{21}^{\prime E} \left({}^{t+\Delta t}e_{xx} - {}^{t}e_{xx}^P \right) + \bar{C}_{22}^{\prime E} \left({}^{t+\Delta t}e_{yy} - {}^{t}e_{yy}^P \right) \end{aligned} \tag{4.4.46}$$

are the elastic solutions for the deviatoric stresses. In writing (4.4.45) we followed the procedure of derivation of (4.4.19) and used (4.4.14). From (4.4.45) and (4.4.22) the solutions for the components ${}^{t+\Delta t}\hat{S}_{xx}$ and ${}^{t+\Delta t}\hat{S}_{yy}$ of the stress radius ${}^{t+\Delta t}\hat{\mathbf{S}}$ are

$$\begin{aligned} {}^{t+\Delta t}\hat{S}_{xx} &= \left({}^{t+\Delta t}\hat{S}_{xx}^E + \Delta\lambda D_{xx} \right) / D_\lambda \\ {}^{t+\Delta t}\hat{S}_{yy} &= \left({}^{t+\Delta t}\hat{S}_{yy}^E + \Delta\lambda D_{yy} \right) / D_\lambda \end{aligned} \tag{4.4.47}$$

where

$$\begin{aligned} D_{xx} &= \left(\bar{C}_{11}^{\prime E} + \hat{C} \right) {}^{t+\Delta t}\hat{S}_{xx}^E - \bar{C}_{12}^{\prime E} \, {}^{t+\Delta t}\hat{S}_{yy}^E \\ D_{yy} &= -\bar{C}_{12}^{\prime E} \, {}^{t+\Delta t}\hat{S}_{xx}^E + \left(\bar{C}_{11}^{\prime E} + \hat{C} \right) {}^{t+\Delta t}\hat{S}_{yy}^E \\ D_\lambda &= 1 + 2\Delta\lambda \left(\bar{C}_{11}^{\prime E} + \hat{C} \right) + (\Delta\lambda)^2 \left[\left(\bar{C}_{11}^{\prime E} + \hat{C} \right)^2 - \left(\bar{C}_{12}^{\prime E} \right)^2 \right] \end{aligned} \tag{4.4.48}$$

and $^{t+\Delta t}\hat{S}_{xx}^E$ and $^{t+\Delta t}\hat{S}_{yy}^E$ are the elastic solutions for the deviatoric stresses in the stress radius according to (4.4.15). In (4.4.48) we have used the symmetry of the matrix $\bar{\mathbf{C}}'^E$ and that $\bar{C}_{22}'^E = \bar{C}_{11}'^E$. The third normal component $^{t+\Delta t}\hat{S}_{zz}$ of the stress radius follows from the deviatoric character of $^{t+\Delta t}\hat{\mathbf{S}}$, i.e.,

$$^{t+\Delta t}\hat{S}_{zz} = -\,^{t+\Delta t}\hat{S}_{xx} - \,^{t+\Delta t}\hat{S}_{yy} \tag{4.4.49}$$

The solutions for the shear components $^{t+\Delta t}\hat{S}_{ij}$ are given in (4.4.23).

The components of the stress radius are expressed in terms of one parameter $\Delta\lambda$. This parameter can be related to $\Delta\bar{e}^P$ by (4.4.24), and, as in the case of three-dimensional deformations, we use the yield curve (4.4.28) and the yield condition (4.4.17) to form the governing equation for $\Delta\bar{e}^P$

$$f\left(\Delta\bar{e}^P\right) = \frac{1}{2}\,^{t+\Delta t}\hat{\mathbf{S}} \cdot \,^{t+\Delta t}\hat{\mathbf{S}} - \frac{1}{3}\,^{t+\Delta t}\hat{\sigma}_y^2 = 0 \tag{4.4.50}$$

Once $\Delta\bar{e}^P$ has been found by solving (4.4.50), we have the solution for $^{t+\Delta t}\hat{\mathbf{S}}$, we can determine $\Delta\mathbf{e}^P$ from (4.4.18), $\Delta\boldsymbol{\alpha}$ from (4.4.20), $^{t+\Delta t}\mathbf{S}$ from (4.4.22), the mean stress $^{t+\Delta t}\sigma_m$ from (4.4.41), and the strain through the thickness as

$$^{t+\Delta t}e_{zz} = \,^{t+\Delta t}e_{zz}^E + \,^{t+\Delta t}e_{zz}^P \tag{4.4.51}$$

where $^{t+\Delta t}e_{zz}^E$ is given in (4.4.40).

The above procedure is given in Kojic and Bathe (1987b) including thermal effects and creep.

Note that a special case of this procedure is the plane stress (membrane) case, in which we have that not only the normal stress σ_{zz} is zero, but also the transverse shear stresses σ_{yz} and σ_{zx} are zero.

4.4.3 Elastic-Plastic Matrix

In this section we present the derivation of the *consistent tangent constitutive relations* (consistent with the stress integration procedure) corresponding to the end of the time step. The derivation presented below is a special case of the general considerations given in Section 4.2 (see (4.2.4) and (4.2.31)). We use the one-index (vector) notation for stresses and strains (see (A1.1) to (A1.6) in Appendix A1), e.g.,

$$[\mathbf{S}]^T = [\ S_1 = S_{xx},\ S_2 = S_{yy},\ S_3 = S_{zz},\ S_4 = S_{xy},\ S_5 = S_{yz},\ S_6 = S_{zx}]$$
$$[\hat{\mathbf{e}}]^T = [\ e_1 = e_{xx},\ e_2 = e_{yy},\ e_3 = e_{zz},\ e_4 = \gamma_{xy},\ e_5 = \gamma_{yz},\ e_6 = \gamma_{zx}]$$

$$(4.4.52)$$

Note that the total strain vector $[\hat{\mathbf{e}}]$ is defined with the engineering shear strain components, as it is common in engineering practice. For other strains, as $^{t+\Delta t}\mathbf{e}'$, $^{t+\Delta t}\mathbf{e}''$, $^{t+\Delta t}\mathbf{e}^P$, we use the tensorial components (e.g., $^{t+\Delta t}e_4' = {}^{t+\Delta t}e_{12}'$). The elastic-plastic tangent matrix $^{t+\Delta t}\mathbf{C}^{EP}$ is

$$^{t+\Delta t}C_{ij}^{EP} = \frac{\partial\ ^{t+\Delta t}\sigma_i}{\partial\ ^{t+\Delta t}e_j}$$

$$(4.4.53)$$

Figure 4.4.5 gives a graphical representation of a component $^{t+\Delta t}C_{ij}^{EP}$.

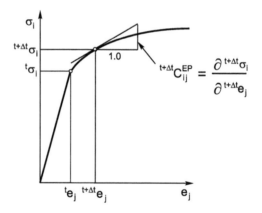

Fig. 4.4.5. Graphical representation of the component $^{t+\Delta t}C_{ij}^{EP}$

The stresses $^{t+\Delta t}\sigma_i$, deviatoric stresses $^{t+\Delta t}S_i$ and the mean stress $^{t+\Delta t}\sigma_m$ can be related as follows

$$
\begin{aligned}
^{t+\Delta t}\sigma_i &= {}^{t+\Delta t}S_i + {}^{t+\Delta t}\sigma_m & i &= 1, 2, 3 \\
^{t+\Delta t}\sigma_i &= {}^{t+\Delta t}S_i & i &= 4, 5, 6
\end{aligned}
$$

$$(4.4.54)$$

We consider the consistent elastic-plastic matrix $^{t+\Delta t}C_{ij}^{EP}$ for general three-dimensional deformations. According to (4.4.6), (4.4.7), (4.4.19), (4.4.12), (4.4.23), (4.4.24) and (4.4.28), we have

$$
{}^{t+\Delta t}\sigma_m = c_m \, {}^{t+\Delta t}e_m \qquad\qquad \text{no sum on m}
$$
$$
{}^{t+\Delta t}S_i = {}^{t+\Delta t}S_i\left({}^{t+\Delta t}\mathbf{e}'', \Delta\bar{e}^P\right) \tag{4.4.55}
$$

where $c_m = 3K$. Hence we have

$$
\boxed{
\begin{aligned}
{}^{t+\Delta t}C_{ij}^{EP} &= {}^{t+\Delta t}C_{ij}' + \frac{1}{3}\, c_m & \quad & i = 1,2,3 \\
& & & j = 1,2,3 \\[2mm]
{}^{t+\Delta t}C_{ij}^{EP} &= {}^{t+\Delta t}C_{ij}' & \quad & i = 1,2,3; \ j = 4,5,6 \\
& & & i = 4,5,6; \ j = 1,2...,6
\end{aligned}
} \tag{4.4.56}
$$

where

$$
{}^{t+\Delta t}C_{ij}' = \frac{\partial \, {}^{t+\Delta t}S_i}{\partial \, {}^{t+\Delta t}e_j} \tag{4.4.57}
$$

It is convenient to express the derivatives ${}^{t+\Delta t}C_{ij}'$ as

$$
\boxed{\;
{}^{t+\Delta t}C_{ij}' = {}^{t+\Delta t}\bar{C}_{ik}\, \frac{\partial \, {}^{t+\Delta t}e_k''}{\partial \, {}^{t+\Delta t}e_j}
\;} \tag{4.4.58}
$$

where

$$
\boxed{\;
{}^{t+\Delta t}\bar{C}_{ij} = \frac{\partial \, {}^{t+\Delta t}S_i}{\partial \, {}^{t+\Delta t}e_j''}
\;} \tag{4.4.59}
$$

and summation on k $(k = 1,2,\ldots,6)$ is implied. According to (4.4.13) and (4.4.52), the derivatives $\partial\, {}^{t+\Delta t}e_i''/\partial^{t+\Delta t}e_j$ can be expressed in the matrix form

$$
\boxed{\;
\left[\frac{\partial \, {}^{t+\Delta t}\mathbf{e}''}{\partial \, {}^{t+\Delta t}\hat{\mathbf{e}}}\right] = \begin{bmatrix} \mathbf{A} & 0 \\ 0 & \frac{1}{2}\mathbf{I}_3 \end{bmatrix}
\;} \tag{4.4.60}
$$

where $\mathbf{A}$ is the matrix defined in (3.2.70), and $\mathbf{I}_3$ is the (3×3) identity matrix. The matrix ${}^{t+\Delta t}\mathbf{C}^{EP}$, given in terms of the constant c_m and the derivatives ${}^{t+\Delta t}\bar{C}_{ij}'$ is presented in Table 4.4.2. Hence, our next task is to derive the expressions for ${}^{t+\Delta t}\bar{C}_{ij}'$.

In accordance with (4.2.4), from (4.4.19) we obtain

$$
\boxed{\;
{}^{t+\Delta t}\bar{C}_{ij}' = {}^{t+\Delta t}S_{i,j} = {}^{t+\Delta t}S_{i,j}^E - 2G\, {}^{t+\Delta t}\hat{S}_i\, \Delta\lambda_{,j} - 2G\, \Delta\lambda\, {}^{t+\Delta t}\hat{S}_{i,j}
\;}
$$
$$
\tag{4.4.61}
$$

Table 4.4.2. Elastic-plastic constitutive matrix (metal plasticity)

$\frac{1}{3}(2\bar{C}'_{11}-\bar{C}'_{12}-\bar{C}'_{13}+c_m)$	$\frac{1}{3}(2\bar{C}'_{12}-\bar{C}'_{11}-\bar{C}'_{13}+c_m)$	$\frac{1}{3}(2\bar{C}'_{13}-\bar{C}'_{11}-\bar{C}'_{12}+c_m)$	$\frac{1}{2}\bar{C}'_{14}$	$\frac{1}{2}\bar{C}'_{15}$	$\frac{1}{2}\bar{C}'_{16}$
$\frac{1}{3}(2\bar{C}'_{21}-\bar{C}'_{22}-\bar{C}'_{23}+c_m)$	$\frac{1}{3}(2\bar{C}'_{22}-\bar{C}'_{21}-\bar{C}'_{23}+c_m)$	$\frac{1}{3}(2\bar{C}'_{23}-\bar{C}'_{21}-\bar{C}'_{22}+c_m)$	$\frac{1}{2}\bar{C}'_{24}$	$\frac{1}{2}\bar{C}'_{25}$	$\frac{1}{2}\bar{C}'_{26}$
$\frac{1}{3}(2\bar{C}'_{31}-\bar{C}'_{32}-\bar{C}'_{33}+c_m)$	$\frac{1}{3}(2\bar{C}'_{32}-\bar{C}'_{31}-\bar{C}'_{33}+c_m)$	$\frac{1}{3}(2\bar{C}'_{33}-\bar{C}'_{31}-\bar{C}'_{32}+c_m)$	$\frac{1}{2}\bar{C}'_{34}$	$\frac{1}{2}\bar{C}'_{35}$	$\frac{1}{2}\bar{C}'_{36}$
$\frac{1}{3}(2\bar{C}'_{41}-\bar{C}'_{42}-\bar{C}'_{43})$	$\frac{1}{3}(2\bar{C}'_{42}-\bar{C}'_{41}-\bar{C}'_{43})$	$\frac{1}{3}(2\bar{C}'_{43}-\bar{C}'_{41}-\bar{C}'_{42})$	$\frac{1}{2}\bar{C}'_{44}$	$\frac{1}{2}\bar{C}'_{45}$	$\frac{1}{2}\bar{C}'_{46}$
$\frac{1}{3}(2\bar{C}'_{51}-\bar{C}'_{52}-\bar{C}'_{53})$	$\frac{1}{3}(2\bar{C}'_{52}-\bar{C}'_{51}-\bar{C}'_{53})$	$\frac{1}{3}(2\bar{C}'_{53}-\bar{C}'_{51}-\bar{C}'_{52})$	$\frac{1}{2}\bar{C}'_{54}$	$\frac{1}{2}\bar{C}'_{55}$	$\frac{1}{2}\bar{C}'_{56}$
$\frac{1}{3}(2\bar{C}'_{61}-\bar{C}'_{62}-\bar{C}'_{63})$	$\frac{1}{3}(2\bar{C}'_{62}-\bar{C}'_{61}-\bar{C}'_{63})$	$\frac{1}{3}(2\bar{C}'_{63}-\bar{C}'_{61}-\bar{C}'_{62})$	$\frac{1}{2}\bar{C}'_{64}$	$\frac{1}{2}\bar{C}'_{65}$	$\frac{1}{2}\bar{C}'_{66}$

All $\bar{C}'_{ij}$ are evaluated at end of time step

where the notation $(\cdot)_{,j} \equiv \partial(\cdot)/\partial^{\,t+\Delta t}e''_j$ is used for simpler writing. The derivatives $^{t+\Delta t}S^E_{i,j}$ follow from (4.4.12),

$$^{t+\Delta t}S^E_{i,j} = 2G\,\delta_{ij} \tag{4.4.62}$$

The governing parameter is $\Delta\bar{e}^P$ and, following (4.2.4), we express $\Delta\lambda_{,j}$ and $^{t+\Delta t}\hat{S}_{i,j}$ in terms of $\Delta\bar{e}^P_{,j}$ by employing (4.4.24) and (4.4.23),

$$\Delta\lambda_{,j} = \ell_P\,\Delta\bar{e}^P_{,j} \tag{4.4.63}$$

$$
^{t+\Delta t}\hat{S}_{i,j} = \frac{^{t+\Delta t}\hat{\sigma}_y}{^{t+\Delta t}\hat{\sigma}^E}\,^{t+\Delta t}S^E_{i,j} + \frac{M\,^{t+\Delta t}\hat{E}_P}{^{t+\Delta t}\hat{\sigma}_y}\,^{t+\Delta t}\hat{S}_i\,\Delta\bar{e}^P_{,j}
$$
$$
- \frac{1}{^{t+\Delta t}\hat{\sigma}^E}\,^{t+\Delta t}\hat{S}_i\,^{t+\Delta t}\hat{\sigma}^E_{,j} \tag{4.4.64}
$$

where

$$\ell_P = \frac{3}{2\,^{t+\Delta t}\hat{\sigma}_y}\left(1 - \frac{M\,^{t+\Delta t}\hat{E}_P}{^{t+\Delta t}\hat{\sigma}_y}\,\Delta\bar{e}^P\right) \tag{4.4.65}$$

and

$$^{t+\Delta t}\hat{E}_P = \frac{\partial\sigma_y}{\partial\bar{e}^P}\bigg|_{\bar{e}^P = M\,^{t+\Delta t}\bar{e}^P} \tag{4.4.66}$$

In deriving (4.4.64) we have used the expression for $^{t+\Delta t}\hat{S}_i$ which follows from (4.4.23), (4.4.26) and (4.4.27), and the definition of $^{t+\Delta t}\hat{\sigma}^E$ in (4.4.29),

$$^{t+\Delta t}\hat{S}_i = \frac{^{t+\Delta t}\hat{\sigma}_y}{^{t+\Delta t}\hat{\sigma}^E}\,^{t+\Delta t}\hat{S}^E_i \tag{4.4.67}$$

We have also used that

$$^{t+\Delta t}\hat{\sigma}_{y,j} = \frac{\partial\left[^{t+\Delta t}\hat{\sigma}_y\left(M\,^{t+\Delta t}\bar{e}^P\right)\right]}{\partial\bar{e}^P}\,\Delta\bar{e}^P_{,j} = M\,^{t+\Delta t}\hat{E}_P\Delta\bar{e}^P_{,j} \tag{4.4.68}$$

From the definition of $^{t+\Delta t}\hat{\sigma}^E$ analogous to (4.4.2), and using (4.4.62), we obtain

$$^{t+\Delta t}\hat{\sigma}^E_{,j} = \frac{3G}{^{t+\Delta t}\hat{\sigma}_y}\,^{t+\Delta t}\hat{S}'_j \tag{4.4.69}$$

where $^{t+\Delta t}\hat{S}'_j = \,^{t+\Delta t}\hat{S}_j$ for the normal terms ($j = 1, 2, 3$), and $^{t+\Delta t}\hat{S}'_j = 2\,^{t+\Delta t}\hat{S}_j$ for the shear terms ($j = 4, 5, 6$).

We now evaluate the expression (4.2.5). Hence, we differentiate the governing equation (4.4.29) with respect to $^{t+\Delta t}e''_j$ to determine $\Delta\bar{e}^P_{,j}$:

$$\frac{\partial f}{\partial (\Delta \bar{e}^P)} \Delta \bar{e}^P_{,j} + \frac{\partial f}{\partial\ {}^{t+\Delta t}e''_j} = 0 \qquad (4.4.70)$$

and then

$$\Delta \bar{e}^P_{,j} = \frac{3G}{\bar{\bar{D}}\ {}^{t+\Delta t}\hat{\sigma}_y}\ {}^{t+\Delta t}\hat{S}'_j \qquad (4.4.71)$$

where

$$\bar{D} = M\ {}^{t+\Delta t}\hat{E}_P + \alpha\left(\frac{\partial\ {}^{t+\Delta t}E_P}{\partial\ {}^{t+\Delta t}\bar{e}^P} - M\frac{\partial\ {}^{t+\Delta t}\hat{E}_P}{\partial\ {}^{t+\Delta t}\bar{e}^P}\right)\Delta\bar{e}^P + \frac{3}{2}(2G + \hat{C}) \quad (4.4.72)$$

In this expression for $\bar{D}$ we have used $\bar{E}_P = (1 - \alpha)\ {}^t E_P + \alpha\ {}^{t+\Delta t}E_P$ and $\hat{E}_P = (1 - \alpha)\ {}^t\hat{E}_P + \alpha\ {}^{t+\Delta t}\hat{E}_P$, for the weighted plastic moduli in $\hat{C}$ (see (4.4.21)), with the parameter $0 \le \alpha \le 1$.

Finally, we substitute ${}^{t+\Delta t}S^E_{i,j}$ from (4.4.62), $\Delta\lambda_{,j}$ from (4.4.63), and ${}^{t+\Delta t}\hat{S}_{i,j}$ from (4.4.64), into (4.4.61), and use (4.4.71) to obtain the coefficients ${}^{t+\Delta t}\bar{C}'_{ij}$ in compact form

$$ {}^{t+\Delta t}\bar{C}'_{ij} = \bar{C}'^m \delta_{ij} - D\ {}^{t+\Delta t}\hat{S}_i\ {}^{t+\Delta t}\hat{S}'_j \qquad (4.4.73)$$

where

$$\bar{C}'^m = 2G(1 - 2G\frac{{}^{t+\Delta t}\hat{\sigma}_y}{{}^{t+\Delta t}\hat{\sigma}^E}\Delta\lambda) \qquad (4.4.74)$$

and

$$D = \frac{6G^2}{{}^{t+\Delta t}\hat{\sigma}_y}\left[\frac{1}{\bar{D}}\left(\ell_P + \Delta\lambda\frac{M\ {}^{t+\Delta t}\hat{E}_P}{{}^{t+\Delta t}\hat{\sigma}_y}\right) - \frac{\Delta\lambda}{{}^{t+\Delta t}\hat{\sigma}^E}\right] \qquad (4.4.75)$$

Note that the matrices ${}^{t+\Delta t}\bar{C}'$ and ${}^{t+\Delta t}\mathbf{C}^{EP}$ are symmetric and that ${}^{t+\Delta t}\mathbf{C}^{EP}$ is in general a full matrix. The matrix ${}^{t+\Delta t}\mathbf{C}^{EP}$ is applicable to general three-dimensional deformations, and also to plane strain and axisymmetric conditions by deleting the corresponding rows and columns. The matrix can be used for shell or plane stress analyses as well, by enforcing the applicable zero stress conditions using static condensation (see Bathe 1996).

The application of the above general procedure to the uniaxial stress conditions is given in Example 4.5.1.

4.4.4 Accuracy Considerations

In this section we briefly consider some aspects of the solution accuracy of the stress integration procedure for the von Mises material model. Similar considerations are applicable to the stress integration using other material models. The approximations adopted in the governing parameter method for stress integration of the von Mises material model pertain to the stress-plastic strain relations (4.4.18) and the constitutive relation for the back stress (4.4.20) (see (3.2.73) and (3.2.78)),

$$
\Delta \mathbf{e}^P = \int_t^{t+\Delta t} \hat{\mathbf{S}} \, d\lambda \approx \Delta \lambda \, {}^{t+\Delta t}\hat{\mathbf{S}} \tag{4.4.76}
$$

and

$$
\Delta \boldsymbol{\alpha} = (1 - M) \int_t^{t+\Delta t} C de^P \approx \frac{2}{3}(\bar{E}_P - M\hat{\bar{E}}_P)\Delta \mathbf{e}^P \tag{4.4.77}
$$

where $\bar{E}_P$ and $\hat{\bar{E}}_P$ are weighted values of the plastic moduli E_P and $\hat{E}_P$. We now want to discuss the influence of these approximations on the solution for the stresses and plastic strains at the end of the time step, for a given strain increment $\Delta \mathbf{e}$.

According to (4.4.23), the problem of the stress integration reduces to finding the stress point D on ${}^tOD_E$ (see Fig. 4.4.6a), such that (4.4.29) is satisfied. The key variables are the normals ${}^t\mathbf{n}$ and ${}^{t+\Delta t}\mathbf{n}$ to the yield surface, defined as

$$
{}^t\mathbf{n} = {}^t\hat{\mathbf{S}} / \left\| {}^t\hat{\mathbf{S}} \right\| \tag{4.4.78}
$$

$$
{}^{t+\Delta t}\mathbf{n} = {}^{t+\Delta t}\hat{\mathbf{S}}^E / \left\| {}^{t+\Delta t}\hat{\mathbf{S}}^E \right\| \tag{4.4.79}
$$

where

$$
{}^t\hat{\mathbf{S}} = 2G \left({}^t\mathbf{e}' - {}^t\mathbf{e}^P \right) - {}^t\boldsymbol{\alpha} \tag{4.4.80}
$$

and

$$
{}^{t+\Delta t}\hat{\mathbf{S}}^E = 2G \left({}^t\mathbf{e}' + \Delta \mathbf{e}' - {}^t\mathbf{e}^P \right) - {}^t\boldsymbol{\alpha} \tag{4.4.81}
$$

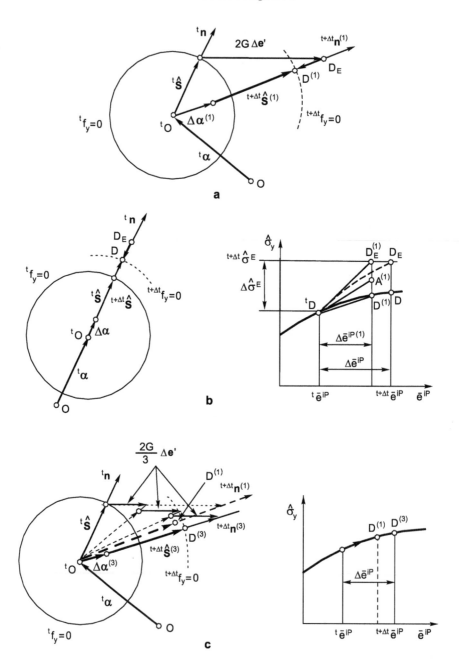

Fig. 4.4.6. Stress integration in case of nonradial and radial loading conditions. **a** One-step solution corresponding to strain increment $\Delta e'$, nonradial loading; **b** Radial loading, solution shown in deviatoric plane and on the yield curve; **c** Solution using three subincrements for the same strain increment as in **a**

Depending on the directions of the normals $^{t+\Delta t}\mathbf{n}$ and $^{t}\mathbf{n}$, we can have two distinct deformation regimes in a time step: *radial loading*, when $^{t+\Delta t}\mathbf{n} = {}^{t}\mathbf{n}$, and *non-radial loading*, when $^{t+\Delta t}\mathbf{n}$ and $^{t}\mathbf{n}$ are different. The importance of the unit normal follows from (4.4.76), which, with use of (4.4.18) and (4.4.24) to (4.4.26), can be written as

$$\Delta \mathbf{e}^P = \frac{3}{2} \int_t^{t+\Delta t} \frac{d\bar{e}^P}{\hat{\sigma}} \hat{\mathbf{S}} = \sqrt{\frac{3}{2}} \int_t^{t+\Delta t} \mathbf{n}\, d\bar{e}^P \qquad (4.4.82)$$

Hence, the increment of plastic strain $\Delta \mathbf{e}^P$ is determined through the magnitude $\|\Delta \mathbf{e}^P\| = \sqrt{1.5}\Delta \bar{e}^P$ and through the change of the normal $\mathbf{n}$ in the time step Δt.

Consider first radial loading, Fig. 4.4.6b. Then we have $\mathbf{n} = {}^{t}\mathbf{n} = {}^{t+\Delta t}\mathbf{n}$ and from (4.4.82)

$$\Delta \mathbf{e}^P = \sqrt{\frac{3}{2}} \Delta \bar{e}^P \; {}^{t}\mathbf{n} \qquad (4.4.83)$$

Also, we have that the stress radius $^{t+\Delta t}\hat{\mathbf{S}}$, the back stress $^{t+\Delta t}\boldsymbol{\alpha}$, and the deviatoric stress $^{t+\Delta t}\mathbf{S}$, can be written as

$$^{t+\Delta t}\hat{\mathbf{S}} = \sqrt{\frac{2}{3}}\, {}^{t+\Delta t}\hat{\sigma}\; {}^{t}\mathbf{n}, \quad {}^{t+\Delta t}\boldsymbol{\alpha} = \sqrt{\frac{2}{3}}\, {}^{t+\Delta t}\bar{\alpha}\; {}^{t}\mathbf{n}, \quad {}^{t+\Delta t}\mathbf{S} = \sqrt{\frac{2}{3}}\, {}^{t+\Delta t}\bar{\sigma}\; {}^{t}\mathbf{n}$$

$$(4.4.84)$$

where $^{t+\Delta t}\bar{\alpha}$ is the effective back stress in the sense of (3.2.41). Since the normal $^{t}\mathbf{n}$ is given, the accuracy of the solution for the stress, plastic strain and back stress components depends on the accuracy of the solution for the effective values in (4.4.83) and (4.4.84). In order to obtain insight into the accuracy of the solution for the effective values, we write (4.4.29) in the form

$$^{t+\Delta t}\hat{\sigma}_y + 3G\Delta \bar{e}^P + (\bar{E}_P - M\hat{\bar{E}}_P)\Delta \bar{e}^P = {}^{t+\Delta t}\hat{\sigma}^E \qquad (4.4.85)$$

where we have used (4.4.21). A graphical representation of the solution with respect to $\Delta \bar{e}^{iP}$ (hence for $\Delta \bar{e}^P$), for a given increment of deviatoric strain $\|\Delta \mathbf{e}'\|$, i.e., for a given $^{t+\Delta t}\hat{\sigma}^E$ and given yield curve $\hat{\sigma}_y(\bar{e}^{iP})$, is shown in Fig. 4.4.6b. Consider the following cases:

- One-step solution. Assume that the weighted plastic moduli $\bar{E}_P = {}^t E_P$ and $\hat{\bar{E}}_P = {}^t \hat{E}_P$ are employed. Then we have that the value ${}^{t+\Delta t}\hat{\sigma}^E$ is obtained by adding two terms linear in $\Delta\bar{e}^{iP}$ to the yield stress ${}^{t+\Delta t}\hat{\sigma}_y$; graphically, the point $D_E^{(1)}$ is reached by adding $D^{(1)}A^{(1)}$ and $A^{(1)}D_E^{(1)}$ to the yield stress ${}^{t+\Delta t}\hat{\sigma}_y^{(1)}$ corresponding to ${}^t\bar{e}^{iP} + \Delta\bar{e}^{iP(1)}$.
- Multistep solution (use of subincrements). Here the total increment $\Delta\hat{\sigma}^E$ is divided into parts and the calculation is repeated for each step, with the weighted plastic moduli $\bar{E}_P$ and $\hat{\bar{E}}_P$ changing from step to step. Hence we have the term $(\bar{E}_P - M\hat{\bar{E}}_P)\Delta\bar{e}^P$ nonlinear with respect to $\Delta\bar{e}^{iP}$, and the final solution for $\Delta\bar{e}^{iP}$ depends on the number of subincrements used. In the graphical representation, we have that the final point D_E (with the corresponding point D on the yield curve), which gives the solution $\Delta\bar{e}^{iP}$, is obtained as the intersection of the curve ${}^t DD_E$ and the line $\hat{\sigma}_y = {}^{t+\Delta t}\hat{\sigma}^E$.

It is obvious therefore that in case of a nonlinear yield curve the solution depends in general on the number of subincrements used (due to the nonlinear term $(\bar{E}_P - M\hat{\bar{E}}_P)\Delta\bar{e}^P$ with respect to $\Delta\bar{e}^{iP}$ in (4.4.85)), and therefore the solution accuracy increases with the increase of number of steps (subincrements). However, when the yield curve is linear in the considered range, all terms on the left-hand side of (4.4.85) are linear with respect to $\Delta\bar{e}^{iP}$ and, for a given increment $\Delta\hat{\sigma}^E$, the one-step solution is giving the exact value. In isotropic hardening ($M = 1$), we have $\bar{E}_P - M\hat{\bar{E}}_P = 0$ and the one-step solution is also exact for any nonlinear yield curve. The analysis of solution accuracy in the case of radial loading, which includes reverse loading conditions, is given in Example 4.5.7.

Consider next the case of nonradial loading. Then the normal **n** changes direction during the step. Hence in evaluating the integral (4.4.82), we have the error in the effective values, as discussed above, and the error due to change of the normal **n**. Both errors decrease when the step size is decreased, or if subincrementation is employed. If subincrements are used, then the algorithm of Table 4.2.2 is applied in each subincrement.

Figure 4.4.6c shows schematically the solution stages using three subincrements of the total deviatoric strain increment $\Delta e'$. The final normal ${}^{t+\Delta t}\mathbf{n}^{(3)}$ and the stress point $D^{(3)}$ in the stress space and on the yield curve are different from the one-step solution normal ${}^{t+\Delta t}\mathbf{n}^{(1)}$ and the stress point $D^{(1)}$. Of course, the solution error increases with the degree of nonradiality and magnitude of the strain increment $\|\Delta e'\|$ (in mixed hardening behavior). In general, the error can only be quantified provided specific conditions are considered. We illustrate these conclusions in Examples 4.5.9 and 4.5.10, where we show, although nonexhaustively, that good accuracy is reached in solutions of nonradial loading conditions.

4.5 Examples

The objective of this section is to present some examples of plasticity solutions. The examples are sufficiently simple that valuable insight into the
solution approach and details is gained. Also, some examples show the calculated material response in typical engineering conditions.

Example 4.5.1. Uniaxial Loading. Write the governing equations and
derive the expression for the tangent constitutive relation in the case of uniaxial loading. Assume a von Mises material with mixed hardening.

The uniaxial stress at the end of the time step is

$$^{t+\Delta t}\sigma = {}^{t+\Delta t}\sigma^E - E\,\Delta e^P \tag{a}$$

where

$$^{t+\Delta t}\sigma^E = E\left({}^{t+\Delta t}e - {}^t e^P\right) \tag{b}$$

Here $^{t+\Delta t}e$ is the total strain at the end of the time step, $^t e^P$ is the plastic
strain at the start of the time step, and Δe^P is the increment of the plastic
strain, all in the loading direction. With use of (4.4.20) and (4.4.21), the yield
condition (4.4.17) can be written in the form

$$^{t+\Delta t}f = 3\,^{t+\Delta t}f_y = \left({}^{t+\Delta t}\tilde{S}\right)^2 - \left({}^{t+\Delta t}\hat{\sigma}_y\right)^2 = 0 \tag{c}$$

where

$$^{t+\Delta t}\tilde{S} = {}^{t+\Delta t}\sigma - \frac{3}{2}\,{}^t\alpha - (\bar{E}_P - M\hat{E}_P)\Delta e^P \tag{d}$$

Here $^t\alpha$ is the back stress component in the loading direction.

Using the increment Δe^P as the governing parameter, we find that the
governing equation (b) of Table 4.2.1 is now (c) (see also (4.4.29)), since all
terms in (c) can be expressed in terms of Δe^P. We then proceed to determine
the unknowns Δe^P, $^{t+\Delta t}\sigma$ and $^{t+\Delta t}\alpha$ as follows. For a trial value Δe^P we
calculate $^{t+\Delta t}\sigma$ from (a), $^{t+\Delta t}\hat{\sigma}_y$ from (4.4.28), and determine $\bar{E}_P$ and $\hat{E}_P$
from the yield curve (see (4.4.21) and (3.2.77)). Then, we evaluate $^{t+\Delta t}\tilde{S}$
from (d) and iterate on Δe^P until (c) is satisfied. The increment $\Delta\alpha$ follows
from (4.4.20).

The tangent constitutive relation is (see (4.4.53))

$$^{t+\Delta t}C^{EP} = \frac{\partial\,^{t+\Delta t}\sigma}{\partial\,^{t+\Delta t}e} \tag{e}$$

From (a) we obtain

$$^{t+\Delta t}C^{EP} = E - E\,\frac{\partial\left(\Delta e^P\right)}{\partial\,^{t+\Delta t}e} \tag{f}$$

We now proceed to calculate $\partial(\Delta e^P)/\partial^{t+\Delta t}e$ by using (4.2.5). Let us first summarize the terms in (4.2.5) that correspond to this example. We have that:

a) The governing parameter p is Δe^P.

b) The governing function is $^{t+\Delta t}f = 3\,^{t+\Delta t}f_y$ given in (c).

c) There is only one equation in (4.2.5), corresponding to the strain $^{t+\Delta t}e$.

d) By inspection of (c), with (a), (b) and (d), we see that the governing function $^{t+\Delta t}f$ depends on $^{t+\Delta t}\sigma$, Δe^P and $^{t+\Delta t}e$. Hence we do not have the terms corresponding to $^{t+\Delta t}e^{IN}$ and $^{t+\Delta t}\beta$ in (4.2.5) (the internal variable $^{t+\Delta t}\alpha$ has already been included in (d), and the inelastic strain $^{t+\Delta t}e^P$, or rather Δe^P, is taken as the governing parameter). Note that (c) can also be written as

$$^{t+\Delta t}f = 3\,^{t+\Delta t}f_y = \left(^{t+\Delta t}\sigma - \frac{3}{2}\,^{t+\Delta t}\alpha\right)^2 - (^{t+\Delta t}\hat{\sigma}_y)^2 = 0$$

In this case we have that the governing function is $^{t+\Delta t}f(^{t+\Delta t}\sigma,\,^{t+\Delta t}\alpha,\,\Delta e^P,\,^{t+\Delta t}e)$ and we would have the nonzero derivatives:

$$\frac{\partial^{t+\Delta t}f}{\partial^{t+\Delta t}\sigma_{11}} = \frac{\partial^{t+\Delta t}f}{\partial^{t+\Delta t}\sigma}, \qquad \frac{\partial^{t+\Delta t}f}{\partial^{t+\Delta t}\beta_{11}} = \frac{\partial^{t+\Delta t}f}{\partial^{t+\Delta t}\alpha}$$

$$\frac{\partial^{t+\Delta t}f}{\partial p} = \frac{\partial^{t+\Delta t}f}{\partial(\Delta e^P)} \quad \text{and} \quad \frac{\partial^{t+\Delta t}f}{\partial^{t+\Delta t}e_{11}} = \frac{\partial^{t+\Delta t}f}{\partial^{t+\Delta t}e}$$

But for this example, it is more direct to use $^{t+\Delta t}f(^{t+\Delta t}\sigma,\,\Delta e^P,\,^{t+\Delta t}e)$ as given in (c). Then the nonzero terms in (4.2.5) are:

$$\frac{\partial^{t+\Delta t}f}{\partial^{t+\Delta t}\sigma}\frac{\partial^{t+\Delta t}\sigma}{\partial(\Delta e^P)}, \qquad \left.\frac{\partial^{t+\Delta t}f}{\partial(\Delta e^P)}\right|_{\sigma=const} \quad \text{and} \quad \left.\frac{\partial^{t+\Delta t}f}{\partial^{t+\Delta t}e}\right|_{(\Delta e^P)=const} \qquad \text{(g)}$$

Further we derive the corresponding expressions for the terms in (g). According to (a) to (d) we first have

$$\frac{\partial^{t+\Delta t}f}{\partial^{t+\Delta t}\sigma}\frac{\partial^{t+\Delta t}\sigma}{\partial(\Delta e^P)} = \frac{\partial^{t+\Delta t}f}{\partial^{t+\Delta t}\tilde{S}}\frac{\partial^{t+\Delta t}\tilde{S}}{\partial^{t+\Delta t}\sigma}\frac{\partial^{t+\Delta t}\sigma}{\partial(\Delta e^P)} = -2\left(^{t+\Delta t}\tilde{S}\right)E \qquad \text{(h)}$$

next,

$$\left.\frac{\partial^{t+\Delta t}f}{\partial(\Delta e^P)}\right|_{\sigma=const} = 2\,^{t+\Delta t}\tilde{S}\left\{-\bar{E}_P + M\hat{\bar{E}}_P - \right.$$

$$\left.\left[\frac{\partial\bar{E}_P}{\partial(\Delta e^P)} - M\frac{\partial\hat{\bar{E}}_P}{\partial(\Delta e^P)}\right]\Delta e^P\right\} - 2\,^{t+\Delta t}\hat{\sigma}_y\,M\,^{t+\Delta t}\hat{E}_P \qquad \text{(i)}$$

and finally,

$$\frac{\partial^{t+\Delta t} f}{\partial^{t+\Delta t} e}\bigg|_{\Delta e^P=const} = \frac{\partial^{t+\Delta t} f}{\partial^{t+\Delta t}\tilde{S}}\frac{\partial^{t+\Delta t}\tilde{S}}{\partial^{t+\Delta t}\sigma}\frac{\partial^{t+\Delta t}\sigma}{\partial^{t+\Delta t} e}\bigg|_{\Delta e^P=const} = 2\left(^{t+\Delta t}\tilde{S}\right)E$$

(j)

Here, $^{t+\Delta t}\hat{E}_P$ is the plastic modulus given in (4.4.66).

Substituting (h) to (j) into (4.2.5), and taking into account that $^{t+\Delta t}\tilde{S} = {}^{t+\Delta t}\hat{\sigma}_y$, which follows from (c), we obtain the solution for $\partial\left(\Delta e^P\right)/\partial^{t+\Delta t} e$ as

$$\frac{\partial\left(\Delta e^P\right)}{\partial^{t+\Delta t} e} = \frac{E}{D_P}$$

(k)

where

$$D_P = E + \bar{E}_P + M\left(^{t+\Delta t}\hat{E}_P - \hat{E}_P\right) + \left[\frac{\partial\bar{E}_P}{\partial\left(\Delta e^P\right)} - M\frac{\partial\hat{E}_P}{\partial\left(\Delta e^P\right)}\right]\Delta e^P$$

(l)

We now substitute $\partial\left(\Delta e^P\right)/\partial^{t+\Delta t} e$ into (f) and obtain $^{t+\Delta t}C^{EP}$. For a bilinear stress-strain relation we have $\partial\bar{E}_P/\partial\left(\Delta e^P\right) = \partial\hat{E}_P/\partial\left(\Delta e^P\right) = 0$, and from (k) and (l) follows

$$\frac{\partial\left(\Delta e^P\right)}{\partial^{t+\Delta t} e} = \frac{E}{E + E_P}$$

(m)

With use of (3.2.5) we obtain from (f)

$$^{t+\Delta t}C^{EP} = E_T$$

(n)

Note that the same result for stress will be obtained when the general algorithm given in Table 4.2.2 is applied, with the governing parameter

$$\beta = \left\|\Delta e^P\right\| = \left(\frac{3}{2}\right)^{1/2}\Delta\bar{e}^P = \left(\frac{3}{2}\right)^{1/2}\left|\Delta e^P\right|$$

(o)

Here we have used the relation (3.2.46). Regarding the calculation of $^{t+\Delta t}C^{EP}$, we can evaluate the expression (4.2.31) as follows. First, instead of (a) we have now

$$^{t+\Delta t}\sigma = {}^{t+\Delta t}\sigma^E - \left(\frac{2}{3}\right)^{1/2}E\left\|\Delta e^P\right\|$$

(p)

and then

$$\Delta\sigma' = -\frac{\partial(^{t+\Delta t}\sigma)}{\partial(\|\Delta e^P\|)} = \left(\frac{2}{3}\right)^{1/2}E$$

(q)

Using $\left\|\Delta e^P\right\|$ as the internal variable, we have

$$\beta' = 1 \tag{r}$$

Also,

$$^{t+\Delta t}\tilde{S}_{,\sigma} = 1 \tag{s}$$

and with β as defined in (o)

$$^{t+\Delta t}\tilde{S}_{,\beta} = \left(\frac{2}{3}\right)^{1/2} \frac{\partial^{t+\Delta t}\tilde{S}}{\partial(\Delta e^P)}, \qquad ^{t+\Delta t}\hat{\sigma}_{y,\beta} = M\left(\frac{2}{3}\right)^{1/2} {}^{t+\Delta t}\hat{E}_P \tag{t}$$

With the above expressions entering (4.2.32) and (4.2.33) we obtain the derivative $\partial(\|\Delta\mathbf{e}^P\|)/\partial^{t+\Delta t}e$ and the coefficient a_σ, and then $^{t+\Delta t}C^{EP}$ from (4.2.31). For a bilinear stress-strain relation we obtain the expression (m) for $^{t+\Delta t}C^{EP}$.

Example 4.5.2. Elastic-Plastic Deformation of Beam and Pipe. Derive the expressions for the stress integration assuming von Mises plasticity with mixed hardening in the case of beam and pipe deformations.

Beam. In the case of beam deformations the stresses different from zero are (see Appendix A1, Fig. A1.2b and (A1.27))

$$S_{xx} = -2S_{yy} = -2S_{zz} = \frac{2}{3}\sigma_{xx} = \frac{2}{3}Ee_{xx}^E \tag{a}$$

The normal deviatoric stress $^{t+\Delta t}S_{xx}$ and the two nonzero shear stresses $^{t+\Delta t}S_{xy}$ and $^{t+\Delta t}S_{xz}$, denoted by $^{t+\Delta t}S_i$, $i = 1, 2, 3$, can be expressed in the form (4.4.11),

$$^{t+\Delta t}S_i = {}^{t+\Delta t}S_i^E - C_{ii}^{\prime E}\Delta e_i^P \qquad \text{no sum on } i \tag{b}$$

where

$$^{t+\Delta t}S_i^E = C_{ii}^{\prime E}\left(^{t+\Delta t}e_i - {}^t e_i^P\right) \tag{c}$$

The elastic coefficients $C_{ii}^{\prime E}$ are the diagonal terms of the matrix (see (A1.28))

$$\mathbf{C}^{\prime E} = \begin{bmatrix} \frac{2}{3}E & & \\ & 2G & \\ & & 2G \end{bmatrix} \tag{d}$$

The derivation of the relations for the stress integration is as for the general three-dimensional deformations. Therefore, the solution for the stress radius $^{t+\Delta t}\hat{\mathbf{S}}$ is analogous to (4.4.23),

$$^{t+\Delta t}\hat{S}_i = \frac{^{t+\Delta t}\hat{S}_i^E}{1 + \left(C_{ii}^{\prime E} + \hat{C}\right)\Delta\lambda} \qquad i = 1, 2, 3 \tag{e}$$

where $\Delta\lambda$ is given by (4.4.24). With use of (a) we have now the governing equation (4.4.50) as

$$f(\Delta\bar{e}^P) = \frac{3}{4}\,{}^{t+\Delta t}\hat{S}_1^2 + {}^{t+\Delta t}\hat{S}_2^2 + {}^{t+\Delta t}\hat{S}_3^2 - \frac{1}{3}\,{}^{t+\Delta t}\hat{\sigma}_y^2 = 0 \qquad (f)$$

Hence, we assume $\Delta\bar{e}^P$, use the yield curve (4.4.28), calculate $\Delta\lambda$ from (4.4.24), determine ${}^{t+\Delta t}\hat{S}_i$ from (e), and iterate until (f) is satisfied. The other computational details follow from Table 4.4.1.

Pipe. In the case of pipe deformations, the non-zero stresses are as shown in Fig. E.4.5-2. We consider a pipe loaded by an internal pressure p so that the hoop stress σ_{cc} can be determined as the static equivalent,

$$\sigma_{cc} = \frac{a_i}{\delta}p \qquad (g)$$

where a_i and δ are the internal radius of the cross-section and the wall thickness, respectively. We have plane stress conditions, with $\sigma_{bb} = 0$, two shear stresses σ_{ab} and σ_{ac}, and the known normal stress σ_{cc}. A detailed analysis of pipe deformations is given in, e.g., Bathe et al. (1983).

We employ the elastic constitutive relation for the axial stress ${}^{t+\Delta t}\sigma_{aa}$,

$$^{t+\Delta t}\sigma_{aa} = \frac{E}{1-\nu^2}({}^{t+\Delta t}e_{aa} - {}^{t+\Delta t}e_{aa}^P + \nu\,{}^{t+\Delta t}e_{cc}^E) \qquad (h)$$

and for the elastic strain ${}^{t+\Delta t}e_{cc}^E$,

$$^{t+\Delta t}e_{cc}^E = \frac{1}{E}({}^{t+\Delta t}\sigma_{cc} - \nu\,{}^{t+\Delta t}\sigma_{aa}) \qquad (i)$$

Substituting (i) into (h) we obtain

$$^{t+\Delta t}\sigma_{aa} = E\,({}^{t+\Delta t}e_{aa} - {}^{t+\Delta t}e_{aa}^P) + \nu\,{}^{t+\Delta t}\sigma_{cc} \qquad (j)$$

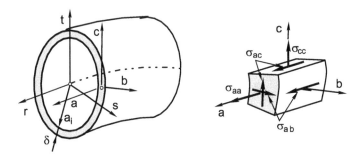

Fig. E.4.5-2. Nonzero stresses at a material point in the pipe wall

The deviatoric stresses $^{t+\Delta t}S_{aa}$, $^{t+\Delta t}S_{cc}$ and $^{t+\Delta t}S_{bb}$ can be expressed as

$$
\begin{aligned}
^{t+\Delta t}S_{aa} &= {}^{t+\Delta t}S^E_{aa} - C'^E_{aa}\,\Delta e^P_{aa} & ^{t+\Delta t}S_{bb} &= -{}^{t+\Delta t}S_{aa} - {}^{t+\Delta t}S_{cc} \\
^{t+\Delta t}S_{cc} &= {}^{t+\Delta t}S^E_{cc} + C'^E_{ac}\,\Delta e^P_{aa}
\end{aligned} \tag{k}
$$

where

$$
\begin{aligned}
^{t+\Delta t}S^E_{aa} &= C'^E_{aa}\left(^{t+\Delta t}e_{aa} - {}^t e^P_{aa}\right) - \frac{1-2\nu}{3}\,^{t+\Delta t}\sigma_{cc} \\[2mm]
^{t+\Delta t}S^E_{cc} &= -C'^E_{ac}\left(^{t+\Delta t}e_{aa} - {}^t e^P_{aa}\right) + \frac{2-\nu}{3}\,^{t+\Delta t}\sigma_{cc}
\end{aligned} \tag{l}
$$

and

$$
C'^E_{aa} = \frac{2}{3}E \qquad C'^E_{ac} = \frac{1}{3}E \tag{m}
$$

We next use the flow rule (4.4.18) and the relation (4.4.22) to express $^{t+\Delta t}S_{aa}$ and $^{t+\Delta t}S_{cc}$. Following the procedure given by (4.4.19) to (4.4.23), we obtain the solutions for the normal components of the stress radius $^{t+\Delta t}\hat{S}_{aa}$, $^{t+\Delta t}\hat{S}_{cc}$ and $^{t+\Delta t}\hat{S}_{bb}$, as

$$
^{t+\Delta t}\hat{S}_{aa} = \frac{^{t+\Delta t}\hat{S}^E_{aa}}{1 + \Delta\lambda\left(C'^E_{aa} + \hat{C}\right)} \qquad
^{t+\Delta t}\hat{S}_{cc} = \frac{^{t+\Delta t}\hat{S}^E_{cc} + \Delta\lambda C'^E_{ac}\,^{t+\Delta t}\hat{S}_{aa}}{1 + \Delta\lambda\hat{C}}
$$

$$
^{t+\Delta t}\hat{S}_{bb} = -{}^{t+\Delta t}\hat{S}_{aa} - {}^{t+\Delta t}\hat{S}_{cc} \tag{n}
$$

The shear components $^{t+\Delta t}\sigma_{ab}$ and $^{t+\Delta t}\sigma_{ac}$ are given by (e) for the beam.

With the components of the stress radius $^{t+\Delta t}\hat{S}$, expressed by (e) and (n), we can follow the above computational steps for the beam. The yield condition (4.4.50) must be satisfied at the end of the time step.

The calculation of the *consistent tangent elastic-plastic matrix* $^{t+\Delta t}C^{EP}$ for the beam or pipe conditions can be performed in two ways: 1) by appropriate differentiations with respect to the strains $^{t+\Delta t}e_i$ of the above governing relations; or, 2) by implementation of the expressions given in Section 4.4.3 for the general three-dimensional deformation, with the corresponding static condensation. Note that in using approach 2), we first have to determine $^{t+\Delta t}\mathbf{S}^E$ according to (4.4.12) and then apply the procedure of Section 4.4.3.

Example 4.5.3. Bending of a Beam. Consider a beam subjected to bending in the x, y plane. Assume that the distributions of the normal strain e_{xx} and the shear strain γ_{xy} over the beam cross-section are

$$
e_{xx} = k_e y \qquad \gamma_{xy} = k_\gamma \tag{a}
$$

where k_e and k_γ are constants. The geometric data of the cross-section, the material data and the strain distributions corresponding to the constants

$$k_e = 1.7453 \times 10^{-4} \quad [1/\text{mm}] \qquad k_\gamma = -1 \times 10^{-3} \tag{b}$$

are given in Fig. E.4.5-3a.

Determine the stress distribution over the cross-section, and the bending moment and the transversal force corresponding to the calculated stresses.

The elastic solution for the effective stress at the top or bottom points is

$$\bar{\sigma}^E_{max} = \sqrt{(\sigma^E_{xx})^2_{max} + 3(\sigma^E_{xy})^2} = 3493.1 \quad \text{MPa} \tag{c}$$

Since $\bar{\sigma}^E_{max} > \sigma_{yv} = 200$, we have elastic-plastic deformations within the cross-section. Using $\Delta\lambda = 0$ and $^{t+\Delta t}\hat{\sigma}_y = \sigma_{yv} = 200$ in (f) of Example 4.5.2, we obtain the equation

$$\frac{E^2}{3}k_e^2 y_0^2 + G^2\gamma_{xy}^2 - \frac{1}{3}\sigma_{yv}^2 = 0 \tag{d}$$

from which we obtain the solution for the elastic-plastic boundary

$$y_0 = \pm 4.2732 \quad \text{mm} \tag{e}$$

In writing (d) we have employed (c) and (d) of Example 4.5.2, and also $^te^P_{xx} = {}^te^P_{xy} = 0$; $\sigma_{xz} = 0$, $e_{xz} = 0$.

For $|y| > |y_0|$ we have elastic-plastic deformations. Since the material behavior in compression is the same as in tension, and the strain distribution is as shown in Fig. E.4.5-3b, we consider only the part $y \geq 0$ of the cross-section. The computational procedure of Example 4.5.2 is implemented and the solution is obtained using one step, with 20 values of the coordinate y: the first is $y = y_0$, the second is $y = 10.0$, and then the increment $\Delta y = 5$ is used up to the value $y = 100$. The governing equation (f) of Example 4.5.2 is solved by a bisection procedure (see Section 2.1.1).

We give the numerical results for the top point $(y = h/2 = 100)$:

$$\hat{\sigma} = 213.17 \qquad \sigma_{xx} = 216.34 \qquad \sigma_{xy} = -4.17$$
$$\hat{S}_{xx} = 142.03 \qquad \hat{S}_{xy} = -4.10$$
$$e^P_{xx} = 1.6372 \times 10^{-2} \qquad \gamma^P_{xy} = -9.4584 \times 10^{-4}$$

Figure E.4.5-3c shows the stress distributions and the distributions of plastic strains for the beam cross section. Note that the normal stress σ_{xx} increases slightly in the plastic region because the hardening of the material is small. The shear stress is much smaller in the plastic than in the elastic region; this is due to the fact that the material is almost perfectly plastic (see, e.g., Prager 1959; Prager and Hodge 1968).

The bending moment and transversal force at the cross-section, calculated from the solution for stresses, are

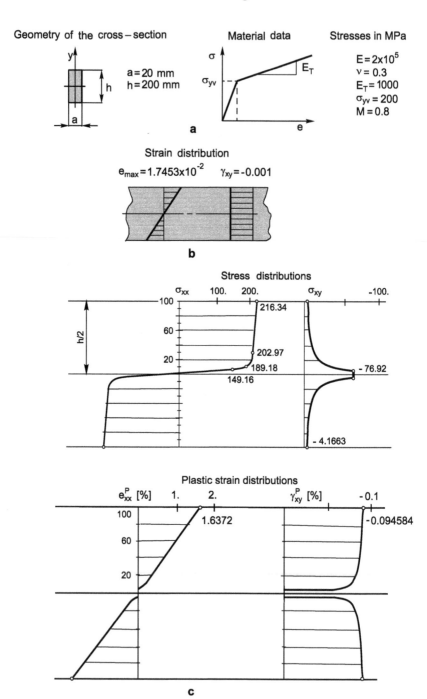

Fig. E.4.5-3. Geometric and material data and solution results for elastic-plastic deformation of a beam cross-section

$$M_z = - \int_{-h/2}^{h/2} y\sigma_{xx}\, dy = -2.0968 \times 10^6 \quad \text{Nmm/mm}$$

$$F_T = \int_{-h/2}^{h/2} \sigma_{xy}\, dy = -3835.99 \quad \text{N/mm}$$

The moment M_z and the transversal force F_T correspond to a unit beam thickness.

Example 4.5.4. Restrained Pipe. A pipe, shown in Fig. E.4.5-4a, restrained at both ends, is subjected to internal pressure. Determine the stresses, strains and the support reactions R_A and R_B using a one-step and then multi-step solutions.

We first define the physical conditions relevant for this problem. They are:

• total strain in axial direction is equal to zero,

$$e_{aa} = 0 \tag{a}$$

• hoop stress is determined as the static equivalent of the internal pressure,

$$\sigma_{cc} = \sigma_{cc}^{(p)} = \frac{a_i}{\delta}p \tag{b}$$

• stress through the pipe thickness and all shear stresses are equal to zero, i.e.,

$$\sigma_{bb} = 0, \qquad \sigma_{ij} = 0 \quad i \neq j \tag{c}$$

We employ the basic equations derived in Example 4.5.2 (see also Example 3.2.3). The expressions for $^{t+\Delta t}\hat{S}_{ij}$ are given by equation (n) of that example. Applying the above condition (a) we obtain (see (l) of Example 4.5.2)

$$^{t+\Delta t}\hat{S}_{aa}^E = -\frac{1-2\nu}{3}\,^{t+\Delta t}\sigma_{cc} - {}^{t}\alpha_{aa} - \frac{2}{3}E\,{}^{t}e_{aa}^P$$

$$^{t+\Delta t}\hat{S}_{cc}^E = \frac{2-\nu}{3}\,^{t+\Delta t}\sigma_{cc} - {}^{t}\alpha_{cc} + \frac{1}{3}E\,{}^{t}e_{aa}^P \tag{d}$$

The support reactions R_A and R_B, with the directions shown in Fig. E.4.5-4a, are

$$R_A = R_B = -\pi\,\delta\,D_m\,\sigma_{aa} + \pi\,a_i^2\,p \tag{e}$$

where $D_m = D - \delta = 98$, $a_i = D/2 - \delta = 48$.

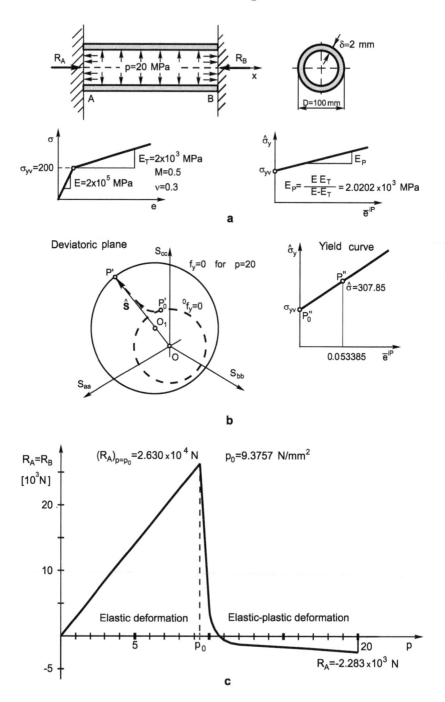

Fig. E.4.5-4. Restrained pipe under internal pressure. **a** Dimensions and material data; **b** Stress states corresponding to $p = p_0 = 9.3757$ and to $p = 20$ MPa; **c** Support reactions in terms of internal pressure

Following the computational procedure of Example 4.5.2, we first solve this example with *one step* and $p = 20$. Besides the relations derived in Example 4.5.2, we employ the expression (4.4.24) for $\Delta\lambda$,

$$\Delta\lambda = \frac{3}{2ME_P}\left(1 - \frac{{}^t\hat\sigma_y}{{}^{t+\Delta t}\hat\sigma}\right) \tag{f}$$

In the one-step solution we use ${}^t\hat\sigma_y = \sigma_{yv} = 200$. The solution results are

$$\sigma_{aa} = 238.17 \quad \hat\sigma = 307.85 \quad \text{N/mm}^2$$
$$e^P_{aa} = -4.7083 \times 10^{-4} \quad e^P_{bb} = -9.2229 \times 10^{-2} \quad e^P_{cc} = 9.2700 \times 10^{-2}$$
$$e_{bb} = -9.3306 \times 10^{-2} \quad e_{cc} = 9.4742 \times 10^{-2} \tag{g}$$
$$\alpha_{aa} = -0.31710 \quad \alpha_{bb} = -62.107 \quad \alpha_{cc} = 62.424 \quad \text{N/mm}^2$$
$$R_A = R_B = -1887.1 \quad \text{N}$$

Figure E.4.5-4**b** shows the initial and the yield surface after deformation, with the stress points P' and P'' on the yield surface and on the yield curve.

We next determine the solution *using 12 steps*: the first step corresponds to start of yielding under pressure $p_0 = 9.3757$, then $p = 10.$, and an additional 10 steps with increments of pressure $\Delta p = 1$ are used. The pressure p_0 is obtained by calculating the hoop stress $(\sigma_{cc})_0$ at the start of yielding and using (b). The point P'_0 in the deviatoric plane has coordinates

$$S_{aa} = \hat S_{aa} = -30.002 \quad S_{bb} = \hat S_{bb} = -97.508$$
$$S_{cc} = \hat S_{cc} = 127.51 \tag{h}$$

The loading path $P'_0 P'$ is shown by a dashed line in Fig. E.4.5-4**b**. We give the results for the last step for several quantities in order to compare them with the one-step solution:

$$\sigma_{aa} = 238.81 \quad R_A = R_B = -2283.3 \tag{i}$$
$$e^P_{aa} = -4.7405 \times 10^{-4} \quad e^P_{bb} = -9.2220 \times 10^{-2} \quad e^P_{cc} = 9.2700 \times 10^{-2}$$

As can be seen, the 1-step solution and the 12-step solution do not differ significantly (the loading is close to radial conditions, see Section 4.4.4).

Figure E.4.5-4**c** shows the dependence of the reaction R_A (with $R_B = R_A$) on the internal pressure p. During the elastic deformation we have that (see (j) of Example 4.5.2)

$$\sigma_{aa} = \nu\sigma_{cc} = \nu\frac{a_i}{\delta}p \tag{j}$$

and then, using (e), we obtain

$$R^E_A = R^E_B = \pi a_i\,(a_i - \nu D_m)\,p = 2804.8\,p \tag{k}$$

showing the linear dependence of the reactions on the pressure. When the deformation is elastic-plastic, the reaction is

$$R_A = R_A^E + \pi \delta D_m E e_{aa}^P \tag{1}$$

Since the plastic strain e_{aa}^P is negative due to plastic flow in the hoop direction ($e_{cc}^P > 0$) and free contraction in the b-direction, we obtain from (1) that R_A decreases during plastic deformation.

Finally, we note that this pipe problem can be solved using only one plane stress finite element restrained in one direction and subjected to extension in the other direction by the hoop stress calculated from (b).

Example 4.5.5. Straight Pipe With Free End. A straight capped pipe is loaded by the axial force F and the internal pressure p. The pipe dimensions and the material data are given in Fig. E.4.5-5a. Assuming that the internal pressure is constant and the force F varies with time (see the figure), determine the pipe deformation using two steps. The von Mises material has the mixed hardening characteristic with the parameter $M = 0.2$.

In this example we know the loading from which the stresses can be calculated, hence the task is to determine the strains from the known stresses.

Let us start by computing the quantities which are constant in further calculations. The pressure static equivalents $\sigma_{aa}^{(p)}$ and $\sigma_{cc}^{(p)}$ are

$$\sigma_{aa}^{(p)} = \frac{a_i^2}{D_m \delta} p = 20.7337 \qquad \sigma_{cc}^{(p)} = \frac{a_i}{\delta} p = 42.4444 \tag{a}$$

where $a_i = D/2 - \delta = 145.542$ is the pipe internal radius. The hoop pressure static equivalent $\sigma_{cc}^{(p)}$ is given by (g) of Example 4.5.2, while the axial pressure static equivalent follows from the equation

$$F_a^{(p)} = \pi a_i^2 p = \pi D_m \delta \sigma_{aa}^{(p)}$$

where $F_a^{(p)}$ is the axial force due to pressure, and $D_m = D - \delta$. All stresses are in (MPa) and lengths in (mm).

Consider the *first step*. The stresses and deviatoric stresses at the end of the first step are

$$\begin{aligned}
{}^1\sigma_{aa} &= {}^1F/A + \sigma_{aa}^{(p)} = 3.3230 \times 10^2 & {}^1\sigma_{cc} &= \sigma_{cc}^{(p)} = 42.4444 \\
{}^1S_{aa} &= 2/3\,{}^1\sigma_{aa} - 1/3\,\sigma_{cc}^{(p)} = 2.0739 \times 10^2 \\
{}^1S_{bb} &= -0.5({}^1S_{aa} + \sigma_{cc}^{(p)}) = -1.2491 \times 10^2 \\
{}^1S_{cc} &= -0.5({}^1S_{aa} - \sigma_{cc}^{(p)}) = -8.2470 \times 10^1
\end{aligned} \tag{b}$$

where $A = D_m \pi \delta$. The shear stresses are equal to zero. Since the effective stress ${}^1\bar{\sigma}$, calculated from (3.2.39) and (b), is greater than σ_{yv}, the material deforms plastically in the load step.

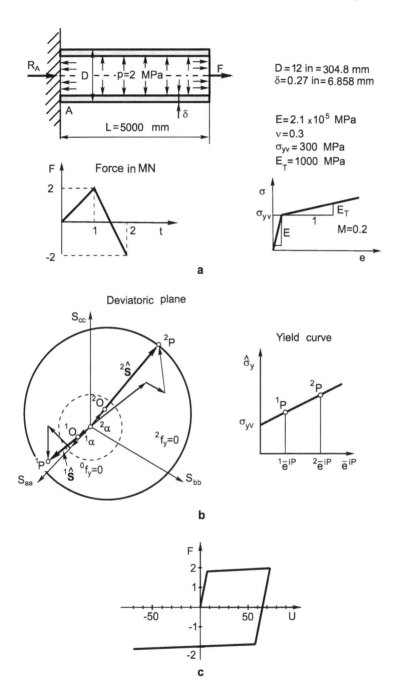

Fig. E.4.5-5. Straight pipe under internal pressure and axial loading. **a** Geometric and material data, and loading conditions; **b** Stress states for steps 1 and 2; **c** Axial force versus axial displacement of the pipe end

In order to determine the plastic strains we use (4.4.22) and obtain the relations

$$
{}^{1}\hat{S}_i = \frac{{}^{1}S_i}{1 + \Delta\lambda\hat{C}} \tag{c}
$$

where $i = 1, 2, 3$ corresponds to the a, b, c axes. Substituting ${}^{1}S_i$ into the yield condition (4.4.27), and using the bilinear stress-strain relation given in Fig. E.4.5-5a, we obtain

$$
\Delta\lambda = \frac{3}{2ME_P}\left(1 - \frac{\sigma_{yv}}{{}^{1}\hat{\sigma}}\right) \tag{d}
$$

$$
{}^{1}\hat{\sigma} = M\,{}^{1}\bar{\sigma} + (1 - M)\sigma_{yv} = 3.0265 \times 10^2 \tag{e}
$$

From (c) to (e) follows

$$
\Delta\lambda = 6.532 \times 10^{-5}
$$
$$
{}^{1}\hat{S}_{aa} = 2.0037 \times 10^2 \quad {}^{1}\hat{S}_{bb} = -1.2069 \times 10^2 \quad {}^{1}\hat{S}_{cc} = -79.681 \tag{f}
$$

The increments of plastic strains can now by obtained from (4.4.18),

$$
\begin{aligned}
{}^{1}e_{aa}^P &= \Delta e_{aa}^P = \Delta\lambda\,{}^{1}\hat{S}_{aa} = 1.3088 \times 10^{-2} \\
{}^{1}e_{bb}^P &= \Delta e_{bb}^P = \Delta\lambda\,{}^{1}\hat{S}_{bb} = -7.8834 \times 10^{-3} \\
{}^{1}e_{cc}^P &= \Delta e_{cc}^P = \Delta\lambda\,{}^{1}\hat{S}_{cc} = -5.2048 \times 10^{-3}
\end{aligned} \tag{g}
$$

We calculate the elastic strains from Hooke's law and obtain the total strains as, for example,

$$
{}^{1}e_{aa} = \frac{1}{E}\left({}^{1}\sigma_{aa} - \nu\sigma_{cc}\right) + {}^{1}e_{aa}^P = 1.4610 \times 10^{-2} \tag{h}
$$

As a check, we can use (n) and (l) of Example 4.5.2 to calculate ${}^{1}\hat{S}_{aa}$,

$$
{}^{1}\hat{S}_{aa} = \frac{2E\,{}^{t}e_{aa} - (1 - 2\nu)\sigma_{cc}}{3 + 2\Delta\lambda\left[E + (1 - M)E_P\right]} = 2.0037 \times 10^2 \tag{i}
$$

which is the same as in (f) (note that the relation of type (i) has not been used in (c) to (e)). The yield surface translation follows from (4.4.20) and (g),

$$
{}^{1}\alpha_{aa} = 7.0138 \qquad {}^{1}\alpha_{bb} = -4.2246 \qquad {}^{1}\alpha_{cc} = -2.7892 \tag{j}
$$

Finally, we calculate the displacement of the free end,

$$
{}^{1}U = L\,{}^{1}e_{aa} = 7.3050 \times 10^1 \tag{k}
$$

We proceed to the calculations for *step 2* for which ${}^{1}\mathbf{e}^P$ and ${}^{1}\boldsymbol{\alpha}$ are used as the initial conditions. Analogous to (c), we obtain the following relations from (4.4.22):

$$^2\hat{S}_i = \frac{^2\hat{S}_i^E}{1 + \Delta\lambda\hat{C}} = \frac{^2S_i - {}^1\alpha_i}{1 + \Delta\lambda\hat{C}} \tag{1}$$

and analogous to (d) and (e) we have

$$\Delta\lambda = \frac{3}{2ME_P}\left(1 - \frac{^1\hat{\sigma}}{^2\hat{\sigma}}\right)$$

$$^2\hat{\sigma} = M\,{}^2\hat{\sigma}^E + (1 - M)\,{}^1\hat{\sigma} = 3.0702 \times 10^2 \tag{m}$$

The calculations now follow the procedure shown for step 1, and some of the results are

$$^2\hat{S}_{aa} = -2.0345 \times 10^2 \quad {}^2\hat{S}_{bb} = 9.2328 \times 10^1 \quad {}^2\hat{S}_{cc} = 1.2113 \times 10^2$$

$$^2e_{aa}^\Gamma = {}^1e_{aa}^\Gamma + \Delta\lambda^2\hat{S}_{aa} = -8.5533 \times 10^{-3} \tag{n}$$

$$^2e_{aa} = \left(^2\sigma_{aa} - \nu\,\sigma_{cc}\right)/E + {}^2e_{aa}^P = -9.9989 \times 10^{-3}$$

The displacement of the free end is

$$^2U = L\,{}^2e_{aa} = -4.9945 \times 10^1 \tag{o}$$

From the yield curve and the result (m) we obtain

$$^2\bar{e}^{iP} = {}^1\bar{e}^{iP} + \Delta\bar{e}^{iP} = \frac{1}{E_P}(^2\hat{\sigma} - \sigma_{yv}) = 0.6990 \times 10^{-3}$$

and then (see (3.2.65) and (3.2.64))

$$^2\bar{e}^P = \frac{1}{M}\,{}^2\bar{e}^{iP} = 3.4951 \times 10^{-2} \quad {}^2\bar{e}^{Pk} = (1 - M)\,{}^2\bar{e}^P = 2.7961 \times 10^{-2}$$

Note that

$$^2\bar{e}^P > \sqrt{\frac{2}{3}\,{}^2\mathbf{e}^P \cdot {}^2\mathbf{e}^P}$$

because the loading is not proportional.

Figure E.4.5-5b shows the results in the deviatoric plane and on the yield curve.

Finally, in order to calculate how the end displacement U varies as a function of the force F, we need to determine the displacements $^1U^E$ and $^2U^E$ corresponding to the elastic limits within the loading regime. From (b) and the yield condition (3.2.56) we obtain

$$^1S_{aa}^E = \sqrt{\frac{4}{9}\sigma_{yv}^2 - \frac{1}{3}\sigma_{cc}^2} = 1.9849 \times 10^2$$

$$^1F^E = (^1\sigma_{aa}^E - \sigma_{aa}^{(p)})\,A = 1.9144 \times 10^6 \tag{p}$$

and using (h),
$$^1U^E = L\,^1e^E_{aa} = 7.2911 \tag{q}$$

For the second step we use (b) and write the yield condition (3.2.57) as

$$(\Delta S^E_{aa} + {}^1\hat{S}_{aa})^2 + (-\frac{1}{2}\Delta S^E_{aa} + {}^1\hat{S}_{bb})^2 + (-\frac{1}{2}\Delta S^E_{aa} + {}^1\hat{S}_{cc})^2 = \frac{2}{3}\,{}^1\hat{\sigma}^2 \tag{r}$$

where
$$\Delta S^E_{aa} = {}^2S^E_{aa} - {}^1S_{aa}$$

is the increment of S_{aa} from ${}^1S_{aa}$ to the elastic limit ${}^2S^E_{aa}$. Equation (r) reduces to

$$\Delta S^E_{aa}\,(\Delta S^E_{aa} + 2\,{}^1\hat{S}_{aa}) = 0$$

The solution $\Delta S^E_{aa} = 0$ corresponds to a further increase of the load, while

$$\Delta S^E_{aa} = -2\,{}^1\hat{S}_{aa}$$

corresponds to the reverse loading. Using this result and the values in (b) and (f), we obtain

$$^2S^E_{aa} = {}^1S_{aa} - 2\,{}^1\hat{S}_{aa} = -1.9336 \times 10^2$$

and then

$$^2F^E = -1.8587 \times 10^6$$
$$^2U^E = L\,({}^2e^E_{aa} + {}^1e^P_{aa}) = 5.8737 \times 10^1 \tag{s}$$

Figure E.4.5-5c shows F as a function of U, which includes the results (p), (q) and (s), as well as ^1U and ^2U for the load steps 1 and 2. A more precise solution can be obtained by increasing the number of steps within the loading regime, since the loading is not radial (see Section 4.4.4).

Note that the support reaction due to internal pressure is equal to zero (see (e) of Example 4.5.4). The reaction only balances the axial force F.

Example 4.5.6. Torsion of Tube. A thin-walled tube is subjected to the torsional moment $M_x = 8 \times 10^5$ Nmm. The dimensions of the tube and the von Mises material characteristics are given in Fig. E.4.5-6a. Determine the stresses, strains and the rotation of the tube using two incremental steps, with $M_x = 3 \times 10^5$ for the first step, and the full Newton iteration procedure.

a) *General Remarks.* The equilibrium equation to be solved for iteration "i" is (see (2.2.22))

$$^{t+\Delta t}K^{(i-1)}_{xx}\,\Delta\theta^{(i)}_x = {}^{t+\Delta t}M_x - {}^{t+\Delta t}M^{(i-1)}_\sigma \tag{a}$$

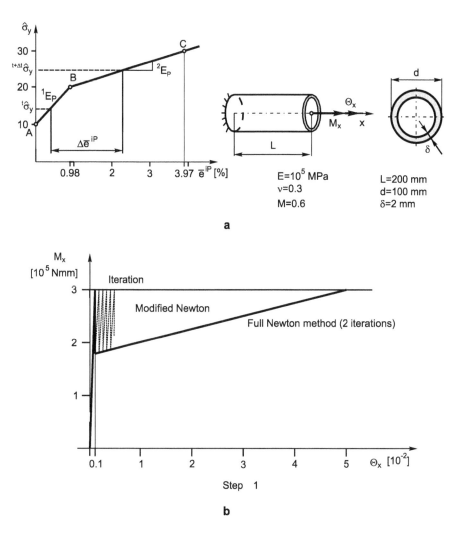

Fig. E.4.5-6. Torsion of tube. **a** Geometric and material data; **b** Moment - rotation relationship during equilibrium iterations

where $t+\Delta t K_{xx}^{(i-1)}$ is the stiffness of the tube, $\Delta\theta_x^{(i)}$ is the increment of rotation, $t+\Delta t M_x$ is the given externally applied moment, and $t+\Delta t M_\sigma^{(i-1)}$ is the moment due to the stress $t+\Delta t \sigma_{xy}^{(i-1)}$. The stiffness $t+\Delta t K_{xx}^{(i-1)}$ can be expressed as follows

$$t+\Delta t K_{xx}^{(i-1)} = \frac{d^{t+\Delta t}M_\sigma^{(i-1)}}{d^{t+\Delta t}\theta_x^{(i-1)}} = C_\theta \frac{d^{t+\Delta t}\sigma_{xy}^{(i-1)}}{d^{t+\Delta t}e_{xy}^{(i-1)}} = C_\theta \, {}^{t+\Delta t}\bar{C}_{44}^{'(i-1)} \qquad \text{(b)}$$

where

$$C_\theta = \frac{\pi \, \delta \, r_m^3}{L} = 3.696 \times 10^3 \quad \text{mm}^3 \tag{c}$$

$^{t+\Delta t}\bar{C}_{44}^{\prime(i-1)}$ is the coefficient in the constitutive matrix $^{t+\Delta t}\bar{\mathbf{C}}'$ defined in (4.4.59), and $r_m = (d - \delta)/2$. According to (4.4.73) and (4.4.74) we have

$$^{t+\Delta t}\bar{C}_{44}^{\prime(i-1)} = 2G \left(1 - 2G \frac{^{t+\Delta t}\hat{\sigma}_y}{^{t+\Delta t}\hat{\sigma}^E} \Delta\lambda\right)^{(i-1)} - 2\,{}^{t+\Delta t}D^{(i-1)} \left[{}^{t+\Delta t}\hat{S}_{xy}^{(i-1)}\right]^2 \tag{d}$$

The coefficient D is defined by (4.4.75), with $^{t+\Delta t}\hat{E}_P$ corresponding to either ^1E_P or ^2E_P shown in Fig. E.4.5-6a. With the yield curve shown in the figure, we have that the coefficient $\bar{D}$ in (4.4.72) is

$$\bar{D} = M\,{}^{t+\Delta t}\hat{E}_P + 3G + (1 - M)\,{}^{t+\Delta t}E_P \tag{e}$$

where we have used $\alpha = 1$ in expressing the weighted modulus $\bar{E}_P$.

We give some of the relations for the stress calculation that follow from Section 4.4.1 :

$$^{t+\Delta t}S_{xy}^E = 2G \left({}^{t+\Delta t}e_{xy} - {}^t e_{xy}^P\right) \qquad ^{t+\Delta t}\hat{S}_{xy}^E = {}^{t+\Delta t}S_{xy}^E - {}^t\alpha_{xy}$$

$$^{t+\Delta t}\hat{\sigma}^E = \sqrt{3} \, \|{}^{t+\Delta t}\hat{S}_{xy}^E\| \qquad \Delta e_{xy}^P = \Delta\lambda\,{}^{t+\Delta t}\hat{S}_{xy} \tag{f}$$

The computational steps are summarized in Table E.4.5-6a.

Table E.4.5-6a. Computational steps for elastic-plastic deformation of tube

---

1. Initial conditions for load step

$^t e_{xy}^P, \; \sigma_{xy}^{(0)} = {}^t\sigma_{xy}, \; ^{t+\Delta t}M_x, \; ^{t+\Delta t}\theta_x^{(0)} = {}^t\theta_x, \; ^{t+\Delta t}K_{xx}^{(0)} = {}^tK_{xx}$

$i = 0$

2. Iteration loop

$\quad i = i + 1$

$\quad ^{t+\Delta t}K_{xx}^{(i-1)} \Delta\theta_x^{(i)} = {}^{t+\Delta t}M_x - 2\pi\delta \, r_m \, {}^{t+\Delta t}\sigma_{xy}^{(i-1)}$

$\quad ^{t+\Delta t}\theta_x^{(i)} = {}^{t+\Delta t}\theta_x^{(i-1)} + \Delta\theta_x^{(i)}$

$\quad ^{t+\Delta t}e_{xy}^{(i)} = 0.5 \, (r_m/L) \, {}^{t+\Delta t}\theta_x^{(i)}$

$\quad\cdot$ Stress calculation using (f) and solving (4.4.29)

$\quad\cdot$ Convergence check. If the convergence criteria are
$\quad\quad$ satisfied, go to step 3

$\quad\cdot$ Determine $^{t+\Delta t}K_{xx}^{(i)}$ and go to beginning of iteration loop

3. Final calculations

$\quad ^{t+\Delta t}e_{xy}^P, \; {}^{t+\Delta t}\alpha_{xy}$

---

b) *Solution for the First Step.* With initial conditions: $^0e_{xy}^P = 0$, $^0\alpha_{xy} = 0$, $^0\sigma_{xy} = 0$, $^0\bar{C}_{44}' = 2G = 7.692 \times 10^4$, we obtain

$$^1K_{xx}^{(0)} = 2.843 \times 10^8 \tag{g}$$

and the solution of (a) gives

$$\Delta\theta_x^{(1)} = 1.055 \times 10^{-3} \tag{h}$$

Then the strain $^{t+\Delta t}e_{xy}^{(1)}$ is

$$^1e_{xy}^{(1)} = \frac{r_m}{2L}\Delta\theta_x^{(1)} = 1.293 \times 10^{-4} \tag{i}$$

With use of (f), we find from (4.4.29) that $\Delta\bar{e}^P = 6.204 \times 10^{-5}$, and then $^1\sigma_{xy}^{(1)} = 5.810$. Note that (4.4.29) is linear with respect to $\Delta\bar{e}^P$ because $E_P = const.$ for each segment of the yield curve. We obtain from (d)

$$^1\bar{C}_{44}'^{(1)} = 6.743 \times 10^2 \tag{j}$$

and then

$$^1K_{xx}^{(1)} = 2.492 \times 10^6 \tag{k}$$

Comparing $^0\bar{C}'_{44}$ and $^1K_{xx}^{(0)}$ with $^1\bar{C}_{44}'^{(1)}$ and $^1K_{xx}^{(1)}$, we see that the decrease of $\bar{C}'_{44}$ and K_{xx} is of the order 10^2.

With $^1K_{xx}^{(1)}$ and $^1M_\sigma^{(1)} = 2r_m\delta\pi\ ^1\sigma_{xy}^{(1)} = 1.753 \times 10^5$, we obtain from (a)

$$\Delta\theta_x^{(2)} = 5.003 \times 10^{-2} \tag{l}$$

and then

$$^1\theta_x^{(2)} = \Delta\theta_x^{(1)} + \Delta\theta_x^{(2)} = 5.109 \times 10^{-2}$$
$$^1e_{xy}^{(2)} = \frac{r_m}{2L}\ ^1\theta_x^{(2)} = 6.258 \times 10^{-3} \tag{m}$$

The stress and the moment are $^1\sigma_{xy}^{(2)} = 9.943$, $^1M_\sigma^{(2)} = 3.000 \times 10^5$. Therefore, the solution has been obtained in two iterations. Note that many (in our run 346) iterations are required if the modified Newton method is employed with $K_{xx}^{(0)}$. Some of the variables calculated during the equilibrium iterations are given in Table E.4.5-6b.

c) *Solution for the Second Step.* We follow the steps given in Table E.4.5-6a, using the solution from the first step, and the plastic modulus ^2E_P in the equilibrium iterations.

As in the case of step 1, the solution is obtained within two iterations. The calculated values are given in Table E.4.5-6b.

Table E.4.5-6b. Results for time step 1 ($^1M_x = 3E05$) and step 2 ($^2M_x = 8E05$)

	Step 1 Iteration		Step 2 Iteration	
	1	2	1	2
e_{xy}	$1.293E\text{-}4$	$6.258E\text{-}3$	$3.083E\text{-}2$	$7.093E\text{-}2$
e^P_{xy}	$5.373E\text{-}5$	$6.129E\text{-}3$	$3.061E\text{-}2$	$7.060E\text{-}2$
α_{xy}	$1.462E\text{-}2$	$1.668E00$	$3.851E00$	$7.417E00$
σ_{xy}	$5.810E00$	$9.943E00$	$3.851E00$	$7.417E00$
$\bar{C}'_{44}$	$6.743E02$	$6.743E02$	$2.223E02$	$2.223E02$
M_σ	$1.753E05$	$3.000E05$	$5.310E05$	$8.000E05$
$\Delta M_x/M_x$	$4.157E\text{-}1$	$1.49E\text{-}15$	$3.362E\text{-}1$	$1.04E\text{-}16$

Example 4.5.7. Two Step Proportional Loading. Assume that a material is subjected to a proportional loading, in accordance with (3.2.29). Considering a von Mises material with mixed hardening, determine the effective plastic strain and the plastic strains. Use two steps with the loading reversed in step 2.

a) Derive the equation for the solution at the end of step 1 for the increment of the effective plastic strain Δe^P_{num} using the governing equations of Section 4.4.1, and for Δe^P_{anal} by imposing the condition that the effective stress at the end of step 1 lies on the yield curve. The deviation of the effective stress corresponding to Δe^P_{num} from the effective stress corresponding to Δe^P_{anal} represents the error in the effective stress. Calculate the error in the effective stress assuming the Ramberg-Osgood yield curve (3.2.7). Also derive the equation for the solution of the effective plastic strain at the end of step 2.

b) Use the derived expressions in a) for the analysis of the plate subjected to the in-plane stresses shown in Fig. E.4.5-7e.

a) Figure E.4.5-7a shows the loading function $f(t)$ for the two steps. The loading of the material is proportional (see (3.2.29)); therefore we can write

$$^tS = f(t)\,^1S \tag{a}$$

where ^1S is the stress deviator at time t_1. This relation is graphically represented in Fig. E.4.5-7b. Substituting (a) into the yield condition (3.2.57), we find the value of the function $f(t_0) = f_0$ at the start of yielding

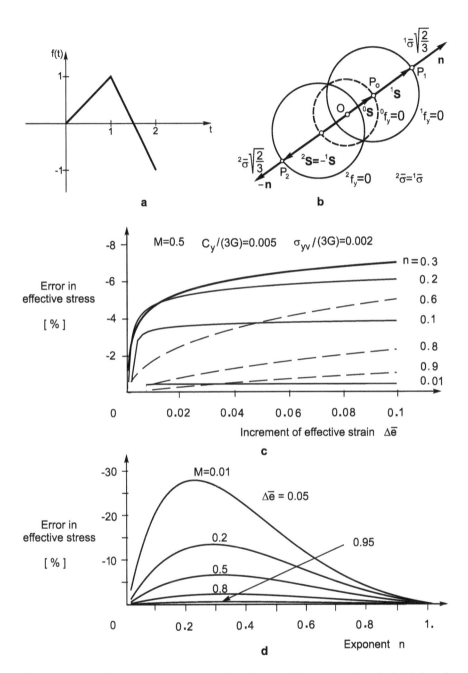

Fig. E.4.5-7. Two step proportional loading. **a** Time function; **b** Initial and yield surfaces for steps 1 and 2; **c** Error in the effective stress in terms of the effective strain increment $\Delta\bar{e}$, for several values of the exponent n; **d** Error in the effective stress in terms of the exponent n, for several values of the mixed hardening parameter M and $\Delta\bar{e} = const. = 0.05$

$$f_0 = \frac{\sigma_{yv}}{{}^1\bar{\sigma}}$$ (b)

where σ_{yv} is the initial yield stress defined in (3.2.7), and ${}^1\bar{\sigma}$ is the effective stress (3.2.39). It follows from (a) that the components of the unit normal vector ${}^t\mathbf{n}$, defined as

$${}^t\mathbf{n} = \frac{{}^t\mathbf{S}}{||{}^t\mathbf{S}||} = \sqrt{\frac{3}{2}} \frac{{}^t\mathbf{S}}{{}^t\bar{\sigma}}$$ (c)

do not change in the time interval $0 \le t < 1$.

The relation (4.4.22), with use of (c), can be written in the form (see (4.4.84))

$${}^{t+\Delta t}\bar{\sigma} \, {}^{t+\Delta t}\mathbf{n} = {}^t\bar{\alpha} \, {}^t\mathbf{n}_\alpha + (1 + \Delta\lambda\hat{C}) \, {}^{t+\Delta t}\hat{\sigma}_y \, {}^{t+\Delta t}\hat{\mathbf{n}}$$ (d)

where ${}^t\bar{\alpha}$ is the effective value of the back stress according to (3.2.39), and ${}^t\mathbf{n}_\alpha = {}^t\boldsymbol{\alpha}/||{}^t\boldsymbol{\alpha}||$, ${}^{t+\Delta t}\hat{\mathbf{n}} = {}^{t+\Delta t}\hat{\mathbf{S}}/{}^{t+\Delta t}||\hat{\mathbf{S}}||$. With use of (c), (4.4.18), (4.4.20), (4.4.21) and (4.4.24), but not using weighted values, we find that the following scalar equation holds

$${}^{t+\Delta t}\bar{\sigma} = {}^t\bar{\alpha} + {}^{t+\Delta t}\hat{\sigma}_y + ({}^{t+\Delta t}E_P - M \, {}^{t+\Delta t}\hat{E}_P)\Delta\bar{e}^P$$ (e)

From the constitutive relation (4.4.11) we obtain

$${}^{t+\Delta t}\bar{\sigma} = 3G({}^{t+\Delta t}\bar{e}'' - \Delta\bar{e}^P)$$ (f)

where ${}^{t+\Delta t}\bar{e}'' = {}^{t+\Delta t}\bar{e} - {}^t\bar{e}^P$. Here ${}^{t+\Delta t}\bar{e} = (2/3 \, {}^{t+\Delta t}\mathbf{e}'' \cdot {}^{t+\Delta t}\mathbf{e}'')^{1/2}$, see (3.2.46) and (4.4.13). If a one step proportional loading is assumed, then the stress point moves from the initial position P_0 to the position P_1 along the unit normal $\mathbf{n}$ in the deviatoric plane, as shown in Fig. E.4.5-7b. The yield surface changes from ${}^0 f_y = 0$ to ${}^1 f_y = 0$.

Consider now the *reverse loading conditions*. The loading is in the negative $\mathbf{n}$-direction, as shown in Fig. E.4.5-7b, and instead of (e) we have now

$${}^{t+\Delta t}\bar{\sigma} = -{}^t\bar{\alpha} + {}^{t+\Delta t}\hat{\sigma}_y + ({}^{t+\Delta t}E_P - M \, {}^{t+\Delta t}\hat{E}_P)\Delta\bar{e}^P$$ (g)

If a one step reverse loading is assumed, the stress point moves from the position P_1 to the position P_2 along the negative unit normal $-\mathbf{n}$, as shown in Fig. E.4.5-7b. The yield surface changes from ${}^1 f_y = 0$ to ${}^2 f_y = 0$. In case the stress integration in the interval 1-2 is performed using more than one step (see (e) in Example 5.4.3), (g) is applicable as long as the back stress ${}^t\boldsymbol{\alpha}$ is in the positive $\mathbf{n}$-direction.

We next analyze the accuracy of the solution for the effective stress, as specified above, in the loading regime $0 - 1$ assuming that the yield curve is given by the Ramberg-Osgood formula (3.2.7). The increment of the effective plastic strain $\Delta\bar{e}^P$ is obtained from (e), which, with use of (3.2.7), (f), and (3.2.77), becomes

$$f_{num} = \Delta\bar{e} - \left[\Delta\bar{e}^P + \frac{C_y}{3G}\left[n + (1-n)M^n\right]\left(\Delta\bar{e}^P\right)^n\right] = 0 \qquad (h)$$

Here we have used the condition $(\bar{e}^E)_{init} - \sigma_{yv}/(3G) = 0$, where $(\bar{e}^E)_{init}$ is the elastic effective strain corresponding to the elastic limit. The basic condition for this equation is that the reduced effective stress $^1\hat{\sigma}$ lies on the yield curve $\sigma_y(M\bar{e}^P)$. On the other hand, using the condition that the effective stress $^1\bar{\sigma}$ lies on the yield curve $\sigma_y(\bar{e}^P)$, we obtain the equation

$$f_{anal} = \Delta\bar{e} - \left[\Delta\bar{e}^P + \frac{C_y}{3G}\left(\Delta\bar{e}^P\right)^n\right] = 0 \qquad (i)$$

The effective stresses that correspond to the solutions $\Delta\bar{e}^P_{num}$ and $\Delta\bar{e}^P_{anal}$ of (h) and (i) are $^1\bar{\sigma}_{num}$ and $^1\bar{\sigma}_{anal}$, and the deviation of $^1\bar{\sigma}_{num}$ from the yield curve can be expressed as

$$(Error)_{\bar{\sigma}} = \frac{^1\bar{\sigma}_{num} - {^1}\bar{\sigma}_{anal}}{^1\bar{\sigma}_{anal}} = \frac{\Delta\bar{e}^P_{anal} - \Delta\bar{e}^P_{num}}{\sigma_{yv}/(3G) + \Delta\bar{e} - \Delta\bar{e}^P_{anal}} \qquad (j)$$

where we have normalized the error with respect to the effective stress $^1\bar{\sigma}_{anal}$.

A graphical representation of the dependence of the error in the effective stress on the effective strain increment $\Delta\bar{e}$ is shown in Fig. E.4.5-7c, for the mixed hardening parameter $M = 0.5$ and for several values of the exponent n. It can be seen that the error is in general increasing as the effective strain increment $\Delta\bar{e}$ increases. Figure E.4.5-7d shows the dependence of the error on the exponent n for the effective strain increment $\Delta\bar{e} = 0.05$ and for several values of the mixed hardening parameter M. The graphs show that the error is equal to zero for $M = 1$ (isotropic hardening) and always for $n = 1$ (bilinear stress-strain relation), for any increment of the effective strain. It can be seen that the error increases as the hardening parameter approaches zero (kinematic hardening behavior). We have used a rather large value of $\Delta\bar{e}$, not usually employed in practical applications, to emphasize the deviation of the effective stress from the yield curve when a large strain increment is used in a load step. If the strain increment is taken as $\Delta\bar{e} = 0.0002$, a value normally used in an engineering analysis, the error (for M=0.01, n=0.22, corresponding to the maximum error shown in Fig. E.4.5-7d) is 0.36%. Also, as pointed out already, we did not use weighted values for the material plastic moduli, which would reduce the error.

The above error analysis is of interest for practical applications, because it shows the dependence of the error in the effective stress, in the case of radial loading and the Ramberg-Osgood yield curve, on the increment of strain (the measure used here is the effective strain) in the time step and on the material parameters of the model. The loading regime assumed in this analysis is typically arising in engineering practice.

b) We next use the results of the above analysis for the analysis of the in-plane loading of the plate shown in Fig. E.4.5-7e. The plate is modeled by a four-node finite element, with a linear displacement field within the element. According to the given boundary conditions, the displacements u_x and u_y as functions of coordinates x and y are

$$u_x = e_{xx}\,x + \gamma_{xy}\,y$$
$$u_y = e_{yy}\,y \tag{k}$$

The components of the strain tensor e_{ij} can be expressed as (see (4.4.4) to (4.4.9))

$$e_{ij} = S_{ij}/(2G) + (\sigma_m/c_m)\delta_{ij} + e_{ij}^P \tag{l}$$

The loading of the material is proportional, with the components of the normal $\mathbf{n}$ in the first step given by (c),

$$n_{xx} = 0.5 \quad n_{yy} = -0.75 \quad n_{zz} = 0.25 \quad n_{xy} = 0.25 \tag{m}$$

Note that the following relation is satisfied :

$$n_{xx}^2 + n_{yy}^2 + n_{zz}^2 + 2n_{xy}^2 = 1 \tag{n}$$

The increment $\Delta\bar{e}_1^P$ represents the solution of (e), which for the Ramberg-Osgood yield curve (3.2.7) is

$$\Delta\bar{e}_1^P = \left[\frac{{}^1\bar{\sigma} - \sigma_{yv}}{C_y[n + (1-n)M^n]}\right]^{(1/n)} \tag{o}$$

and the components ${}^1e_{ij}^P$ are

$$^1e_{ij}^P = \sqrt{\frac{3}{2}}\,\Delta\bar{e}_1^P\,n_{ij} \tag{p}$$

The values of plastic strains and displacements for isotropic, kinematic, and mixed ($M = 0.5$) hardening are given in Table E.4.5-7.

The loading in the second step is the reverse radial loading, hence we solve numerically (g) with respect to $\Delta\bar{e}_2^P$, and calculate the plastic strains ${}^2e_{ij}^P$ for the second step as

$$^2e_{ij}^P = {}^1e_{ij}^P - \sqrt{\frac{3}{2}}\,\Delta\bar{e}_2^P\,n_{ij} \tag{q}$$

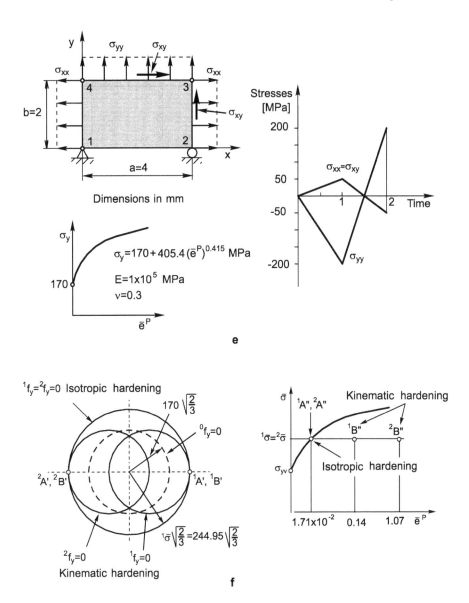

Fig. E.4.5-7 (continued). e Boundary conditions, loading and material data for plate element; **f** Yield surfaces and representative points in the deviatoric plane and in the coordinate system $\bar{\sigma}$, $\bar{e}^P$

The total strains and displacements are calculated using (k) and (l). Some of the calculated values are given in Table E.4.5-7.

Table E.4.5-7. Results for isotropic, kinematic and mixed hardening

step1/step2	Isotropic hardening		
$e_{xx}^P, e_{yy}^P, \gamma_{xy}^P$	$1.048E\text{-}2$	$-1.572E\text{-}2$	$1.048E\text{-}2$
	$1.048E\text{-}2$	$-1.572E\text{-}2$	$1.048E\text{-}2$
$U_{x2}, U_{x4}, U_{y3} = U_{y4}$	$4.633E\text{-}2$	$2.357E\text{-}2$	$-3.575E\text{-}2$
	$3.753E\text{-}2$	$1.837E\text{-}2$	$-2.715E\text{-}2$
	Kinematic hardening		
$e_{xx}^P, e_{yy}^P, \gamma_{xy}^P$	$8.726E\text{-}2$	$-1.309E\text{-}1$	$8.726E\text{-}2$
	$-4.800E\text{-}1$	$7.200E\text{-}1$	$-4.800E\text{-}1$
$U_{x2}, U_{x4}, U_{y3} = U_{y4}$	$3.534E\text{-}1$	$1.771E\text{-}1$	$-2.661E\text{-}1$
	$-1.924E00$	$-9.626E\text{-}1$	$1.444E00$
	Mixed hardening		
$e_{xx}^P, e_{yy}^P, \gamma_{xy}^P$	$1.534E\text{-}2$	$-2.302E\text{-}2$	$1.534E\text{-}2$
	$6.208E\text{-}3$	$-9.312E\text{-}3$	$6.208E\text{-}3$
$U_{x2}, U_{x4}, U_{y3} = U_{y4}$	$6.577E\text{-}2$	$3.329E\text{-}2$	$-5.033E\text{-}2$
	$2.043E\text{-}2$	$9.816E\text{-}3$	$-1.432E\text{-}2$

Figure E.4.5-7f shows the yield surfaces and the corresponding points in the deviatoric plane and in the $\bar{\sigma}, \bar{e}^P$ coordinate system, for isotropic and kinematic hardening. The yield surfaces and the representative points for mixed hardening lie between those for isotropic and kinematic hardening. Note that the points $^1B'$, $^2B''$ corresponding to kinematic hardening do not lie on the yield curve because the yield condition of the form $\bar{\sigma}_y(\bar{e}^P)$ is not used. This yield condition is used only in isotropic hardening when the points $^1A''$ and $^2A''$ are on the yield curve. Plastic flow occurs during the reverse loading in kinematic or mixed hardening, while in isotropic hardening the stress point moves from $^1A'$ to $^2A'$ within the yield surface $^1f_y = 0$ and the deformation is elastic.

Note that very large values of plastic strains are used in this example merely to emphasize the influence of the model characteristics on the predicted material response (but a small strain analysis was of course performed).

Example 4.5.8. Plane Strain Plastic Deformation. A material element shown in Fig. E.4.5-8a is loaded by compressive forces R_x and R_y. The material is in plane strain conditions (strain $e_{zz} = 0$) and it is of the von

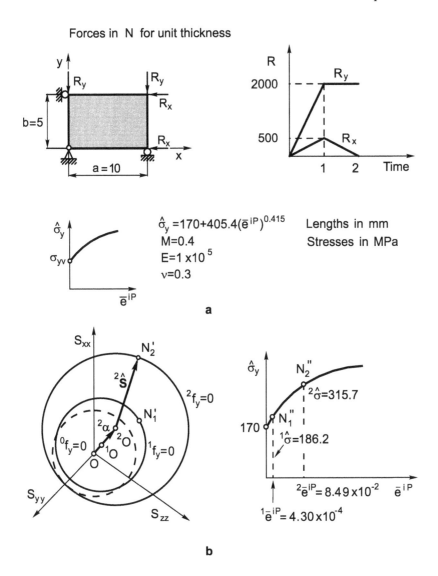

Forces in N for unit thickness

Fig. E.4.5-8. Plane strain element under compression. **a** Boundary and loading conditions; **b** Stress states in the deviatoric plane and on the yield curve

Mises type with mixed hardening (parameter $M = 0.4$). The yield curve is defined by the Ramberg-Osgood formula (3.2.7), as given in the figure.

Assuming a uniform stress/strain state in the element and the load change as given in the figure, determine the displacements using a two-step solution.

We start with the principle of virtual work and form the incremental-iterative equation (see (2.3.13) and Bathe 1996),

$$\int_A (\delta e_{xx})(^{t+\Delta t}C^{EP}_{xx} e_{xx} + {}^{t+\Delta t}C^{EP}_{xy} e_{yy})\, dA +$$

$$\int_A (\delta e_{yy})(^{t+\Delta t}C^{EP}_{yx} e_{xx} + {}^{t+\Delta t}C^{EP}_{yy} e_{yy})\, dA =$$

$$-2(\delta{}^{t+\Delta t}U_x)^{t+\Delta t}R_x - 2(\delta{}^{t+\Delta t}U_y)^{t+\Delta t}R_y \qquad \text{(a)}$$

$$-\int_A (\delta e_{xx})\,{}^t\sigma_{xx}\, dA - \int_A (\delta e_{yy})\,{}^t\sigma_{yy}\, dA$$

Here we have neglected nonlinear strain terms, considering that the displacements are small. Also, we have taken into account that the only non-zero strains are e_{xx} and e_{yy}. Under the assumption that the strains are uniform within the element, we have the relations

$$e_{xx} = U_x/a \qquad e_{yy} = U_y/b \qquad \text{(b)}$$

The components of the elastic-plastic matrix $^{t+\Delta t}C^{EP}_{xx}$, $^{t+\Delta t}C^{EP}_{yy}$, $^{t+\Delta t}C^{EP}_{xy} = {}^{t+\Delta t}C^{EP}_{yx}$, are constant over the element. With these conditions we obtain two equilibrium equations

$$^{t+\Delta t}K^{(i-1)}_{xx}\Delta U^{(i)}_x + {}^{t+\Delta t}K^{(i-1)}_{xy}\Delta U^{(i)}_x = 2\,{}^{t+\Delta t}R_x - {}^{t+\Delta t}F^{(i-1)}_x$$

$$^{t+\Delta t}K^{(i-1)}_{xy}\Delta U^{(i)}_x + {}^{t+\Delta t}K^{(i-1)}_{yy}\Delta U^{(i)}_y = 2\,{}^{t+\Delta t}R_y - {}^{t+\Delta t}F^{(i-1)}_y \qquad \text{(c)}$$

where

$$^{t+\Delta t}K^{(i-1)}_{xx} = b/a\,{}^{t+\Delta t}C^{EP(i-1)}_{xx} \qquad {}^{t+\Delta t}K^{(i-1)}_{yy} = a/b\,{}^{t+\Delta t}C^{EP(i-1)}_{yy}$$

$$^{t+\Delta t}K^{(i-1)}_{xy} = {}^{t+\Delta t}C^{EP(i-1)}_{xy} \qquad \text{(d)}$$

$$^{t+\Delta t}F^{(i-1)}_x = b\,{}^{t+\Delta t}\sigma^{(i-1)}_{xx} \qquad {}^{t+\Delta t}F^{(i-1)}_y = a\,{}^{t+\Delta t}\sigma^{(i-1)}_{yy}$$

With use of the Ramberg-Osgood formula (3.2.7), the governing equation (4.4.29) becomes

$$f(\Delta \bar{e}^P) = \sigma_{yv} + C_y\, M^n\, (^t\bar{e}^P + \Delta \bar{e}^P)^n$$

$$+ \frac{3}{2}(2G + \hat{C})\, \Delta \bar{e}^P - {}^{t+\Delta t}\hat{\sigma}^E = 0 \qquad \text{(e)}$$

where (see (4.4.21) and (3.2.80))

$$\hat{C} = 2/3\, nC_y(1 - M^n)(^t\bar{e}^P + \Delta \bar{e}^P)^{n-1} \qquad \text{(f)}$$

To solve (e) we employ the Newton iteration, hence

$$\Delta \bar{e}^{P(k)} = \Delta \bar{e}^{P(k-1)} - f^{(k-1)}/f'^{(k-1)} \qquad \text{(g)}$$

where

$$f'(\Delta \bar{e}^P) = nM^n C_y \left({}^t e^P + \Delta \bar{e}^P\right)^{(n-1)} +$$

$$\frac{3}{2}(2G + \hat{C}) + n(n-1)C_y(1 - M^n) \left({}^t e^P + \Delta \bar{e}^P\right)^{(n-2)} \qquad \text{(h)}$$

The coefficients of the elastic-plastic matrix in (d) can be determined according to (4.4.56). The derivatives of the stress radius follow from (4.4.64), and the derivatives of $\Delta \bar{e}^P$ with respect to the strains ${}^{t+\Delta t}e_j$ can be obtained by differentiation of (e) or by use of (4.4.70). In this procedure we can use the derivatives given by (h), and also

$$\frac{\partial^{\,t+\Delta t}\hat{\sigma}}{\partial^{\,t+\Delta t}\bar{e}^P} = n\, C_y\, M^n \left({}^{t+\Delta t}\bar{e}^P\right)^{n-1} \qquad \text{(i)}$$

Table E.4.5-8 gives values of the relative unbalanced force and the relative unbalanced energy during equilibrium iterations of the system (c). The convergence is quadratic due to the tangent character of the elastic-plastic matrix $\mathbf{C}^{EP}$. The solution of (e) was obtained with a maximum of five trials. The quantities in the table are calculated as follows

$$\Delta \mathbf{F}^{(i)} = {}^{t+\Delta t}\mathbf{R} - {}^{t+\Delta t}\mathbf{F}^{(i)}$$

$$\Delta E^{(i)} = \Delta F_x^{(i-1)}\,\Delta U_x^{(i)} + \Delta F_y^{(i-1)}\,\Delta U_y^{(i)}$$

$$F_{ref} = 2\sqrt{({}^1 R_x)^2 + ({}^1 R_y)^2} = 0.412 \times 10^4$$

Table E.4.5-8. Unbalanced force and unbalanced energy during iterations

| Iteration | Step 1 | | Step 2 | |
	$\dfrac{\Delta E^{(i)}}{\Delta E^{(0)}}$	$\left\|\dfrac{\Delta F^{(i)}}{F_{ref}}\right\|$	$\dfrac{\Delta E^{(i)}}{\Delta E^{(0)}}$	$\left\|\dfrac{\Delta F^{(i)}}{F_{ref}}\right\|$
1	1.37E-2	1.12E-4	3.38E00	2.46E-2
2	1.72E-3	1.29E-5	7.40E00	6.87E-2
3	2.96E-5	2.39E-7	2.46E00	1.80E-2
4	8.61E-9	7.2E-11	7.01E-2	5.09E-4
5	7.2E-16	6.1E-18	3.51E-5	2.55E-7
6			8.2E-12	6.0E-14

$$\Delta E^{(0)} = 5.98E01$$

Figure E.4.5-8b gives a graphical representation of the solution. It can be seen from the positions of the stress points N_1' and N_2' that the loading is very nonradial (see Section 4.4.4). The points N_1'' and N_2'', corresponding to the stress points N_1' and N_2', are exactly on the yield curve.

Example 4.5.9. Radial, Nonradial and Reverse Radial Loading.
Consider the stress conditions given in Fig. E.4.5-9a. The stresses σ_{xx}, σ_{yy}, σ_{zz} and σ_{xy} change linearly to the given values in the three loading (time) intervals, providing radial, non-radial and reverse radial loading regimes. The von Mises material has the mixed hardening characteristic (with the param-

			Loading regime	
Stresses [MPa]		(1) Radial	(2) Nonradial	(3) Reverse radial
σ_{xx}		100.00	40.00	-100.00
σ_{yy}		60.00	-30.00	-12.08
σ_{zz}		-100.00	-130.00	102.80
σ_{xy}		60.00	100.00	-125.30

a

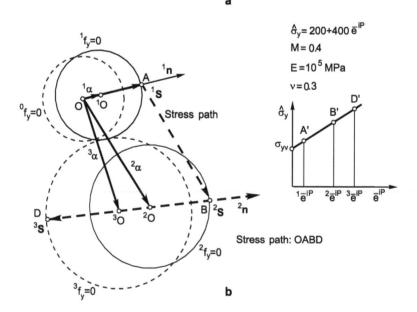

b

Fig. E.4.5-9. Radial, nonradial and reverse radial loading regimes at a material point. **a** Stresses at end of loading regimes; **b** Yield surfaces and yield curve

eter $M = 0.4$). The yield curve is defined by the Ramberg-Osgood formula (3.2.7), with the constants given in Fig. E.4.5-9.

Employ various number of time steps in the three regimes to calculate the strains and compare the results.

We write the governing equations for this problem assuming a nonlinear yield curve, and give the numerical solutions using the bilinear stress-strain relation with the material constants given in Fig. E.4.5-9. The stresses are given, and from (4.4.22) and (4.4.24) we obtain

$$\sqrt{\frac{2}{3}}\,^{t+\Delta t}\hat{\sigma} + \sqrt{\frac{3}{2}}\,\hat{C}\,\Delta\bar{e}^P - \|\,^{t+\Delta t}\mathbf{S} - \,^t\boldsymbol{\alpha}\| = 0 \tag{a}$$

and, further, with use of (3.2.7), (4.4.21), and (3.2.80)

$$f(\Delta\bar{e}^P) = \sigma_{yv} + C_y\,M^n(\,^t\bar{e}^P + \Delta\bar{e}^P)^n +$$

$$nC_y(1 - M^n)(\,^t\bar{e}^P + \Delta\bar{e}^P)^{n-1}\Delta\bar{e}^P - \sqrt{\frac{3}{2}}\,\|\,^{t+\Delta t}\mathbf{S} - \,^t\boldsymbol{\alpha}\| = 0 \tag{b}$$

This equation is nonlinear with respect to $\Delta\bar{e}^P$ when the exponent $n \neq 1$, but it is linear for $n = 1$. With $\Delta\bar{e}^P$ determined, we calculate $^{t+\Delta t}\hat{\sigma}$ from the yield curve $^{t+\Delta t}\hat{\sigma}(M\,^{t+\Delta t}\bar{e}^P)$, $\Delta\lambda$ from (4.4.24), the stress radius $^{t+\Delta t}\hat{\mathbf{S}}$ from (4.4.22),

$$^{t+\Delta t}\hat{\mathbf{S}} = (\,^{t+\Delta t}\mathbf{S} - \,^t\boldsymbol{\alpha})/(1 + \hat{C}\Delta\lambda) \tag{c}$$

and the increments of plastic strain $\Delta\mathbf{e}^P$ from (4.4.18). With the known plastic strains $^{t+\Delta t}\mathbf{e}^P$ we obtain the total strains (see (4.4.4) and (4.4.5)), i.e.,

$$^{t+\Delta t}\hat{\mathbf{e}} = (\mathbf{C}^E)^{-1}\,^{t+\Delta t}\boldsymbol{\sigma} + \,^{t+\Delta t}\hat{\mathbf{e}}^P \tag{d}$$

The stresses are linearly increased and the stress point moves along the normal $^1\mathbf{n}$, from point O to point A in Fig. E.4.5-9b. We use one step and ten steps (the first step in the ten step loading corresponds to start of yielding), and obtain that the plastic and total strains are the same. The results are given in Table E.4.5-9. If the yield curve were nonlinear, the solution would depend on the number of time steps (see Example 4.5.7).

The stress path in the *second regime* is orthogonal to the normal $^1\mathbf{n}$, and we calculate plastic strains using one and ten steps. Since the loading is nonradial (at the beginning the nonradiality is at a maximum), the results depend on the number of load steps. Note that the results do not differ significantly, especially for the effective quantities $\hat{\sigma}$ and $\bar{e}^P$.

In the *third regime* we have the reverse radial loading conditions, with the stress path along the normal $^2\mathbf{n}$ calculated from the ten-step solution in the second regime. The stress is increased linearly between the stress points B and D. We have used the results of the ten-step solution in the second

Table E.4.5-9. Results in the three loading regimes

	Radial 1 and 10 steps	Nonradial 1 step	Nonradial 10 steps	Reverse radial 1 and 10 steps
	\multicolumn{4}{c}{Loading regime}			
e_{xx}	$1.637E$-2	$3.920E$-2	$4.017E$-2	$-7.253E$-2
e_{yy}	$8.226E$-3	$1.021E$-2	$1.300E$-2	$1.042E$-3
e_{zz}	$-2.436E$-2	$-4.989E$-2	$-5.365E$-2	$7.146E$-2
γ_{xy}	$2.444E$-2	$8.388E$-2	$7.916E$-2	$-2.083E$-1
α_{xx}	$2.440E00$	$6.131E00$	$6.286E00$	$1.140E01$
α_{yy}	$1.220E00$	$1.638E00$	$2.085E00$	$1.874E$-1
α_{xy}	$1.830E00$	$6.502E00$	$6.125E00$	$-1.640E01$
$\hat{\sigma}$	$2.043E02$	$2.113E02$	$2.114E02$	$2.452E02$
$\bar{e}^P$	$2.678E$-2	$7.078E$-2	$7.106E$-2	$2.824E$-1

regime as the initial conditions for the third loading regime. As stated in Table E.4.5-9, the same results are obtained when using one or ten steps. The first stress increment in the ten-step solution corresponds to the start of yielding in this regime, while in the other nine steps they are equal. Again, if the yield curve were nonlinear, the solution would depend on the number of time steps used, as discussed in Example 4.5.7.

The above results demonstrate the solution accuracy that is obtained. Note that the large strain values are merely used to better indicate the solution differences that can be expected; we of course always used the materially-nonlinear-only (MNO) formulation.

Example 4.5.10. Nonradial Loading of Perfectly Plastic Material. Compare the numerical solutions obtained using various number of steps with the analytical solution, for a plane strain deformation and perfect plasticity. The initial stress state on the yield surface is defined by $S_{xx} = -S_{yy}$, $S_{zz} = 0$ (stress point 1 in Fig. E.4.5-10a), and the deviatoric strain rate is tangential to the yield surface at the stress point 1. The total strain increments from the state 1 are $\Delta e_{xx} = \Delta e_{yy} = 0.01$. The material data are given in the figure.

The analytical solution is (see Krieg RD and Krieg DB 1977)

$$^{t+\Delta t}\mathbf{S} = \alpha_a(^1\mathbf{S} + \beta_a \Delta \mathbf{S}_e) \tag{a}$$

where

$$\alpha_a = \frac{2C_a}{1 + C_a^2 + (1 - C_a^2)\cos\psi_c} \tag{b}$$

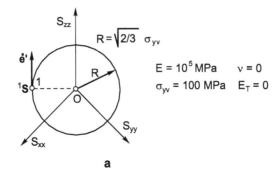

a

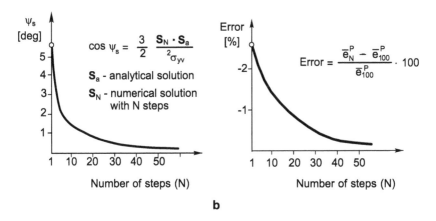

b

Fig. E.4.5-10. Accuracy analysis in case of perfect plasticity plane strain deformation; **a** Material data; **b** Accuracy of solutions

and

$$C_a = \exp(-\|\Delta \mathbf{S}_e\|/R)$$
$$\beta_a = \frac{\left[1 - C_a^2 + (1 - C_a)^2 \cos\psi_c\right] R}{2C_a\|\Delta \mathbf{S}_e\|} \tag{c}$$

Here R is the radius of the yield surface, and $\Delta \mathbf{S}_e$ is the increment of deviatoric stress calculated as (see (4.4.9))

$$\Delta \mathbf{S}_e = 2G\,\Delta \mathbf{e}' \tag{d}$$

where $\Delta \mathbf{e}'$ is the increment of deviatoric strain tensor. The value $\cos\psi_c$ is defined as

$$\cos\psi_c = \frac{{}^1\mathbf{S}\cdot\Delta \mathbf{S}_e}{R\,\|\Delta \mathbf{S}_e\|} \tag{e}$$

where $^1\mathbf{S}$ is the deviatoric stress at the start of yielding.

The numerical solution is obtained according to Table 4.4.1, using

$$^1S_{xx} = -\,^1S_{yy} = \sigma_{yv}/\sqrt{3}; \quad ^1S_{zz} = 0 \tag{f}$$

and

$$^1e'_{xx} = -\,^1e'_{yy} = a_E\,S_{xx} = 0.57735 \times 10^{-3} \tag{g}$$

where $a_E = 0.5/G$. Since $\dot{\mathbf{e}}'$ is tangential to the yield surface at stress point 1, we have

$$\cos\psi_c = 0 \tag{h}$$

From the conditions (f) and (h) we obtain

$$\Delta e_{xx} = \Delta e_{yy} \qquad \Delta e'_{xx} = \Delta e'_{yy} = \Delta e_{xx} \tag{i}$$

The analytical solution, and the numerical results for 1, 20, 50 and 100 steps, are given in Table E.4.5-10 (see Kojic and Bathe 1987a). Figure E.4.5-10b shows the error in the deviatoric stress $^2\mathbf{S}$ direction, and in the effective plastic strain $^2\bar{e}^P$ with respect to the hundred-step solution. It can be seen that the solution accuracy increases with the number of steps used. Also, the error is not very large when only one step is used. Note that the nonradiality is at its maximum at the start of the plastic deformation, and that the error decreases as the loading approaches the radial loading condition.

Table E.4.5-10. Analytical and numerical results for plane strain perfect plasticity

	Analytical solution	Numerical solutions - number of steps		
		1	20	50
S_{xx}	$3.334E01$	$3.891E01$	$3.336E01$	$3.334E01$
S_{yy}	$3.333E01$	$2.742E01$	$3.330E01$	$3.332E01$
e^P_{xx}	–	$3.522E\text{-}3$	$3.577E\text{-}3$	$3.577E\text{-}3$
e^P_{zz}	–	$-6.003E\text{-}3$	$-6.000E\text{-}3$	$-6.000E\text{-}3$
$\bar{e}^P$	–	$6.033E\text{-}3$	$6.166E\text{-}3$	$6.186E\text{-}3$

Example 4.5.11. Gurson Material Model. Derive the relations for implicit stress integration in the case of the Gurson material model for metals.

In this example we a) describe briefly the Gurson model, b) implement the governing parameter method of Section 4.2 to develop the implicit stress integration algorithm for general 3-D and shell (plane stress) conditions, and c) show the numerical results in a simple example.

a) *Model description.* It has been observed in experiments that void nucleation and void growth occur in the case of large local plastic flow of metals (usual in ductile fracture processes). Then the hydrostatic stress independent plasticity material models are no longer adequate to describe the behavior of the metals.

The basic plasticity model for a porous metal was proposed by Gurson (1977). The yield condition of this model was modified by Tvergaard (1981), (1987) into the following form

$$f_y = \frac{1}{2} \mathbf{S} \cdot \mathbf{S} + \frac{1}{3} \left[2f^* q_1 \cosh \left(\frac{3q_2 \sigma_m}{2\sigma_y} \right) - 1 - q_3^2 f^{*2} \right] \sigma_y^2 = 0 \qquad (a)$$

where σ_y is the yield stress, $\mathbf{S}$ and σ_m are the stress deviator and mean stress, f^* is a function of porosity (volume fraction) f ; and q_1, q_2 and q_3 are material constants. The function f^* is given as

$$f^* = \begin{vmatrix} f & \text{for } f \le f_c \\ f_c + K_f (f - f_c) & \text{for } f > f_c \end{vmatrix} \qquad (b)$$

where f_c is a critical porosity when the onset of rapid volume coalescence begins, and

$$K_f = \frac{1/q_1 - f_c}{f_f - f_c} \qquad (c)$$

Here f_f is the value of f at the material failure.

The increment of porosity Δf in the time step Δt can be decomposed into the parts corresponding to the void growth, nucleation and coalescence (Brunet and Sabourin 1996), and can in general be expressed in terms of the volumetric plastic strain increment Δe_V^P and the effective plastic strain increment $\Delta \bar{e}^P$,

$$\Delta f = (1 - f) \Delta e_V^P + A \Delta \bar{e}^P \qquad (d)$$

where A may be constant, or dependent on the effective plastic strain and porosity.

It is assumed that the flow rule (3.2.71) is applicable (Gurson 1977), as well as the equivalence of plastic work (3.2.32) which now has the form (Worswick and Pick 1991)

$$\boldsymbol{\sigma} \cdot \dot{\mathbf{e}}^P = (1 - f) \sigma_y \dot{\bar{e}}^P \qquad (e)$$

b) *Stress Integration Procedure.* We first derive the basic relations for a general three-dimensional deformation, and then for the shell (plane stress) conditions.

The constitutive relations (4.4.6) and (4.4.11) have now the following form

$$^{t+\Delta t}\sigma_m = {}^{t+\Delta t}\sigma_m^E - c_m \Delta e_m^P \qquad \text{no sum on } m \qquad (f)$$

and

$$t+\Delta t \mathbf{S} = {}^{t+\Delta t}\mathbf{S}^E - 2G\Delta\mathbf{e}'^P \tag{g}$$

where Δe_m^P and $\Delta\mathbf{e}'^P$ are the increments of the mean and deviatoric plastic strains. The elastic mean stress ${}^{t+\Delta t}\sigma_m^E$ is

$$^{t+\Delta t}\sigma_m^E = c_m \left({}^{t+\Delta t}e_m - {}^t e_m^P\right) \tag{h}$$

and the elastic deviatoric stress ${}^{t+\Delta t}\mathbf{S}^E$ is given in (4.4.12). The increments Δe_m^P and $\Delta\mathbf{e}'^P$ follow from (4.2.10) and (a),

$$\Delta e_m^P = \frac{\Delta\lambda}{3}\, {}^{t+\Delta t}f_y' \tag{i}$$

and

$$\Delta\mathbf{e}'^P = \Delta\lambda\, {}^{t+\Delta t}\mathbf{S} \tag{j}$$

where

$$^{t+\Delta t}f_y' = \frac{\partial^{t+\Delta t}f_y}{\partial^{t+\Delta t}\sigma_m} = q_1 q_2\, {}^{t+\Delta t}\sigma_y\, {}^{t+\Delta t}f^* \sinh\left(\frac{3q_2}{2}\frac{{}^{t+\Delta t}\sigma_m}{{}^{t+\Delta t}\sigma_y}\right) \tag{k}$$

Substituting (j) into (g) we solve for ${}^{t+\Delta t}\mathbf{S}$ as

$$^{t+\Delta t}\mathbf{S} = \frac{{}^{t+\Delta t}\mathbf{S}^E}{1 + 2G\Delta\lambda} \tag{l}$$

From the evolution equation (d) for porosity, using ${}^{t+\Delta t}f$ for f, we obtain

$$\Delta f = \left[3(1 - {}^t f)\,\Delta e_m^P + A\,\Delta\bar{e}^P\right] / \left(1 + 3\Delta e_m^P\right) \tag{m}$$

With use of (i) and (j) we obtain (e) corresponding to the end of the time step in the form

$$^{t+\Delta t}f_e = \Delta\lambda\, {}^{t+\Delta t}\mathbf{S} \cdot {}^{t+\Delta t}\mathbf{S} + 3\, {}^{t+\Delta t}\sigma_m\,\Delta e_m^P - (1 - {}^{t+\Delta t}f)\, {}^{t+\Delta t}\sigma_y\Delta\bar{e}^P = 0 \tag{n}$$

Finally, the yield condition (a) at the end of time step,

$$^{t+\Delta t}f_y\left({}^{t+\Delta t}\mathbf{S}, {}^{t+\Delta t}\sigma_m, {}^{t+\Delta t}\sigma_y, {}^{t+\Delta t}f\right) = 0 \tag{o}$$

must be satisfied.

The computational steps are in accordance with Table 4.2.1. The governing parameter is Δe_m^P and the plasticity calculations are as follows: for an assumed Δe_m^P we determine ${}^{t+\Delta t}\sigma_m$ from (f) and then iterate on $\Delta\bar{e}^P$ (by calculating ${}^{t+\Delta t}\sigma_y({}^t\bar{e}^P + \Delta\bar{e}^P)$ from the yield curve, Δf from (m), ${}^{t+\Delta t}f^*$

from (b), $^{t+\Delta t}f'_y$ from (k), $\Delta\lambda$ from (i), and $^{t+\Delta t}\mathbf{S}$ from (l)) until (n) is satisfied within a selected numerical tolerance. The iterations on Δe^P_m continue until (o) is satisfied.

The consistent tangent elastic-plastic matrix can be obtained by implementing the approach described in Sections 4.2 and 4.4.3. The computational details of the above stress integration procedure and the elastic-plastic matrix calculation are given in Kojic et al. (2002b).

Consider next the above procedure when applied to the *shell (plane stress)* conditions (Kojic et al. 2004). Following the concept of Section 4.4.2, we obtain from the fundamental condition (4.4.38) that the normal components of the stress deviator are

$$
\begin{aligned}
{}^{t+\Delta t}S_{xx} &= \left(p_1 e''_{xx} + p_2 e''_{yy}\right) / \left(p_1^2 - p_2^2\right) \\
{}^{t+\Delta t}S_{yy} &= \left(p_2 e''_{xx} + p_1 e''_{yy}\right) / \left(p_1^2 - p_2^2\right) \\
{}^{t+\Delta t}S_{zz} &= -\,{}^{t+\Delta t}\sigma_m = -\,{}^{t+\Delta t}S_{xx} - {}^{t+\Delta t}S_{yy}
\end{aligned}
\tag{p}
$$

where

$$
\begin{aligned}
p_1 &= c_1\Delta\lambda + a_E \qquad p_2 = c_\nu\Delta\lambda \\
e''_{xx} &= c_1\left({}^{t+\Delta t}e_{xx} - {}^t e^P_{xx}\right) - c_\nu\left({}^{t+\Delta t}e_{yy} - {}^t e^P_{yy}\right) \\
e''_{xx} &= -c_\nu\left({}^{t+\Delta t}e_{xx} - {}^t e^P_{xx}\right) + c_1\left({}^{t+\Delta t}e_{yy} - {}^t e^P_{yy}\right)
\end{aligned}
\tag{q}
$$

with

$$
c_\nu = \frac{1-\nu}{3\left(1-\nu\right)} \qquad c_1 = 1 - c_\nu
$$

and $a_E = 1/(2G)$. The relations for the shear stress components are given in (l).

The computational steps are similar to those for the general 3-D deformations given above. The governing parameter is Δe^P_m. We initialize Δe^P_m and $\Delta\bar{e}^P$ and then iterate on Δe^P_m. With a selected Δe^P_m we calculate Δf from (m) and find $\Delta\bar{e}^P$ numerically by satisfying equation (n). We employ the governing relations (p) for the shell conditions within the numerical solution of (m). The iterations on Δe^P_m continue until the yield condition (o) is satisfied.

The computational algorithm can be extended to large strain deformations and we will give one large strain numerical example in Chapter 7 (Example 7.3.5).

c) *Numerical Example.* We consider a block of material in tension as shown schematically in Fig. E.4.5-11a. The material data are given in the figure.

The dependence of the stress on the strain is calculated for several values of the initial porosity ^0f. It can be seen that the stiffness of the material decreases with an increase of the initial porosity, and when the initial porosity is small the response of the material is as for the von Mises material.

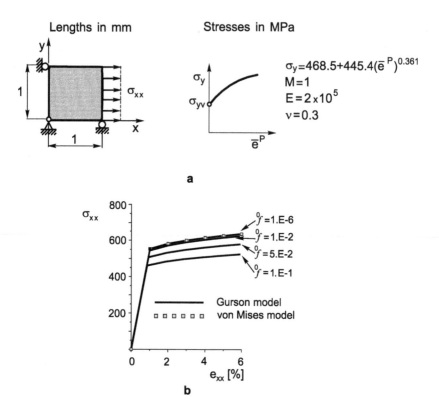

Fig. E.4.5-11. Uniaxial loading of material modeled using the Gurson model. Plane strain conditions. **a** Finite element model and material data; **b** Stress-strain dependence for several values of initial porosity ^0f and the von Mises model

4.6 Orthotropic Material Model - Stress Integration

In this section we present the stress integration procedure for Hill's orthotropic model of Section 3.3.1. The procedure represents a generalization of the algorithm for an initially isotropic metal, described in Section 4.4. The basic relations that we derive here reduce — as the orthotropic model itself — to the isotropic relations of the perfectly plastic von Mises material as a special case. We consider general three-dimensional elastic-plastic deformations and the shell (plane stress) conditions.

The stress integration for the sheet metal models of Section 3.3.2 can be performed by applying the relations derived in Section 4.2.

4.6.1 Stress Integration

The elastic constitutive relations for the mean stress $^{t+\Delta t}\sigma_m$ and the deviatoric stresses $^{t+\Delta t}\mathbf{S}$ corresponding to the end of the time step are (see

Appendix A1)

$$
{}^{t+\Delta t}\sigma_m = \sum_{i=1}^{3} C_{ii}^{mE}\ {}^{t+\Delta t}e_i^E \tag{4.6.1}
$$

$$
{}^{t+\Delta t}\mathbf{S} = \mathbf{C}'^E\ {}^{t+\Delta t}\hat{\mathbf{e}}^E \tag{4.6.2}
$$

where ${}^{t+\Delta t}\hat{\mathbf{e}}^E$ is the elastic strain vector defined according to (4.4.52), and C_{ii}^{mE} and $\mathbf{C}'^E$ are elastic constants and the elastic deviatoric matrix, respectively. The coefficients C_{ii}^{mE} can be obtained from the definition of the mean stress (3.2.12) and the elastic constitutive relations for the orthotropic material as

$$
C_{ii}^{mE} = \frac{1}{3}\sum_{j=1}^{3} C_{ij}^E \tag{4.6.3}
$$

where the C_{ij}^E are the entries in the elasticity matrix $\mathbf{C}^E$ (A1.41). According to the definition of the deviatoric stress (3.2.14), we obtain the coefficients $C_{ij}'^E$ as

$$
C_{ij}'^E = A_{ik}C_{kj}^E \qquad i,j,k = 1,2,3 \tag{4.6.4}
$$

and

$$
C_{ij}'^E = C_{ii}^E \delta_{ij} \qquad \begin{matrix} i = 4,5,6 & \text{no sum on } i \\ j = 1,2,\dots,6 \end{matrix} \tag{4.6.5}
$$

where the matrix $\mathbf{A}$ is given in (3.2.70). Summation on the dummy index k is implied in (4.6.4).

In the case of shell conditions we employ the constitutive matrix $\bar{\mathbf{C}}^E$ whose terms $\bar{C}_{ij}^E$ are given in (A1.44), and instead of (4.6.1) and (4.6.2), we have

$$
{}^{t+\Delta t}\sigma_m = \bar{C}_{11}^{mE}\ {}^{t+\Delta t}e_1^E + \bar{C}_{22}^{mE}\ {}^{t+\Delta t}e_2^E \tag{4.6.6}
$$

and

$$
{}^{t+\Delta t}\mathbf{S} = \bar{\mathbf{C}}'^E\ {}^{t+\Delta t}\hat{\mathbf{e}}^E \tag{4.6.7}
$$

where

$$
\bar{C}_{11}^{mE} = \frac{1}{3}\left(\bar{C}_{11}'^E + \bar{C}_{12}^E\right)
$$

$$\bar{C}_{22}^{mE} = \frac{1}{3}\left(\bar{C}_{22}^E + \bar{C}_{12}^E\right) \tag{4.6.8}$$

The matrix $\bar{\mathbf{C}}'^E$ of order (5×5) contains a (2×2) nonzero submatrix for the normal components, with

$$\bar{C}_{11}'^E = \frac{1}{3}\left(2\bar{C}_{11}^E - \bar{C}_{12}^E\right) \qquad \bar{C}_{12}'^E = \frac{1}{3}\left(2\bar{C}_{12}^E - \bar{C}_{22}^E\right)$$

$$\bar{C}_{21}'^E = \frac{1}{3}\left(2\bar{C}_{12}^E - \bar{C}_{11}^E\right) \qquad \bar{C}_{22}'^E = \frac{1}{3}\left(2\bar{C}_{22}^E - \bar{C}_{12}^E\right) \tag{4.6.9}$$

The part of $\bar{\mathbf{C}}'^E$ corresponding to the shear stresses is the same as for the three-dimensional deformations, given in (4.6.5). Also, regarding the shear stresses and strains, we note that in plane stress deformations (membrane conditions) we have only the in-plane shear stress and strain components.

For the elastic-plastic response, in analogy with (4.4.11), we have

$$\boxed{{}^{t+\Delta t}\mathbf{S} = \mathbf{C}'^E\left({}^{t+\Delta t}\hat{\mathbf{e}}'' - \Delta\hat{\mathbf{e}}^P\right) = {}^{t+\Delta t}\mathbf{S}^E - \mathbf{C}'^E\Delta\hat{\mathbf{e}}^P} \tag{4.6.10}$$

where

$$\tag{4.6.11}{}^{t+\Delta t}\hat{\mathbf{e}}'' = {}^{t+\Delta t}\hat{\mathbf{e}} - {}^{t}\hat{\mathbf{e}}^P$$

and ${}^{t+\Delta t}\mathbf{S}^E = \mathbf{C}'^E\ {}^{t+\Delta t}\hat{\mathbf{e}}''$ is the elastic solution for the deviatoric stresses. Note that we use here the strain vectors with the engineering shear strains, while for the von Mises models we used tensorial shear strains in order to have the simple relation (4.4.9). By employing (4.2.8) and (3.3.10), we express $\Delta\hat{\mathbf{e}}^P$ in the form

$$\boxed{\Delta\hat{\mathbf{e}}^P = \Delta\lambda\,\hat{\mathbf{N}}\ {}^{t+\Delta t}\mathbf{S}} \tag{4.6.12}$$

where $\hat{\mathbf{N}}$ is the matrix of material constants defined in (3.3.5) but with twice the coefficients corresponding to shear.

We are searching for ${}^{t+\Delta t}\mathbf{S}$ on the yield surface $f_y = 0$. Figure 4.6.1 shows schematically the return mapping procedure in the deviatoric stress space. Note that the stress path from the elastic stress point D_E to the yield surface is not in the normal direction to the yield surface. Using (4.6.10) and (4.6.12) the solution for ${}^{t+\Delta t}\mathbf{S}$ is

$$\boxed{{}^{t+\Delta t}\mathbf{S} = (\mathbf{I}_6 + \Delta\lambda\mathbf{C}'^E\hat{\mathbf{N}})^{-1}\ {}^{t+\Delta t}\mathbf{S}^E} \tag{4.6.13}$$

where $\mathbf{I}_6$ is the (6×6) identity matrix.

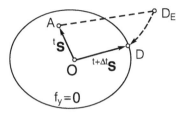

Fig. 4.6.1. Stress states at time t and at time $t + \Delta t$ in the deviatoric plane with the return mapping path; orthotropic plasticity

Considering the above relations (4.6.10) to (4.6.13) we have that all unknowns are functions of one parameter, $\Delta\lambda$. Hence, according to the governing parameter method described in Table 4.2.1, we can establish the governing equation, here the yield condition $f_y = 0$, in the form (3.3.3)

$$f_y = \frac{1}{2} \, {}^{t+\Delta t}\mathbf{S}^T \, \mathbf{N} \, {}^{t+\Delta t}\mathbf{S}' - \frac{1}{3}\sigma_y^2 = 0 \qquad (4.6.14)$$

and solve for $\Delta\lambda$. Table 4.6.1 summarizes the computational steps for the stress calculation.

Table 4.6.1. Computational steps for stress integration, orthotropic plasticity

---

1. Assume $\Delta\lambda$

2. Calculate ${}^{t+\Delta t}\mathbf{S}$ from (4.6.13)

3. Check for $|f_y| \le \varepsilon$ according to (4.6.14)
   If the condition is not satisfied, go to Step 1,
   with a new trial value $\Delta\lambda$

4. Calculate final quantities for time step

   ${}^{t+\Delta t}\mathbf{e}^P$, ${}^{t+\Delta t}\boldsymbol{\sigma}$

We note that, unlike for the isotropic material, the mean stress ${}^{t+\Delta t}\sigma_m$ depends on the plastic strains (the elastic strains in (4.6.1) can be evaluated after the plastic strains have been determined using (4.6.12)). The final stresses ${}^{t+\Delta t}\sigma_{ij}$ are

$$\boxed{{}^{t+\Delta t}\sigma_{ij} = {}^{t+\Delta t}S_{ij} + {}^{t+\Delta t}\sigma_m \delta_{ij}}$$

(4.6.15)

It is computationally more efficient to solve explicitly for ${}^{t+\Delta t}\mathbf{S}$ instead of inverting a matrix as indicated in (4.6.13). Hence we transform (4.6.13) into the form (4.4.47),

$$
\begin{aligned}
{}^{t+\Delta t}S_{xx} &= \left({}^{t+\Delta t}S^E_{xx} + \Delta\lambda D_{xx}\right)/D_\lambda \\
{}^{t+\Delta t}S_{yy} &= \left({}^{t+\Delta t}S^E_{yy} + \Delta\lambda D_{yy}\right)/D_\lambda \\
{}^{t+\Delta t}S_{zz} &= -\,{}^{t+\Delta t}S^E_{xx} - {}^{t+\Delta t}S^E_{yy} \\
{}^{t+\Delta t}S_i &= {}^{t+\Delta t}S^E_i/\left(1+2G_{ii}\Delta\lambda\, N_{ii}\right) \quad \begin{matrix} i=4,5,6 \\ \text{no sum on } i \end{matrix}
\end{aligned}
$$

(4.6.16)

where

$$D_{xx} = C_{yy}\,{}^{t+\Delta t}S^E_{xx} + C_{xy}\,{}^{t+\Delta t}S^E_{yy}$$

$$D_{yy} = C_{yx}\,{}^{t+\Delta t}S^E_{xx} + C_{xx}\,{}^{t+\Delta t}S^E_{yy}$$

$$D_\lambda = 1 + \Delta\lambda\left(C_{xx}+C_{yy}\right) + (\Delta\lambda)^2\left(C_{xx}C_{yy}-C_{xy}C_{yx}\right)$$

(4.6.17)

and

$$C_{xx} = C'^E_{1k}\left(N_{k1}-N_{k3}\right)$$

$$C_{xy} = -C'^E_{1k}\left(N_{k2}-N_{k3}\right)$$

$$C_{yx} = -C'^E_{2k}\left(N_{k1}-N_{k3}\right)$$

$$C_{yy} = C'^E_{2k}\left(N_{k2}-N_{k3}\right)$$

(4.6.18)

$$k = 1,2$$

with summation on k. To establish these relations we have taken into account the deviatoric character of ${}^{t+\Delta t}\mathbf{S}$.

The expressions (4.6.13) or (4.6.16) are also applicable to shell analysis conditions, but using (4.6.9) and $C'^E_{i3} = 0$, $i = 1,2,3$. They can also be reduced to the pipe or beam conditions, see Examples 4.6.1 to 4.6.4. For plane strain or axisymmetric deformations, the appropriate strain components must be set to zero.

The above computational procedure reduces of course to the procedure given for isotropic conditions. Then the matrices $\hat{\mathbf{N}}$ and $\mathbf{C}'^E$ (see (3.3.8), (3.3.9)) to be used in (4.6.13) are

$$\hat{\mathbf{N}} = \begin{bmatrix} \mathbf{A} & 0 \\ 0 & \mathbf{I}_3 \end{bmatrix} \tag{4.6.19}$$

and

$$\mathbf{C}'^E = 2G\,\hat{\mathbf{N}} \tag{4.6.20}$$

where $\mathbf{A}$ is the matrix defined in (3.2.70). Of course, the computational procedure of Section 4.4.1 is then numerically more efficient.

The above derived computational procedure assumes a perfectly plastic orthotropic material. This procedure can be extended to orthotropic materials with hardening (Kojic et al. 1996b). If, for example, the hardening is defined by the uniaxial yield curve in one of the principal material directions (for example the rolling direction for sheet metals, see Section 3.3.2), then an approximate expression for (3.3.14) is

$$\Delta\lambda = \frac{3}{2}\frac{\Delta\bar{e}^P}{^{t+\Delta t}\sigma_y} \tag{4.6.21}$$

This expression corresponds to the von Mises material with isotropic hardening. Here $\Delta\bar{e}^P$ is the increment of the effective plastic strain, and $^{t+\Delta t}\sigma_y$ is the yield stress on the uniaxial yield curve $\sigma_y = \sigma_y(\bar{e}^P)$. The governing parameter for stress integration becomes $\Delta\bar{e}^P$. Hence in step 1 of Table 4.6.1 we would assume $\Delta\bar{e}^P$, then obtain $^{t+\Delta t}\sigma_y$ from the yield curve, and calculate $\Delta\lambda$ from (4.6.21). The other steps in the table remain the same (of course, iterations are performed on $\Delta\bar{e}^P$).

4.6.2 Elastic-Plastic Matrix

The general approach for the derivation of the consistent elastic-plastic tangent matrix is the same as for the isotropic material model in Section 4.4.3. Here we give a few details specific for the orthotropic model.

First, the derivatives of the deviatoric stresses $\partial\,^{t+\Delta t}S_i/\partial\,^{t+\Delta t}e_j = {}^{t+\Delta t}S_{i,j} = {}^{t+\Delta t}C'_{ij}$ follow directly from (4.6.16) and can be written in compact form

$$^{t+\Delta t}C'_{ij} = P_{ij} + Q_i\Delta\lambda_{,j} \tag{4.6.22}$$

where P_{ij} and Q_i are the coefficients expressed in terms of known variables at the end of the step. As in the case of isotropic plasticity, in accordance with (4.2.5) and (4.4.70), we obtain $\Delta\lambda_{,j}$ by differentiation of (4.6.14),

$$\Delta\lambda_{,j} = -W_j/f'_y \tag{4.6.23}$$

where

$$W_j = N_{ik}\,{}^{t+\Delta t}S'_i\,P_{kj} \tag{4.6.24}$$

and $f'_y = \partial^{t+\Delta t}f_y/\partial(\Delta\lambda)$. Substituting $\Delta\lambda_{,j}$, the coefficients ${}^{t+\Delta t}C'_{ij}$ are obtained.

Second, by differentiation of (4.6.1) and with use of (4.6.12), we obtain the derivatives of the mean stress ${}^{t+\Delta t}\sigma_{m,j}$ as

$$
\begin{array}{ll}
{}^{t+\Delta t}\sigma_{m,j} = C_{jj}^{mE} - \Delta\lambda P_j^m + Q_j^m & \begin{array}{l} j = 1,2,3 \\ \text{no sum on } j \end{array} \\[2ex]
{}^{t+\Delta t}\sigma_{m,j} = \dfrac{1}{2}Q_j^m & j = 4,5,6
\end{array}
\tag{4.6.25}
$$

where P_j^m and Q_j^m are the known coefficients.

Details of the above derivations are given in Kojic et al. (1996b). We note that the matrix ${}^{t+\Delta t}\mathbf{C}^{EP}$ reduces to the one of an isotropic material if the constants N_{ij} have the isotropic material values (3.3.9).

4.6.3 Examples

Example 4.6.1. Orthotropic Composite Beam. Derive the basic relations for stress integration for the orthotropic composite beam shown in Fig. E.4.6-1. The beam is composed of layers with material axes inclined symmetrically with respect to the longitudinal axis, see also Kojic et al. (1995b).

We consider that at a given material point the strains e_{rr}, γ_{rs} and γ_{rt} are known from the displacements of the cross-section center and the rotations of the cross-section. The non-zero stresses are σ_{rr}, σ_{rs} and σ_{rt}. The layers

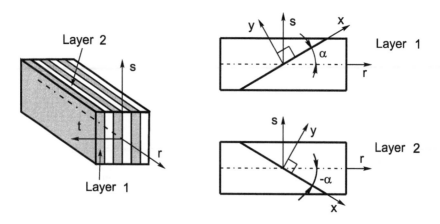

Fig. E.4.6-1. Composite beam

are parallel to the s-r plane, and we use x, y, z as the material axes, with the first axis x inclined at an angle α, the second one inclined at $(-\alpha)$, the third one again inclined at α, and so on, as shown in the figure.

In order to determine the stresses from the given strains, we assume that the elastic constitutive relations for the beam coordinate system r, s, t are

$$
\begin{aligned}
{}^{t+\Delta t}e^E_{rr} &= F^E_{11}{}^{t+\Delta t}\sigma_{rr} + F^E_{14}{}^{t+\Delta t}\sigma_{rs} + F^E_{16}{}^{t+\Delta t}\sigma_{rt} \\
{}^{t+\Delta t}\gamma^E_{rs} &= F^E_{41}{}^{t+\Delta t}\sigma_{rr} + F^E_{44}{}^{t+\Delta t}\sigma_{rs} + F^E_{46}{}^{t+\Delta t}\sigma_{rt} \\
{}^{t+\Delta t}\gamma^E_{rt} &= F^E_{61}{}^{t+\Delta t}\sigma_{rr} + F^E_{64}{}^{t+\Delta t}\sigma_{rs} + F^E_{66}{}^{t+\Delta t}\sigma_{rt}
\end{aligned}
\tag{a}
$$

where the F^E_{ij} are the components of the elastic *compliance matrix* corresponding to the beam coordinate system. To calculate F^E_{ij} from the compliance matrix $\bar{\mathbf{F}}^E = (\bar{\mathbf{C}}^E)^{-1}$, given in (A1.38), we define for each layer the transformation matrix $\mathbf{T}$,

$$
\begin{array}{ccc}
 & x & y & z
\end{array}
$$

$$
\mathbf{T} = \begin{array}{c} r \\ s \\ t \end{array} \begin{bmatrix} c & -s & 0 \\ s & c & 0 \\ 0 & 0 & 1 \end{bmatrix}
\tag{b}
$$

where we use $c = \cos\alpha$, $s = \sin\alpha$ for layer 1, and obtain (see (A1.53) and (A1.62))

$$
\mathbf{F}^E = \mathbf{T}^\epsilon\, \bar{\mathbf{F}}^E\, (\mathbf{T}^\epsilon)^T
\tag{c}
$$

where the matrix $\mathbf{T}^\epsilon$ is given in (A1.56). For layer 2 we proceed in the same manner but now use $(-\alpha)$. From (b) and (c) we obtain the nonzero terms $F^E_{11}, F^E_{14} = F^E_{41}, F^E_{44}$ and F^E_{66}. With the elastic constants F^E_{ij} we write (a) in the form

$$
\begin{aligned}
F^E_{11}{}^{t+\Delta t}\sigma_{rr} + F^E_{14}{}^{t+\Delta t}\sigma_{rs} &= {}^{t+\Delta t}e''_{rr} - \Delta e^P_{rr} \\
F^E_{14}{}^{t+\Delta t}\sigma_{rr} + F^E_{44}{}^{t+\Delta t}\sigma_{rs} &= {}^{t+\Delta t}e''_{rs} - 2\Delta e^P_{rs} \\
F^E_{66}{}^{t+\Delta t}\sigma_{rt} &= {}^{t+\Delta t}e''_{rt} - 2\Delta e^P_{rt}
\end{aligned}
\tag{d}
$$

where

$$
\begin{aligned}
{}^{t+\Delta t}e''_{rr} &= {}^{t+\Delta t}e_{rr} - {}^t e^P_{rr} & {}^{t+\Delta t}e''_{rs} &= {}^{t+\Delta t}\gamma_{rs} - {}^t\gamma^P_{rs} \\
{}^{t+\Delta t}e''_{rt} &= {}^{t+\Delta t}\gamma_{rt} - {}^t\gamma^P_{rt}
\end{aligned}
\tag{e}
$$

The next step is to express the increments $\Delta e^P_{rr}, \Delta e^P_{rs}$ and Δe^P_{rt} in terms of the stresses ${}^{t+\Delta t}\sigma_{rr}, {}^{t+\Delta t}\sigma_{rs}$ and ${}^{t+\Delta t}\sigma_{rt}$. We first write these strain increments as (see (A1.52))

$$\Delta e_{rr}^P = T_{1k}^\epsilon \Delta \bar{e}_k^P = c^2 \Delta e_{xx}^P + s^2 \Delta e_{yy}^P - 2sc\Delta e_{xy}^P$$

$$\Delta e_{rs}^P = \frac{1}{2}T_{4k}^\epsilon \Delta \bar{e}_k^P = sc(\Delta e_{xx}^P - \Delta e_{yy}^P) + (c^2 - s^2)\Delta e_{xy}^P \qquad \text{(f)}$$

$$\Delta e_{rt}^P = \frac{1}{2}T_{6k}^\epsilon \Delta \bar{e}_k^P = -s\Delta e_{yz}^P + c\Delta e_{xz}^P$$

where $\Delta \bar{e}_k^P$ are the components in the material coordinate system of the plastic strain increment tensor Δe^P. Then we transform the beam deviatoric stresses $^{t+\Delta t}S_k$ to $^{t+\Delta t}\bar{S}_i$ corresponding to the material coordinate system,

$$^{t+\Delta t}\bar{S}_i = T_{ki}^\epsilon \, ^{t+\Delta t}S_k \qquad \text{(g)}$$

and obtain $^{t|\Delta t}S_{xx}, ..., \, ^{t+\Delta t}S_{xz}$ in terms of the stresses $^{t+\Delta t}\sigma_{rr}, .., \, ^{t+\Delta t}\sigma_{rt}$. We further use the stress-plastic strain relations (4.6.12) and the expressions (g) for the deviatoric stresses, to obtain after some calculations

$$\Delta e_{rr}^P = \Delta\lambda \, (N_{rr}\, ^{t+\Delta t}\sigma_{rr} + N_{rs}\, ^{t+\Delta t}\sigma_{rs})$$

$$\Delta e_{rs}^P = \Delta\lambda \, (N_{sr}\, ^{t+\Delta t}\sigma_{rr} + N_{ss}\, ^{t+\Delta t}\sigma_{rs}) \qquad \Delta e_{rt}^P = \Delta\lambda \, N_{rt}\, ^{t+\Delta t}\sigma_{rt} \qquad \text{(h)}$$

where

$$N_{rr} = \bar{N}_{xx}(c^2 - \frac{1}{3}) + \bar{N}_{yy}(s^2 - \frac{1}{3}) + 2s^2c^2 N_{xy}$$

$$N_{rs} = 2sc\left[\bar{N}_{xx} - \bar{N}_{yy} - (c^2 - s^2)N_{xy}\right]$$

$$N_{sr} = sc\left[(2N_1 + 2N_2 - N_3)(c^2 - \frac{1}{3})\right.$$

$$\left. + (-2N_1 + N_2 - 2N_3)(s^2 - \frac{1}{3}) - (c^2 - s^2)N_{xy}\right] \qquad \text{(i)}$$

$$N_{ss} = 2s^2c^2(4N_1 + N_2 + N_3) + (c^2 - s^2)^2 \, N_{xy}$$

$$N_{rt} = c^2 N_{xy} + s^2 N_{yz}$$

and

$$\bar{N}_{xx} = c^2(N_1 + 2N_2) - s^2(N_1 - N_3)$$

$$\bar{N}_{yy} = -c^2(N_1 - N_2) + s^2(N_1 + 2N_3) \qquad \text{(j)}$$

Finally, substituting (h) into (d) we solve, with some calculations, for the stresses

$$^{t+\Delta t}\sigma_{rr} = \frac{1}{D_\lambda}\left[F_{44}^E \, ^{t+\Delta t}e_{rr}'' - F_{14}^E \, ^{t+\Delta t}e_{rs}''\right.$$

$$\left. + \Delta\lambda \, (2N_{ss}\, ^{t+\Delta t}e_{rr}'' - N_{rs}\, ^{t+\Delta t}e_{rs}'')\right]$$

$$^{t+\Delta t}\sigma_{rs} = \frac{1}{D_\lambda} \left[-F_{14}^{E}\,{}^{t+\Delta t}e_{rr}'' + F_{11}^{E}\,{}^{t+\Delta t}e_{rs}'' \right.$$
$$\left. + \Delta\lambda(-2N_{sr}\,{}^{t+\Delta t}e_{rr}'' + N_{rr}\,{}^{t+\Delta t}e_{rr}'') \right] \qquad (k)$$

$$^{t+\Delta t}\sigma_{rt} = \frac{{}^{t+\Delta t}e_{rt}''}{F_{66}^{E} + 2\Delta\lambda\,N_{rt}}$$

where

$$D_\lambda = (F_{11}^{E} + \Delta\lambda N_{rr})(F_{44}^{E} + 2\Delta\lambda N_{ss}) - (F_{14}^{E} + 2\Delta\lambda N_{sr})(F_{14}^{E} + \Delta\lambda N_{rs})$$

In the expressions (k) we have the stresses in terms of the unknown $\Delta\lambda$. Therefore we can employ the computational steps in Table 4.6.1 to determine $\Delta\lambda$. Or, we can use the approximation (4.6.21) and take into account the material hardening.

The elastic-plastic matrix $^{t+\Delta t}\mathbf{C}^{EP}$ can be determined in two ways:

- by calculation of $^{t+\Delta t}\bar{\mathbf{C}}^{EP}$ in the material axes, as in Section 4.6.2 with the static condensation; or
- by differentiation of the relations (k) with respect to $^{t+\Delta t}e_{rr}$, $^{t+\Delta t}\gamma_{rs}$ and $^{t+\Delta t}\gamma_{rt}$.

The second approach is more efficient.

Example 4.6.2. Composite Pipe. Derive the basic relations for the stress integration for the composite pipe shown in Fig. E.4.6-2 when the pipe is subjected to internal pressure (see also Kojic et al. 1995a). The pipe wall is composed of layers of orthotropic material.

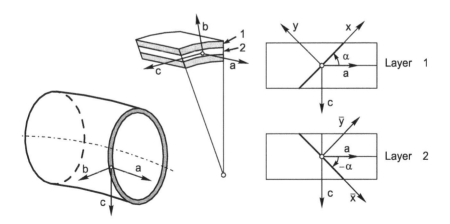

Fig. E.4.6-2. Multilayered pipe made of orthotropic material

As in the case of the analysis of the beam (Example 4.6.1), we employ the compliance matrix to appropriately take into account the stress conditions in the pipe wall. These conditions are that the stress σ_{cc} is known from the pipe internal pressure (see Example 4.5.2) and $\sigma_{bb} = 0$, $\sigma_{bc} = 0$. Also, the strain $\gamma_{bc} = 0$. Using these conditions we can write the following constitutive relations:

$$
\begin{aligned}
{}^{t+\Delta t}e_{aa} - {}^{t}e_{aa}^{P} - \Delta e_{aa}^{P} &= F_{11}^{E}\,{}^{t+\Delta t}\sigma_{aa} + F_{13}^{E}\,{}^{t+\Delta t}\sigma_{cc} + \\
&\quad F_{14}^{E}\,{}^{t+\Delta t}\sigma_{ab} + F_{16}^{E}\,{}^{t+\Delta t}\sigma_{ac} \\
{}^{t+\Delta t}\gamma_{ac} - {}^{t}\gamma_{ac}^{P} - 2\Delta e_{ac}^{P} &= F_{16}^{E}\,{}^{t+\Delta t}\sigma_{aa} + F_{36}^{E}\,{}^{t+\Delta t}\sigma_{cc} + \\
&\quad F_{46}^{E}\,{}^{t+\Delta t}\sigma_{ab} + F_{66}^{E}\,{}^{t+\Delta t}\sigma_{ac} \\
{}^{t+\Delta t}\gamma_{ab} - {}^{t}\gamma_{ab}^{P} - 2\Delta e_{ab}^{P} &= F_{14}^{E}\,{}^{t+\Delta t}\sigma_{aa} + F_{34}^{E}\,{}^{t+\Delta t}\sigma_{cc} + \\
&\quad F_{44}^{E}\,{}^{t+\Delta t}\sigma_{ab} + F_{46}^{E}\,{}^{t+\Delta t}\sigma_{ac}
\end{aligned}
\tag{a}
$$

The elastic constants F_{ij}^{E} can be expressed by the transformation of the compliance matrix $\bar{\mathbf{F}}^{E} = (\bar{\mathbf{C}}^{E})^{-1}$ (see (A1.38) and (A1.62)), which corresponds to the material coordinate system x, y, z. Therefore, we specify the transformation matrix $\mathbf{T}$ for each layer as

$$
\begin{array}{ccc}
 & x & y & z \\
\mathbf{T} = \begin{array}{c} a \\ b \\ c \end{array} & \left[\begin{array}{ccc} c & -s & 0 \\ 0 & 0 & 1 \\ -s & -c & 0 \end{array} \right]
\end{array}
\tag{b}
$$

where we use $c = \cos\alpha$, $s = \sin\alpha$ for layer 1. For layer 2 we proceed in the same manner but now use the angle $(-\alpha)$. We employ the transformation (c) of Example 4.6.1 and obtain F_{ij}^{E}. For example,

$$
\begin{aligned}
F_{11}^{E} &= c^{4}\bar{F}_{11}^{E} + s^{4}\bar{F}_{22}^{E} + s^{2}c^{2}(2\bar{F}_{12}^{E} + \bar{F}_{44}^{E}) \qquad F_{14}^{E} = F_{34}^{E} = F_{46}^{E} = 0 \\
F_{36}^{E} &= 2sc\left[c^{2}\bar{F}_{22}^{E} - s^{2}\bar{F}_{11}^{E} - (c^{2} - s^{2})(\bar{F}_{12}^{E} + \tfrac{1}{2}\bar{F}_{44}^{E})\right] \\
F_{66}^{E} &= 4s^{2}c^{2}(\bar{F}_{11}^{E} + \bar{F}_{22}^{E} - 2\bar{F}_{12}^{E}) + (c^{2} - s^{2})^{2}\bar{F}_{44}^{E}
\end{aligned}
\tag{c}
$$

With these elastic constants we rewrite the constitutive relations (a) in the form

$$
\begin{aligned}
F_{11}^{E}\,{}^{t+\Delta t}\sigma_{aa} + F_{16}^{E}\,{}^{t+\Delta t}\sigma_{ac} &= {}^{t+\Delta t}e_{aa}'' - \Delta e_{aa}^{P} \\
F_{16}^{E}\,{}^{t+\Delta t}\sigma_{aa} + F_{66}^{E}\,{}^{t+\Delta t}\sigma_{ac} &= {}^{t+\Delta t}e_{ac}'' - 2\Delta e_{ac}^{P} \\
F_{44}^{E}\,{}^{t+\Delta t}\sigma_{ab} &= {}^{t+\Delta t}e_{ab}'' - 2\Delta e_{ab}^{P}
\end{aligned}
\tag{d}
$$

where

$$
\begin{aligned}
{}^{t+\Delta t}e_{aa}'' &= {}^{t+\Delta t}e_{aa} - {}^{t}e_{aa}^{P} - F_{13}^{E}\,{}^{t+\Delta t}\sigma_{cc} \\
{}^{t+\Delta t}e_{ac}'' &= {}^{t+\Delta t}\gamma_{ac} - {}^{t}\gamma_{ac}^{P} - F_{36}^{E}\,{}^{t+\Delta t}\sigma_{cc} \\
{}^{t+\Delta t}e_{ab}'' &= {}^{t+\Delta t}\gamma_{ab} - {}^{t}\gamma_{ab}^{P}
\end{aligned}
\tag{e}
$$

are the known strains.

As in Example 4.6.1, we express the increments of plastic strains in (d) in terms of increments of plastic strains $\Delta \bar{e}_k^P$ in the material axes (see (f) of Example 4.6.1), and also the deviatoric stresses $^{t+\Delta t}S_{xx}, ..., {}^{t+\Delta t}S_{xz}$ in terms of the stresses $^{t+\Delta t}\sigma_{aa}, ..., {}^{t+\Delta t}\sigma_{cc}$. With use of the stress-plastic strain relations (4.6.12) we obtain

$$
\begin{aligned}
\Delta e_{aa}^P &= \Delta\lambda \left(\bar{N}_{11}\,{}^{t+\Delta t}\sigma_{aa} + \bar{N}_{16}\,{}^{t+\Delta t}\sigma_{ac} + \bar{N}_{13}\,{}^{t+\Delta t}\sigma_{cc} \right) \\
\Delta e_{ac}^P &= \Delta\lambda \left(\bar{N}_{61}\,{}^{t+\Delta t}\sigma_{aa} + \bar{N}_{66}\,{}^{t+\Delta t}\sigma_{ac} + \bar{N}_{63}\,{}^{t+\Delta t}\sigma_{cc} \right) \qquad \text{(f)} \\
\Delta e_{ab}^P &= \Delta\lambda \left(s^2 \bar{N}_{yz} + c^2 \bar{N}_{zx} \right){}^{t+\Delta t}\sigma_{aa}
\end{aligned}
$$

where $\bar{N}_{ij}$ are expressed in terms of the material coefficients $N_1, ..., N_{zx}$, and $\cos\alpha$ and $\sin\alpha$.

We finally substitute expressions (f) into (d) and solve for the stresses,

$$
\begin{aligned}
{}^{t+\Delta t}\sigma_{aa} &= \left[\left({}^{t+\Delta t}e_{aa}'' - \Delta\lambda\bar{N}_{13}\,{}^{t+\Delta t}\sigma_{cc} \right) \left(F_{66}^E + 2\Delta\lambda\bar{N}_{66} \right) \right. \\
&\quad \left. - \left({}^{t+\Delta t}e_{ac}'' - 2\Delta\lambda\bar{N}_{63}\,{}^{t+\Delta t}\sigma_{cc} \right) \left(F_{16}^E + \Delta\lambda\bar{N}_{16} \right) \right] / D_\lambda \\
{}^{t+\Delta t}\sigma_{ac} &= \left[\left({}^{t+\Delta t}e_{ac}'' - 2\Delta\lambda\bar{N}_{63}\,{}^{t+\Delta t}\sigma_{cc} \right) \left(F_{11}^E + \Delta\lambda\bar{N}_{11} \right) \right. \\
&\quad \left. - \left({}^{t+\Delta t}e_{aa}'' - \Delta\lambda\bar{N}_{13}\,{}^{t+\Delta t}\sigma_{cc} \right) \left(F_{16}^E + 2\Delta\lambda\bar{N}_{61} \right) \right] / D_\lambda \\
{}^{t+\Delta t}\sigma_{ab} &= \frac{{}^{t+\Delta t}e_{ab}''}{F_{44}^E + 2\Delta\lambda \left(s^2 N_{yz} + c^2 N_{xz} \right)} \qquad\qquad \text{(g)}
\end{aligned}
$$

where

$$
D_\lambda = (F_{11}^E + \Delta\lambda\bar{N}_{11})(F_{66}^E + \Delta\lambda\bar{N}_{66}) - (F_{16}^E + 2\Delta\lambda\bar{N}_{61})(F_{16}^E + \Delta\lambda\bar{N}_{16})
$$

The computational procedure for the calculation of $\Delta\lambda$, deviatoric stresses, increments of plastic strains, and stresses, is the same as in Example 4.6.1. Also, the tangent elastic-plastic matrix can be obtained by the appropriate differentiations of the above expressions for stresses with respect to the strains. The calculated matrix $^{t+\Delta t}\mathbf{C}^{EP}$ is symmetric (see Kojic et al. 1995a).

Example 4.6.3. Plate Loaded by in-Plane Loads. A plate, modeled by a 4-node finite element, is strained as shown in Fig. E.4.6-3. The material is first strained to the elastic limit in the direction 1-2, then fully strained further to point 2, and then subjected to shear along the paths 2-3 and 3-4. The material data are as follows (Hill's orthotropic material, Section 3.3.1):

$$
\begin{aligned}
E_{\bar{x}} &= 2E05 & E_{\bar{y}} &= E_{\bar{z}} = 1E05 \\
\nu_{\bar{x}\bar{y}} &= 0.3 & \nu_{\bar{y}\bar{z}} &= 0.2 & \nu_{\bar{z}\bar{x}} &= 0.15 \\
G_{\bar{x}\bar{y}} &= 6E04 & G_{\bar{y}\bar{z}} &= 4.17E04 & G_{\bar{z}\bar{x}} &= 6E04 \\
\bar{X} &= 200 & \bar{Y} &= \bar{Z} = 40 \\
Y_{\bar{x}\bar{y}} &= Y_{\bar{y}\bar{z}} = Y_{\bar{z}\bar{x}} = 80
\end{aligned}
$$

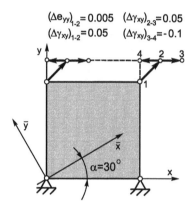

Fig. E.4.6-3. Plastic deformation of an orthotropic plate subjected to prescribed straining

where the moduli and yield stresses are in MPa. The material axes $\bar{x}, \bar{y}$ are at an angle $\alpha = 30^0$ with respect to the x, y coordinate system, as shown in the figure.

The stress calculation is performed according to Table 4.6.1. The final solutions obtained by using one step in the path 1-2, and ten steps in each of the paths 2-3 and 3-4 are:

$$\sigma_{xx} = -104.4 \qquad \sigma_{yy} = -70.90 \qquad \sigma_{xy} = -95.79$$
$$e_{xx}^P = 3.580E\text{-}4 \qquad e_{yy}^P = 5.237E\text{-}3 \qquad e_{zz}^P = -5.595E\text{-}3$$
$$\gamma_{xy}^P = 1.322E\text{-}3$$

5. Creep and Viscoplasticity

In this chapter we consider time-dependent inelastic deformations of isotropic metals and include thermal effects. We consider creep, thermoplastic and viscoplastic models. After the introduction in Section 5.1, we give the fundamental characteristics of creep and thermoplastic models in Section 5.2 and of a viscoplastic model in Section 5.3.

In Sections 5.4 and 5.5 we develop computational procedures for stress integration for the material models described in Sections 5.2 and 5.3. The computational algorithms are implicit and represent applications of the governing parameter method of Section 4.2. General 3-D deformation conditions are considered, as well as the special conditions of shell, beam and pipe deformations. A number of solved examples show the computed material response in typical engineering conditions.

5.1 Introduction

In Chapters 3 and 4 we considered plastic deformations of metals; that is time-independent inelastic deformations. The main assumption was that the deformation occurs *instantaneously* with the application of the load. The load (strain) level was the measure of the external action on the material, and time was a fictitious parameter. However, it is known from experiments that after the initial plastic deformations of a material, the *plastic flow can continue* with the development of time-dependent inelastic strains. These effects are to some degree present in many materials, but they may or may not be significant, depending on the physical conditions under which the material is loaded.

It is considered that there are generally two types of time-dependent (mathematical model) inelastic deformations, *creep* and *viscoplastic deformations*. These two types of inelastic deformations are governed by different constitutive laws. However, some materials, for example metals at elevated temperatures, exhibit creep and viscoplastic phenomena simultaneously. It is not possible to experimentally distinguish between these two types of inelastic deformations, and their separation has been an analytical convenience.

Both creep and viscoplastic deformations are important in the cases of long-term material loading and are pronounced at elevated temperatures.

Many structural parts are exposed to high temperatures and loading over long time intervals (e.g., elements of power plants, gas turbines, pressure vessels, etc.). It is necessary to formulate material models to adequately describe the material characteristics in these conditions, and to have numerical procedures to predict the material response.

There is, however, a fundamental difference in the character between creep and viscoplastic flows. Namely, under constant stresses (above a level that depends on the material and temperature), creep deformations *progress* while viscoplastic deformations *diminish* with time. The creep models assume that the creep deformations increase with time leading to a possible material rupture. On the other hand, the viscoplastic models are based on an overstress assumption, according to which the stress point in the stress space is outside the yield surface and moves toward the yield surface with time. The viscoplastic flow continues until the stress point reaches the yield surface. A stress relaxation occurs in both types of deformations and has the same character. Note that a viscoplastic constitutive law may be reduced to a creep law, if the yield stress is taken to be zero (see Section 5.3.2).

We will describe these two types of inelastic deformations in the next two sections. The constitutive descriptions are macroscopic, phenomenological in character, as for the description of the time independent plastic deformations in Chapter 3.

The methods of analysis that include thermo-elastic-plastic and creep deformations were first of an analytical type and were applicable to simple deformation and loading conditions, Finnie and Heller (1959); Penny and Marriott (1971); Odquist (1974); Kraus (1980). The traditional creep models have been further modified by introducing anisotropic creep behavior, based on the potential theory of creep, Brown (1970); Rees (1983). Later, with the progress of numerical methods, especially the finite element method, numerical procedures have been developed that can be used to realistically predict the response of very complex structures (Bathe 1999, 2001a). Included in these response predictions are also phenomena related to damage, fatigue due to cyclic loading, and fracture, which, however, we do not consider in this book.

5.2 Creep and Thermoplastic Material Models

In this section we first describe some basic temperature effects on material behavior, and then introduce commonly used material models for creep and thermoplastic deformations of metals (e.g., Penny and Marriott 1971; Kraus 1980).

Figure 5.2.1 shows schematically the influence of temperature on a uniaxial stress-strain relation. It can be seen that Young's modulus and the yield stress of the material *decrease* with the temperature *increase*. A detailed de-

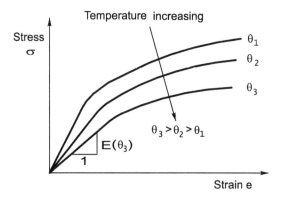

Fig. 5.2.1. Effect of temperature on uniaxial stress-strain curve

scription of the uniaxial stress-strain relation is given in Section 3.2.1 (see Figs. 3.2.1 and 3.2.2).

Creep can be thought of as a time-dependent inelastic deformation of material grains resulting from the relative sliding at the grain boundaries (e.g., McClintock and Argon 1966), where permanent strains which progress with time are developed. These inelastic strains are known as *creep strains*. At room temperature creep strains of metals are generally very small, but they increase with the temperature. A typical displacement-time curve is shown in Fig. 5.2.2 for uniaxial loading of the material at an elevated temperature. At the instant of the load application, an initial elongation u_0 is developed. This instantaneous deformation can be elastic or elastic-plastic, depending on the load level (see Section 3.2). The deformation progresses with time and can lead to material rupture (at time t_R).

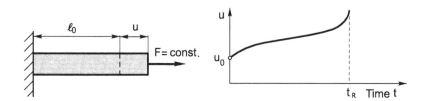

Fig. 5.2.2. Elongation of a bar during time due to creep

5.2.1 Uniaxial Creep

Consider the uniaxial load conditions on a specimen resulting in the typical creep curve, corresponding to a constant stress and constant temperature, shown in Fig. 5.2.3. The strain starts with the initial, instantaneous strain

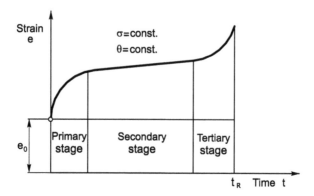

Fig. 5.2.3. Schematics of a creep curve

e_0 (which can be elastic or elastic-plastic) and increases with time. There are in general three stages of creep: the *primary*, *secondary* and *tertiary* creep stages, as indicated in the figure. In the primary stage the creep strain rate is first high and then decreases to an approximately constant value, leading to the so-called *stationary creep rate* during the secondary creep stage. In the tertiary stage the creep strain rate increases again and the strain progresses until the material rupture is reached at time t_R. The length of each stage on the creep curve depends on the material. Some materials hardly exhibit secondary creep, whereas other materials, for example, have a very small tertiary stage.

The creep curves are strongly dependent on the stress level and the temperature, as schematically shown in Fig. 5.2.4. We see that at a given temperature, the creep strain increases with stress, and for a given stress, the

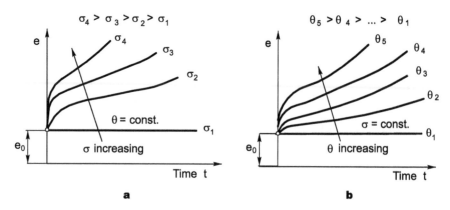

Fig. 5.2.4. Schematics of creep curves for various stress and temperature levels. **a** Effect of stress level on creep curve; **b** Effect of temperature on creep curve

creep increases with temperature. Note that there is a stress level at a given temperature below which the creep is negligible (stress σ_1 in the figure). Also, for a given stress there is a temperature below which the creep strain is very small (temperature θ_1 in the figure).

Consider next the material behavior when a load is maintained for a certain period of time, causing creep, and then removed. Figure 5.2.5 shows schematically the material response under these conditions. At the moment when the load is released, the strain drops instantaneously for the elastic strain e^E, and then an additional decrease e_R^C is developed over a certain period of time. This additional strain decrease is called *creep recovery*, with e_R^C representing the creep recovery strain. We see that the final permanent strain e_F is

$$e_F = {}^1e - e^E - e_R^C \tag{5.2.1}$$

where ${}^1e$ is the strain at time t_1 prior to the stress release.

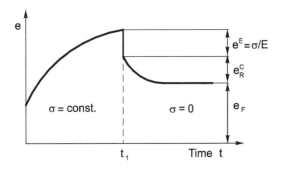

Fig. 5.2.5. Creep recovery

Another important phenomenon related to material behavior in creep is the so-called *creep relaxation*. Namely, if the material is loaded over a period of time, with the total strain held constant, the stress will decrease with time. For example, if we extend a bar by an elongation U, as shown in Fig. 5.2.6, and keep U constant, the stress will decrease with time as shown in the figure. The relaxation curves are drawn for several stress levels. As can be seen, there is a threshold for the stress below which the relaxation is negligible.

We considered so far uniaxial creep in tension. It is generally assumed that creep curves in uniaxial compression are identical to those in uniaxial tension. However, if the material is subjected to cyclic loading, the material response is more complex.

Let us consider first the uniaxial creep behavior under *variable loading*. The main task of the mathematical theory of creep is to establish the relations that describe creep under arbitrary variable loading conditions. The most common and generally accepted approach results into the so-called *equation-of-state* method. This approach is based on the assumption that the creep

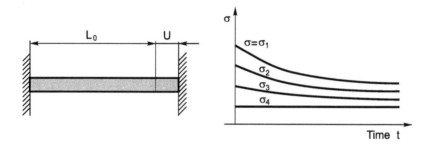

Fig. 5.2.6. Relaxation curves for different stress levels

response of a material at a certain time t depends solely on the magnitude of the variables at that time. Another approach is based on the *material memory*: the material response at a given time depends on the current parameters and also on the history of deformation (e.g., Malvern 1969). We will further use the equation-of-state approach because experimental data are available for this description of creep and the approach is generally accepted for engineering applications.

The creep curve according to the equation-of-state approach is in general written in the form

$$e^C = \sum_{i=1}^{n} \phi_i(\sigma)\, g_i(t)\, h_i(\theta) \tag{5.2.2}$$

where e^C is the creep strain; and ϕ_i, g_i and h_i are functions of stress, σ, time, t, and temperature, θ, respectively. We use a simple form of (5.2.2),

$$\boxed{e^C = f_1(\sigma)\, f_2(t)\, f_3(\theta)} \tag{5.2.3}$$

with f_1, f_2 and f_3 representing the functions of stress, time and temperature. The relation (5.2.3) defines the so-called *creep law*. Analogous to plasticity (Section 3.2), where the basic material behavior was given by the yield curve $\sigma = \sigma(e^P)$ (see (3.2.3)), we have now that the main characteristic of the material response is the creep law. We list here three creep laws, known as the power creep law, exponential creep law, and the eight-parameter creep law (the expressions in (5.2.4), (5.2.5) and (5.2.6), respectively):

$$\boxed{e^C = a_0 \sigma^{a_1} t^{a_2}} \tag{5.2.4}$$

$$\boxed{e^C = f(1 - e^{-rt}) + gt} \tag{5.2.5}$$

with

$$f = a_0 e^{a_1 \sigma} \qquad r = a_2 \left(\frac{\sigma}{a_3}\right)^{a_4} \qquad g = a_5 e^{a_6 \sigma}$$

and

$$e^C = a_0 \sigma^{a_1} \left(t^{a_2} + a_3 t^{a_4} + a_5 t^{a_6}\right) e^{-a_7/(\theta+273.16)} \qquad (5.2.6)$$

Here a_0, a_1, ... , a_7 are the material constants, stress, time and temperature independent, and θ is the temperature in $^{\circ}C$.

In order to apply the creep law to variable loading and variable temperature conditions, we need to determine the *creep strain rate* $\dot{e}^C$ at a given time. From (5.2.3) we obtain

$$\dot{e}^C = \frac{\partial e^C}{\partial t} = f_1(\sigma)\, \dot{f}_2(t)\, f_3(\theta) \qquad (5.2.7)$$

where $\dot{f}_2 = df_2/dt$ is the time derivative of the function $f_2(t)$. In deriving (5.2.7) we have neglected the derivatives of the stress and temperature with respect to time, which strictly holds only in constant stress and temperature conditions. Generally, creep effects are important when the stresses and temperature are held at certain levels over a relatively long time period and then change relatively rapidly, like schematically shown in Fig. 5.2.7. Under these conditions the relation (5.2.7) can be considered adequate. The expression (5.2.7) represents one of the *fundamental relations* used to model creep effects.

Let us apply the relations (5.2.3) and (5.2.7) to a simple example. A material is loaded uniaxially by a constant stress over certain time periods, as

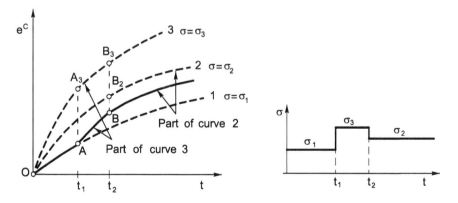

Fig. 5.2.7. Creep strain history according to time hardening

shown in Fig. 5.2.7. To the constant stresses σ_1, σ_2, σ_3 correspond different creep curves, represented schematically in the figure. There are basically two approaches to determine the creep strain in this case: the *time hardening* approach and the *strain hardening* approach. To demonstrate these approaches, we apply both to this example.

In the *time hardening* assumption we proceed as follows. In the time period 0 to t_1 we follow the creep curve corresponding to the stress σ_1. Then, at time t_1 we consider that creep continues according to the creep curve 3 – but using the curve from time $t = t_1$ onward (hence the segment A_3B_3 is translated to AB), and so on.

In the *strain hardening* approach, on the other hand, we use the creep law to express time in terms of the creep strain e^C. This results in the creep strain rate $\dot{e}^C$ given by

$$\dot{e}^C = \dot{e}^C(e^C, \sigma, \theta) \tag{5.2.8}$$

Hence, we have that the creep strain rate is a function of the current values of stress, temperature and creep strain (instead of time). For the creep law of the form (5.2.3) we have

$$\boxed{\dot{e}^C = f_1(\sigma)\,\phi(e^C, \sigma, \theta)\,f_3(\theta)} \tag{5.2.9}$$

where

$$\phi = \dot{f}_2\left\{f_2^{-1}\left[\frac{e^C}{f_1(\sigma)f_3(\theta)}\right]\right\} \tag{5.2.10}$$

Here f_2^{-1} stands for the inverse of the function f_2. For example, if creep is defined by the power creep law (5.2.4), then

$$\boxed{\dot{e}^C = a_2 a_0^{1/a_2}\sigma^{a_1/a_2}(e^C)^{(1-1/a_2)}} \tag{5.2.11}$$

For a complex creep law, it may of course not be possible to obtain the explicit form (5.2.8) or (5.2.9) for the creep strain rate. Then the dependence of the creep strain rate on the creep strain must be calculated numerically (see (5.4.15) in Section 5.4.1).

Let us illustrate the application of the strain hardening approach for the example in Fig. 5.2.7. As shown in Fig. 5.2.8, we follow the creep curve 1 to point A corresponding to time t_1, at which time the creep strain reaches the value $^1e^C$. Then we use the creep curve 3, from the point A_3 onwards, where A_3 is defined by the creep strain $^1e^C$. This use of the creep curve 3 results in translating the segment A_3B_3 to become the segment AB, and so on.

Although both approaches are based on the same fundamental relations (5.2.3) and (5.2.7), the strain hardening approach has been favored because

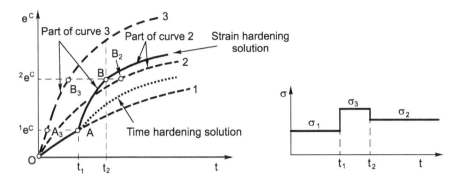

Fig. 5.2.8. Creep strain history according to strain hardening

the results obtained are closer to experimental results (Kraus 1980). The strain hardening approach, in particular, models cyclic loading conditions more accurately. Of course, in general, time hardening and strain hardening solutions are not identical.

Let us consider next uniaxial creep in *cyclic loading* using the strain hardening approach. It is generally assumed that the creep curves in compression are the reflected creep curves in tension, as shown in Fig. 5.2.9. The modeling we discuss corresponds to the so-called *modified hardening rules*, developed at the Oak Ridge National Laboratory (Pugh et al. 1972, 1974; Pugh 1975; Pugh and Robinson 1978).

For the discussion we use the cyclic loading shown in Fig. 5.2.9**a**: tension by stress σ_1, then compression by stress σ_2, and finally tension by stress σ_3 (Kraus 1980). In the first load period the creep occurs according to the creep curve OA in Fig. 5.2.9**c**, which is the part OA_1 of the creep curve 1 in Fig. 5.2.9**b**. From time t_1 onwards, creep continues according to the curve AB which represents the segment OB_2 of the creep curve $\sigma = -\sigma_2$. Hence, the hardening which we would have at strain e_1^C (that is at time t_1) on the creep curve 2 if the stress σ_2 were tensile is not used, and the creep in compression starts from point O on the curve $\sigma = -\sigma_2$. Next, at time t_2 when the stress becomes again positive, $\sigma = \sigma_3$, the creep strain progresses with time as follows. Since the accumulated positive creep strain at time t_2 is equal to $(e_1^C - e_2^C)$, where e_2^C is the magnitude of the negative creep strain generated along the segment OB_2 of the reflected creep curve 2, creep continues following the creep curve 3 but from the point B_3 defined by the creep strain $(e_1^C - e_2^C)$. The creep curve BC represents the segment B_3C_3 of the curve 3. If e_2^C were greater in magnitude than e_1^C, the point B_3 would have been at the origin O (the hardening in tension would have been lost).

To make the O.R.N.L. rule simple for applications, it is effective to introduce the so-called *modified creep strain* e^H as a measure for the strain hardening. The creep strain rate obtains the form

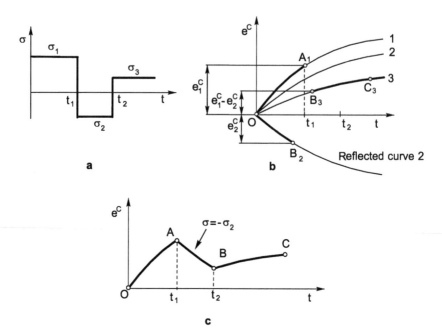

Fig. 5.2.9. Creep response in cyclic uniaxial loading. **a** Stress history; **b** Uniaxial creep curves; **c** Creep response according to O.R.N.L. hardening rule

$$\dot{e}^C = \dot{e}^C(e^H, \sigma, \theta) \qquad (5.2.12)$$

If the material is subjected to tension in a time interval $t_n - t_{n+1}$, the modified creep strain is defined as

$$e^H = e^C - e^+ \qquad (5.2.13)$$

where e^+ is the *"origin"* in tension. The origin e^+ represents the minimum creep strain reached in compression up to the start of the time interval, i.e.,

$$e^+ = \min e^C \qquad 0 \le t \le t_n$$
$$e^+ \le 0 \qquad (5.2.14)$$

Of course, if there was no loading in compression, the origin e^+ is equal to zero. The creep strain e^H at the start of the interval, i.e., at time $t = t_n$, represents the accumulated creep strain in tension corrected for compression (as is the creep strain $e_1^C - e_2^C$ in Fig. 5.2.9). Analogously, if the material is subjected to compression in a time interval $t_m - t_{m+1}$, the modified creep strain is defined as

$$e^H = e^C - e^- \qquad (5.2.15)$$

where the origin e^- represents the maximum creep strain reached in tension up to the start of the interval t_m

$$e^- = \max e^C \qquad 0 \leq t \leq t_m$$
$$e^- \geq 0 \tag{5.2.16}$$

It follows from the definitions (5.2.13) and (5.2.15) that the modified creep strain e^H represents the "distance" from the corresponding origin, as given by the accumulated creep strain in tension or compression, respectively.

The use of the O.R.N.L. hardening rule is summarized in Table 5.2.1 and illustrated in Fig. 5.2.10. The material is subjected to uniaxial cyclic loading (Fig. 5.2.10a). The creep strain changes according to the creep curve in time periods of tension, while in compression it follows the reflected creep curve, as shown in Fig. 5.2.10b. It can be seen from Fig. 5.2.10c that, for the loading in tension the origin $e^+ = 0$ corresponds to the load periods 1 and 3, while for periods 5 and 7 the origin is $e^+ = e_D^+$. Hence, the segment BC on the creep curve $B'C = \bar{O}\bar{C}$ in period 3 starts from the point B; on the creep curve corresponding to period 5 creep starts from the origin, and so on. In the case of the compressive stress the origin for the curve AB is e_A^-, while for the periods 4 and 6 the origin is reset to e_C^-. The creep strain in periods 2 and 4 starts from the origin $\bar{O}$ of the reflected creep curve, while in the period 6 it starts from point E on the creep curve $E'F = \bar{O}\bar{F}$. If the stresses in the load periods have different magnitudes, the creep strain change in a time period follows the creep curve (or reflected curve) corresponding to the stress in that period (see Fig. 5.2.9).

Table 5.2.1. Computational steps according to the O.R.N.L. hardening rule for uniaxial cyclic loading

---

1. Set origins

$e^+ = 0$
$e^- = 0$

2. Stress positive

$e^H = e^C - e^+$

$e^C = \int\limits_{t_n}^{t_{n+1}} \dot{e}^C(e^H, \sigma, t)\,dt$

If $^{n+1}e^C > c^-$, reset e^- to $^{n+1}e^C$

3. Stress negative

$e^H = e^C - e^-$

$e^C = \int\limits_{t_n}^{t_{n+1}} \dot{e}^C(e^H, \sigma, t)\,dt$

If $^{n+1}e^C < e^+$, reset e^+ to $^{n+1}e^C$

4. Next step of creep strain calculation

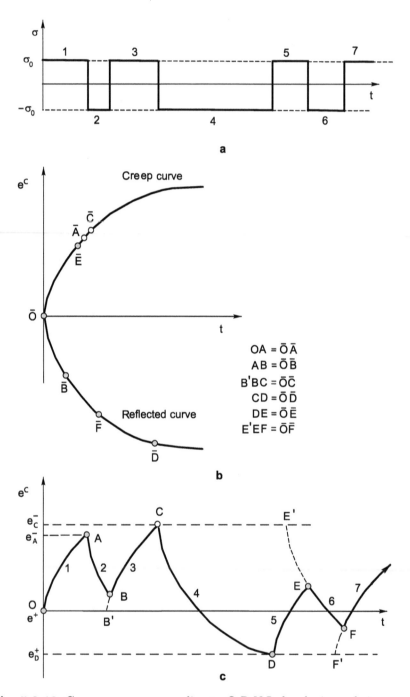

Fig. 5.2.10. Creep response according to O.R.N.L. hardening rule in case of uniaxial cyclic loading. **a** Stress change with time; **b** Creep curve and reflected creep curve; **c** Creep strain as function of time

5.2.2 Multiaxial Creep Model

In this section we describe a creep material model applicable to general, three-dimensional loading of an isotropic metal. The relations given here represent a generalization of the uniaxial creep conditions for which the two fundamental relations are the creep law (5.2.3), $e^C = e^C(\sigma, t, \theta)$, and the creep strain rate (5.2.7), $\dot{e}^C = \partial e^C / \partial t$. The generalization is based on experimental observations.

Several important facts or requirements have to be considered in formulating the basic relations for creep (Kraus 1980):

1) The creep deformation is incompressible.
2) The mean stress has no effect on the creep strains.
3) The principal directions of the creep strain rate and stress coincide.
4) The general relations must reduce to the uniaxial relations for uniaxial loading conditions.

These requirements are analogous to those in plasticity (Section 3.2) where the uniaxial plasticity relations are generalized to the three-dimensional stress conditions.

We start with the creep constitutive relations of the form

$$\dot{\mathbf{e}}^C = \gamma\, \mathbf{S} \qquad (5.2.17a)$$

or, using the components,

$$\dot{e}^C_{ij} = \gamma\, S_{ij} \qquad (5.2.17b)$$

Here $\dot{\mathbf{e}}^C$ is the *creep strain rate tensor* (with the shear components equal to half of the engineering strains), $\mathbf{S}$ is the deviatoric stress tensor (A1.17); and γ is a proportionality factor. The creep constitutive relations (5.2.17) satisfy the requirements 1 to 3 listed above. We next consider the determination of the proportionality factor γ.

In analogy with the theory of plasticity (relation (3.2.46)), we introduce the *effective creep strain rate* $\dot{\bar{e}}^C$ as

$$\dot{\bar{e}}^C = \left(\frac{2}{3}\, \dot{\mathbf{e}}^C \cdot \dot{\mathbf{e}}^C \right)^{1/2} \qquad (5.2.18)$$

where, in uniaxial creep,

$$\dot{\bar{e}}^C = \dot{e}^C_{11}$$

$$\dot{e}^C_{33} = \dot{e}^C_{22} = -\frac{1}{2}\dot{e}^C_{11}$$

$$\dot{e}^C_{ij} = 0 \qquad i \neq j \tag{5.2.19}$$

and $\dot{e}^C_{11}$ is the uniaxial creep strain rate in the x_1- direction. Then, taking the scalar product (see (A2.29)) on both sides in (5.2.17) we obtain

$$\dot{\bar{e}}^C = \frac{2}{3}\gamma\bar{\sigma} \tag{5.2.20}$$

where $\bar{\sigma}$ is the effective stress (see (3.2.39))

$$\bar{\sigma} = (\frac{3}{2}\mathbf{S} \cdot \mathbf{S})^{1/2} \tag{5.2.21}$$

and

$$\gamma = \frac{3}{2}\frac{\dot{\bar{e}}^C}{\bar{\sigma}} \tag{5.2.22}$$

The creep variable γ is determined from the uniaxial creep curve. Using (5.2.8) we obtain

$$\gamma = \gamma(\bar{e}^C, \bar{\sigma}, \theta) \tag{5.2.23}$$

where we have substituted the effective creep strain $\bar{e}^C$ and the effective stress $\bar{\sigma}$ for the uniaxial creep strain e^C and uniaxial stress σ. For example, for the power creep law we obtain from (5.2.22) and (5.2.11)

$$\gamma = \frac{3}{2}a_2 a_0^{1/a_2} \bar{\sigma}^{(a_1/a_2 - 1)} (\bar{e}^C)^{(1-1/a_2)} \tag{5.2.24}$$

Finally, we give the basic ideas for the generalization of the O.R.N.L. hardening rule to multiaxial conditions. We define the two "origins" $\mathbf{e}^+$ and $\mathbf{e}^-$ as schematically shown in Fig. 5.2.11, and the modified creep strains $\hat{\mathbf{e}}^+$ and $\hat{\mathbf{e}}^-$ as

$$\hat{\mathbf{e}}^+ = \mathbf{e}^C - \mathbf{e}^+$$

$$\hat{\mathbf{e}}^- = \mathbf{e}^C - \mathbf{e}^- \tag{5.2.25}$$

where $\mathbf{e}^C$ is the creep strain tensor.

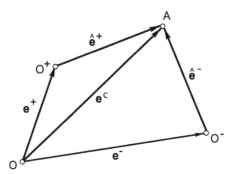

Fig. 5.2.11. Schematic representation of origins for creep according to O.R.N.L. hardening rule

Then, we introduce the distances $\bar{e}^+$ and $\bar{e}^-$ from the origins as

$$\bar{e}^+ = \left(\frac{2}{3} \hat{e}^+ \cdot \hat{e}^+ \right)^{1/2}$$

$$\bar{e}^- = \left(\frac{2}{3} \hat{e}^- \cdot \hat{e}^- \right)^{1/2} \tag{5.2.26}$$

and the distance $\hat{e}$ between the origins,

$$\hat{e} = \left[\frac{2}{3} \left(\mathbf{e}^+ - \mathbf{e}^- \right) \cdot \left(\mathbf{e}^+ - \mathbf{e}^- \right) \right]^{1/2} \tag{5.2.27}$$

From Fig. 5.2.11 we see that $\bar{e}^+$, $\bar{e}^-$ and $\hat{e}$ are proportional to the distances O^+A, O^-A and O^-O^+, respectively.

The creep strain rate is given by (5.2.17), but now

$$\gamma = \gamma \left(\bar{e}^H, \bar{\sigma}, \theta \right) \tag{5.2.28}$$

where $\bar{e}^H$ is the *modified effective creep strain*. We have that $\bar{e}^H$ is either $\bar{e}^+$ or $\bar{e}^-$ depending on the loading conditions at the considered time t. Basically, depending on the scalar product between $\mathbf{S}$, and $\mathbf{e}^+$ and $\mathbf{e}^-$, the conclusion on whether the load changes character can be drawn (see Snyder and Bathe 1980 for details).

It can be seen from the above discussion that the general 3-D relations reduce to the uniaxial relations for uniaxial loading conditions (the requirement 4).

A generalization of the isotropic creep model (the O.R.N.L. hardening rule is not assumed) to orthotropic material behavior is based on the potential theory of creep (Brown 1970; Rees 1983).

5.2.3 Thermoplasticity and Creep

Considering a thermoelastic material model the elastic coefficients depend on temperature, and the thermal strains are proportional to the temperature. Hence, we have (see (A1.29) and (A1.31))

$$
\begin{aligned}
E &= E(\theta) \\
\nu &= \nu(\theta) \\
\alpha &= \alpha(\theta)
\end{aligned}
\tag{5.2.29}
$$

and the thermal strains are modeled using

$$
e_{ij}^{TH} = \alpha(\theta - \theta_{ref})\,\delta_{ij}
\tag{5.2.30}
$$

where e_{ij}^{TH} is the thermal strain; E, ν and α are Young's modulus, Poisson's ratio and the coefficient of thermal expansion, respectively; θ is the current temperature, and θ_{ref} is the reference temperature.

If the stress and temperature levels are high, plastic deformations also result, where it was found experimentally that the size and position of the yield surface depend on temperature. A temperature increase leads to a reduction of the yield stress and hence the yield surface size (see Fig. 5.2.1).

The yield condition (3.2.56) is now

$$
f_y(\theta) = \frac{1}{2}(\mathbf{S} - \boldsymbol{\alpha}) \cdot (\mathbf{S} - \boldsymbol{\alpha}) - \frac{1}{3}\hat{\sigma}_y^2(\theta) = 0
\tag{5.2.31}
$$

where the yield stress depends on the temperature, $\hat{\sigma}_y(\theta)$. Also, the modulus C in the relation (3.2.73) is temperature dependent, since the plastic modului E_P and $\hat{E}_P$ depend on the temperature. Therefore, we have (see (3.2.78))

$$
C(\theta) = \frac{2}{3}\left[E_P(\theta) - M\hat{E}_P(\theta)\right]/(1 - M)
\tag{5.2.32}
$$

When both plastic and creep effects are considered, we add the plastic and creep strains, $\mathbf{e}^P$ and $\mathbf{e}^C$, to the elastic and thermal strains $\mathbf{e}^E$ and $\mathbf{e}^{TH}$. Therefore the total strain tensor $\mathbf{e}$ is

$$
\mathbf{e} = \mathbf{e}^E + \mathbf{e}^P + \mathbf{e}^C + \mathbf{e}^{TH}
\tag{5.2.33}
$$

This is a *fundamental relation used in modeling thermoelastic-plastic-creep* behavior of materials. The plastic and creep strain rates are defined by (3.2.71) and (5.2.17) respectively. Note that this formulation can directly be used to model creep rupture (e.g., Walczak et al. 1983) and phase changes (e.g., Rammerstorfer et al. 1981).

5.3 Viscoplastic Material Models

As stated briefly in Section 5.1, the viscoplasticity theory is based on the *overstress assumption.* This means that viscoplastic flow is assumed to occur when the stress point lies outside the yield surface. We will use the *Perzyna model and summarize the associated form of the viscoplastic strain-stress relations.* These relations are established primarily for metals but can also be developed for other materials and in particular geomaterials with quite complex behavior (Cristescu and Goida 1994).

In the following we first present a simple one-dimensional model and describe the basic characteristics of viscoplastic flow by simple examples. Then we generalize this model to three-dimensional deformations in a manner analogous to the generalizations given in Sections 3.2 and 5.2 for plasticity and creep.

5.3.1 One-Dimensional Model

Consider the model shown in Fig. 5.3.1 that consists of a spring, dashpot and a Coulomb friction slider (e.g., Hinton and Owen 1980). Assume that the total stress carried by the model is σ. The stress changes in time enforcing the same relative displacements (strains) in the dashpot and the slider. The elastic

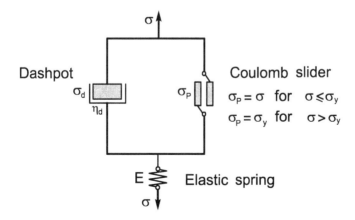

Fig. 5.3.1. One-dimensional viscoplastic model carrying stress σ

spring has the elastic constant E, the dashpot characteristic is represented by the viscosity coefficient η_d, and the slider is characterized by the yield stress σ_y. The stress in the slider, σ_P, is defined as

$$\sigma_P = \sigma \qquad \text{for } \sigma \leq \sigma_y$$
$$\sigma_P = \sigma_y \qquad \text{for } \sigma > \sigma_y \tag{5.3.1}$$

and the stress in the dashpot σ_d is

$$\sigma_d = 0 \qquad \text{for } \sigma \leq \sigma_y$$
$$\sigma_d = \sigma - \sigma_y \qquad \text{for } \sigma > \sigma_y \tag{5.3.2}$$

The presence of the dashpot allows the stress to instantaneously exceed the value σ_y predicted by plasticity theory. The stress σ carried by the system tends to σ_y (or vice versa) as time increases and the system approaches steady state conditions. The viscoplastic flow occurs due to the *overstress* $\sigma_d = \sigma - \sigma_y$.

Two strains are present in the model: the *elastic strain* e^E and the *viscoplastic strain* e^{VP}, which in sum give the total strain e, i.e.,

$$e = e^E + e^{VP} \tag{5.3.3}$$

The *viscoplastic constitutive law* is defined as

$$\sigma_d = \eta_d \dot{e}^{VP} \tag{5.3.4}$$

where $\dot{e}^{VP}$ is the viscoplastic strain rate. The yield stress σ_y is a function of the viscoplastic strain

$$\sigma_y = \sigma_y(e^{VP}) \tag{5.3.5a}$$

and we have the yield condition

$$f_y(\sigma_y, e^{VP}) = 0 \tag{5.3.5b}$$

Based on the above fundamental relations, we can derive the following differential equation for the stress carried by the system in Fig. 5.3.1 during viscoplastic flow,

$$\sigma = \sigma_y + \eta_d \dot{e} - \frac{\eta_d}{E} \dot{\sigma} \tag{5.3.6}$$

In order to illustrate the main characteristics of the viscoplastic material behavior, we consider two special cases of material response:

a) the case of a constant applied stress, and
b) the case of stress relaxation.

For simplicity we use a bilinear hardening law

$$\sigma_y = \sigma_{yv} + E_P e^{VP} \tag{5.3.7}$$

where σ_{yv} and E_P are the initial yield stress and the plastic modulus, respectively. This relation is a special case of the Ramberg-Osgood expression (3.2.7). In the case of a constant stress $\sigma = \sigma_A$, with $\sigma_A > \sigma_{yv}$, the solution of (5.3.6) is

$$e = \frac{\sigma_A}{E} + \frac{\sigma_A - \sigma_{yv}}{E_P} [1 - \exp(-\gamma_d E_P t)] \tag{5.3.8}$$

and in the case of perfect plasticity $(E_P = 0)$,

$$e - \frac{\sigma_A}{E} + \gamma_d (\sigma_A \quad \sigma_{yv}) t \tag{5.3.9}$$

where $\gamma_d = 1/\eta_d$ is the *fluidity parameter*. Figures 5.3.2a,b show graphically these solutions. The viscoplastic flow diminishes in hardening as the yield stress approaches the stress point (Figs. 5.3.2a,c), see also Example 5.5.2. In the case of a material with no hardening the viscoplastic flow continues indefinitely with a constant rate. Note that the initial strain is the elastic strain $e^E = \sigma_A/E$.

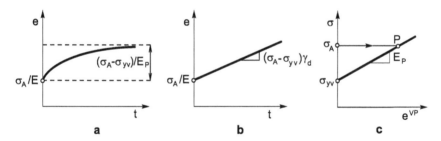

Fig. 5.3.2. Viscoplastic deformation under constant stress σ_A. **a** Strain as a function of time for a material with hardening; **b** Strain as a function of time in the case of no hardening; **c** Representation in the $\sigma - e^{VP}$ plane

In the case of stress relaxation we suppose that the material is subjected to a constant strain e_0, hence $\dot{e} = 0$ in (5.3.6). The solution of (5.3.6) is then

$$\sigma = (E e_0 - \sigma_\infty) e^{-t/\tau} + \sigma_\infty \tag{5.3.10}$$

where

$$\sigma_\infty = \frac{E}{E + E_P} (E_P e_0 + \sigma_{yv}) \tag{5.3.11}$$

and

$$\tau = \eta_d/(E + E_P) \tag{5.3.12}$$

is the so-called *relaxation time*. Note that the stress σ tends to the value of σ_∞, where we have in the case of a material with no hardening, $\sigma_\infty = \sigma_{yv}$. Also, the stress relaxation process is controlled by the ratio between the real time t and the relaxation time τ. Figure 5.3.3 shows the relaxation curve (5.3.10) (see also Example 5.5.2).

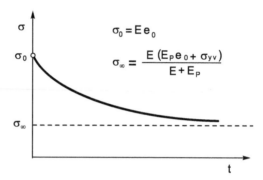

Fig. 5.3.3. Stress relaxation for one-dimensional viscoplasticity

The above two cases illustrate typical phenomena in viscoplasticity. We next generalize the above model to arbitrary multiaxial loading conditions.

5.3.2 General Three-Dimensional Viscoplastic Models

We start with a rather general viscoplastic constitutive law

$$\dot{\mathbf{e}}^{VP} = \gamma < \psi(f_y) > \frac{\partial f}{\partial \sigma} \tag{5.3.13}$$

where $\dot{\mathbf{e}}^{VP}$ is the viscoplastic strain rate tensor, σ is the stress, $\psi(f_y)$ is a function of the yield function f_y, f is a function governing the flow rule, γ is (as in (5.3.9)) a fluidity parameter, and $<>$ represents the Macauley bracket[1]. We will further consider the associated viscoplasticity, in which case the constitutive law (5.3.13) is

---

[1] The Macauley bracket of a function f is defined as

$$< f > = f \qquad \text{for } f \geq 0$$
$$< f > = 0 \qquad \text{for } f < 0$$

$$\dot{e}^{VP} = \gamma < \psi(f_y) > \frac{\partial f_y}{\partial \boldsymbol{\sigma}^*} \qquad (5.3.14)$$

where $\boldsymbol{\sigma}^*$ is the conjugate stress, represented by the stress point P^* on the yield surface as the "closest-point projection" of the stress point P. Figure 5.3.4 shows the stress points and also the unit normal $\mathbf{n}$ to the yield surface.

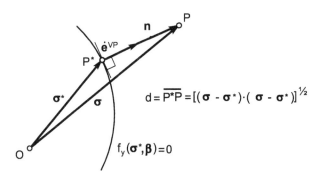

$$d = \overline{P^*P} = [(\boldsymbol{\sigma} - \boldsymbol{\sigma}^*)\cdot(\boldsymbol{\sigma} - \boldsymbol{\sigma}^*)]^{\frac{1}{2}}$$

$$f_y(\boldsymbol{\sigma}^*, \boldsymbol{\beta}) = 0$$

Fig. 5.3.4. Definition of the conjugate stress $\boldsymbol{\sigma}^*$ and "distance" d

The function in the Macauley bracket depends on the "distance" of the stress point P from the conjugate stress point P^* (see Bathe 1996; Simo and Hughes 1998). An example of the viscoplastic constitutive law for a von Mises material with isotropic hardening is (Perzyna 1966)

$$\dot{e}^{VP} = \gamma < \phi(\bar{\sigma}) > \mathbf{n} \qquad (5.3.15)$$

where

$$\phi(\bar{\sigma}) = \sqrt{3/2} \left(\frac{\bar{\sigma} - \bar{\sigma}^*}{\bar{\sigma}^*} \right)^N \qquad (5.3.16)$$

and N is a material constant. Note that $(\bar{\sigma} - \bar{\sigma}^*)$ in (5.3.16) is equal to $\sqrt{3/2}\,d$, where d is the "distance" in the deviatoric stress space.

In accordance with (5.3.14) and (5.3.15) the viscoplastic constitutive law can be written as (see also Kojic 1996a, Bathe 1996)

$$\dot{e}^{VP} = \frac{d}{\eta\,\eta_F\,(d, \boldsymbol{\sigma}^*, \boldsymbol{\beta}^*)}\,\mathbf{n} \qquad (5.3.17)$$

where d is defined in Fig. 5.3.4, and

$$\frac{d}{\eta\,\eta_F\,(d, \boldsymbol{\sigma}^*, \boldsymbol{\beta}^*)} = \gamma < \psi(f_y) > \left\| \frac{\partial f_y}{\partial \boldsymbol{\sigma}^*} \right\| \qquad (5.3.18)$$

Here η is a viscosity coefficient and $\eta_F\,(d, \boldsymbol{\sigma}^*, \boldsymbol{\beta}^*)$ is a viscosity function which depends on the "distance", d, and also on the conjugate stress $\boldsymbol{\sigma}^*$ and internal variables $\boldsymbol{\beta}^*$ of the yield function. We should note that η introduced here and η_d in (5.3.4) may have different units. Since $\mathbf{n} = (\boldsymbol{\sigma} - \boldsymbol{\sigma}^*)/d$, we obtain

$$\dot{\mathbf{e}}^{VP} = \frac{\boldsymbol{\sigma} - \boldsymbol{\sigma}^*}{\eta\,\eta_F\,(d, \boldsymbol{\sigma}^*, \boldsymbol{\beta}^*)} \tag{5.3.19}$$

It follows from this general form of the constitutive law that the viscoplastic flow diminishes as the stress point approaches the yield surface, i.e., we have the same material behavior as described by the uniaxial viscoplastic model. The viscosity coefficient and the viscosity function represent material characteristics and would be determined by experimental investigations. Illustrations of using viscosity coefficients and viscosity functions for metal viscoplasticity are given in Examples 5.5.2 and 5.5.3.

The material response governed by the viscoplastic law (5.3.19) depends on the viscosity coefficient, the viscosity function and yield function. We write the yield condition in a general form (3.2.83), i.e.,

$$f_y\,(\boldsymbol{\sigma}^*, \boldsymbol{\beta}^*) = 0 \tag{5.3.20}$$

When the yield condition is independent of the mean stress, for an isotropic material we can write the yield condition as (see (3.2.15))

$$f_y\,(J_{2D}^*, \boldsymbol{\beta}^*) = 0 \tag{5.3.21}$$

where J_{2D}^* is the second invariant of the conjugate stress $\boldsymbol{\sigma}^*$. Then, the viscoplastic constitutive law (5.3.14) can be written as

$$\dot{\mathbf{e}}^{VP} = \gamma < \psi(f_y\,(J_{2D}^*, \boldsymbol{\beta}^*)) > \frac{\partial f_y}{\partial J_{2D}^*}\,\mathbf{S}^* \tag{5.3.22}$$

where (3.2.13) has been used. Hence the viscoplastic stains have the deviatoric character. Further, instead of (5.3.19) we have now

$$\dot{\mathbf{e}}^{VP} = \frac{\mathbf{S} - \mathbf{S}^*}{\eta\,\eta_F\,(d, \mathbf{S}^*, \boldsymbol{\beta}^*)} \tag{5.3.23}$$

where $\mathbf{S}^*$ is the conjugate deviatoric stress.

On inspecting the above viscoplastic constitutive laws it is noted that the creep of material, described in Section 5.2, may be considered as a *special case* of viscoplasticity. By taking the stress $\mathbf{S}^* = \mathbf{0}$ and the viscosity function η_F as the appropriate function of stresses, the viscoplastic constitutive law (5.3.23) for the case of a von Mises material reduces to the creep constitutive relation (5.2.17).

5.4 Stress Integration for Creep and Thermoplastic Models

In this section we develop the numerical procedures for stress integration of the thermoplasticity and creep models described in Section 5.2. The algorithms represent a generalization of the procedures used for isotropic plasticity. We consider general three-dimensional deformations in Section 5.4.1 and then shell conditions in Section 5.4.2. We also give some basic expressions for the derivation of the consistent tangent elastic-plastic-creep constitutive matrix, see Section 5.4.3, and finally we illustrate the application of the algorithms in a number of solved examples in Section 5.4.4. The basic procedure presented in this section was, in effect, originally introduced by Kojic and Bathe as the effective-stress-function algorithm (Bathe et al. 1984; Kojic and Bathe 1987a,b).

5.4.1 General Three-Dimensional Deformations

Using the stress integration procedure, the fundamental relations to be satisfied for the thermoelastic-plastic-creep model are:

- The thermoelastic constitutive relations at the end of the time step,
- The stress-plastic strain relations based on the normality rule at the end of the time step,
- The yield condition at the end of the time step, and
- The creep constitutive relations for the time step using the Euler backward method.

Since the plastic and creep deformations are incompressible, the mean stress $^{t+\Delta t}\sigma_m$ is independent of the plastic and creep strains, and $^{t+\Delta t}\sigma_m$ is determined by (see (A1.12))

$$^{t+\Delta t}\sigma_m = {}^{t+\Delta t}c_m({}^{t+\Delta t}e_m - {}^{t+\Delta t}e^{TH}) \qquad \text{no sum on m} \qquad (5.4.1)$$

where $^{t+\Delta t}c_m = {}^{t+\Delta t}E/(1 - 2\,{}^{t+\Delta t}\nu)$ and corresponds to temperature $^{t+\Delta t}\theta$. The thermal strain $^{t+\Delta t}e^{TH}$ is given by (5.2.30) (see also (A1.31)),

$$t+\Delta t \, e^{TH} = \, t+\Delta t \, \alpha \left(t+\Delta t \, \theta - \theta_{ref} \right) \tag{5.4.2}$$

As before, we use the superscript $t + \Delta t$ on the material constants to indicate that they correspond to time $t + \Delta t$ and temperature $t+\Delta t \theta$.

The stress-plastic strain relations have the form (4.4.18). To include the creep effects we use the Euler backward scheme to express the increment of creep strain tensor Δe^C (see (5.2.17))

$$\Delta e^C = \Delta t \, ^{t+\Delta t} \gamma \, ^{t+\Delta t} \mathbf{S} \tag{5.4.3}$$

where $^{t+\Delta t} \gamma$ is evaluated from (5.2.23) for the effective stress $^{t+\Delta t} \bar{\sigma}$, time $t + \Delta t$ (or pseudo-time, for the strain-hardening assumption, see (5.4.15)) and temperature $^{t+\Delta t} \theta$. The constitutive relation for the deviatoric stresses is (see (A1.19), (4.4.18) and (5.2.33))

$$^{t+\Delta t}\mathbf{S} = \, ^{t+\Delta t}\mathbf{S}^E - 2 \, ^{t+\Delta t}G(\Delta\lambda \, ^{t+\Delta t}\hat{\mathbf{S}} + \Delta t \, ^{t+\Delta t} \gamma \, ^{t+\Delta t}\mathbf{S}) \tag{5.4.4}$$

where $^{t+\Delta t}\mathbf{S}^E$ is the elastic solution given by (4.4.12), with

$$^{t+\Delta t}\mathbf{e}'' = \, ^{t+\Delta t}\mathbf{e}' - \, ^t\mathbf{e}^P - \, ^t\mathbf{e}^C \tag{5.4.5}$$

We next employ the relation (4.4.22) and, with use of the constitutive relations (5.4.4), we solve for the deviatoric stress $^{t+\Delta t}\mathbf{S}$ and for the stress radius $^{t+\Delta t}\hat{\mathbf{S}}$,

$$^{t+\Delta t}\mathbf{S} = \frac{^{t+\Delta t}\mathbf{S}^\lambda}{1 + a_\gamma + a_\lambda} \tag{5.4.6}$$

and

$$^{t+\Delta t}\hat{\mathbf{S}} = \frac{^{t+\Delta t}\hat{\mathbf{S}}^\gamma}{1 + a_\gamma + \left[2 \, ^{t+\Delta t}G + (1 + a_\gamma)\,\hat{C}\right]\Delta\lambda} \tag{5.4.7}$$

where

$$a_\gamma = 2 \, ^{t+\Delta t}G\Delta t \, ^{t+\Delta t}\gamma$$
$$a_\lambda = \frac{2 \, ^{t+\Delta t}G}{1 + \Delta\lambda\hat{C}}\Delta\lambda \tag{5.4.8}$$

and

$$^{t+\Delta t}\mathbf{S}^\lambda = {}^{t+\Delta t}\mathbf{S}^E + a_\lambda\,{}^t\boldsymbol{\alpha} \tag{5.4.9}$$

$$^{t+\Delta t}\hat{\mathbf{S}}^\gamma = {}^{t+\Delta t}\hat{\mathbf{S}}^E - a_\gamma\,{}^t\boldsymbol{\alpha} \tag{5.4.10}$$

Here $^{t+\Delta t}\hat{\mathbf{S}}^E$ is the elastic solution for $^{t+\Delta t}\hat{\mathbf{S}}$, corresponding to $\Delta\bar{e}^P = 0$ (see (4.4.15)).

Following the governing parameter method of Section 4.2, we select the effective stress $^{t+\Delta t}\bar{\sigma}$ as the governing parameter. By taking the Euclidean norm (see (A2.31)) on both sides of (5.4.6), we obtain

$$f_1(^{t+\Delta t}\bar{\sigma}) = \frac{^{t+\Delta t}\bar{\sigma}_\lambda}{1 + a_\gamma + a_\lambda} - {}^{t+\Delta t}\bar{\sigma} = 0 \tag{5.4.11}$$

where $^{t+\Delta t}\bar{\sigma}_\lambda$ is the effective stress corresponding to $^{t+\Delta t}\mathbf{S}^\lambda$. We solve (5.4.11) by using one of the numerical procedures of Section 2.1. Therefore we select a value of $^{t+\Delta t}\bar{\sigma}$ and calculate $^{t+\Delta t}\gamma$ from the creep law as described in Section 5.2.2, then determine a_γ from (5.4.8) and $^{t+\Delta t}\hat{\mathbf{S}}^\gamma$ from (5.4.10). Next, we calculate the effective stress $^{t+\Delta t}\hat{\sigma}_\gamma$ corresponding to $^{t+\Delta t}\hat{\mathbf{S}}^\gamma/(1 + a_\gamma)$, and check whether $^{t+\Delta t}\hat{\sigma}_\gamma$ exceeds the yield stress $^t\hat{\sigma}_y$. If $^{t+\Delta t}\hat{\sigma}_\gamma \leq {}^t\hat{\sigma}_y$, there is no plastic deformation, we use $\Delta\lambda = 0$, $a_\lambda = 0$, $^{t+\Delta t}\mathbf{S}^\lambda = {}^{t+\Delta t}\mathbf{S}^E$ from (5.4.9), and calculate $^{t+\Delta t}\bar{\sigma}_\lambda^E$ corresponding to $^{t+\Delta t}\mathbf{S}^E$; then we check whether (5.4.11) is satisfied, if not, we select a new trial value for $^{t+\Delta t}\bar{\sigma}$. In case $^{t+\Delta t}\hat{\sigma}_\gamma > {}^t\hat{\sigma}_y$, then plastic deformation occurs for the selected $^{t+\Delta t}\bar{\sigma}$, and we determine $\Delta\lambda$ from (4.4.24), with use of the yield curve and by solving the nonlinear equation

$$f_2(\Delta\bar{e}^P) = \frac{^{t+\Delta t}\hat{\sigma}_\gamma}{(1 + a_\gamma)\,^{t+\Delta t}\hat{\sigma}_y + \frac{3}{2}\left[2\,^{t+\Delta t}G + (1 + a_\gamma)\,\hat{C}\right]\Delta\bar{e}^P} - 1 = 0 \tag{5.4.12}$$

This equation is obtained by taking the Euclidean norm on both sides of (5.4.7). We have indicated that the function $f_2(\Delta\bar{e}^P)$ depends on the parameter $\Delta\bar{e}^P$, the increment of effective plastic strain, since $^{t+\Delta t}\hat{\sigma}_y$ and $\hat{C}$ depend on $\Delta\bar{e}^P$.

Table 5.4.1 summarizes the computational steps. It can be shown, as in the case of the function $f(\Delta\bar{e}^P)$ given in (4.4.29), that both functions (5.4.11) and (5.4.12) are monotonic, for $\Delta\lambda = const.$, and for $^{t+\Delta t}\gamma = const.$, respectively.

Figure 5.4.1 shows graphically the stress integration in time step Δt. As can be seen from the figure, we start with the elastic solution corresponding

Table 5.4.1. Computational steps for stress integration in case of thermoplastic-creep material model

---

1. Assume a value of $^{t+\Delta t}\bar\sigma$

Calculate $^{t+\Delta t}\gamma$ from the creep law; then a_γ and $^{t+\Delta t}\hat\sigma_\gamma$

2. Check for yielding

If $^{t+\Delta t}\hat\sigma_\gamma \le {^t\hat\sigma_y}$, no plastic deformation, use $^{t+\Delta t}\bar\sigma_\lambda = {^{t+\Delta t}\bar\sigma_\lambda^E}$, go to step 4

3. Plasticity calculations

Solve $f_2(\Delta\bar e^P) = 0$, equation (5.4.12)
Calculate a_λ and $^{t+\Delta t}\bar\sigma_\lambda$

4. Check if (5.4.11) is satisfied (solution for $^{t+\Delta t}\bar\sigma$)

If $f_1(^{t+\Delta t}\bar\sigma) \le \varepsilon$ go to step 5
Repeat calculations, go to step 1

5. Final calculations

Determine:

$^{t+\Delta t}\mathbf{S}$	from	(5.4.6)
$^{t+\Delta t}\hat{\mathbf{S}}$	from	(5.4.7)
$\Delta\mathbf{e}^P$	from	(4.4.18)
$\Delta\mathbf{e}^C$	from	(5.4.3)
$^{t+\Delta t}\bar e^C$	from	(5.4.14)

---

to the stress point D_E in the deviatoric plane and search along the lines OD_λ and $^tOD_\gamma$ to find the final point D on the surfaces $^{t+\Delta t}f_1 = 0$ and $^{t+\Delta t}f_2 = 0$. This search is governed by the changes of the effective stress $\bar\sigma$ and the increments of the effective plastic strain $\Delta\bar e^P$ on the creep and yield curves (points D' and D''). Note that the stress path D_ED does not lie on the vector $^{t+\Delta t}\mathbf{S}^E$ or on $^{t+\Delta t}\hat{\mathbf{S}}^E$, because of the terms $a_\lambda {^t\boldsymbol\alpha}$ and $-a_\gamma {^t\boldsymbol\alpha}$ in the expressions (5.4.9) and (5.4.10). The stress path D_ED lies on $^{t+\Delta t}\mathbf{S}^E$ in the case of isotropic hardening, or creep without plastic deformation. In the case of thermoplasticity with no creep D_ED lies on $^{t+\Delta t}\hat{\mathbf{S}}^E$.

The creep strain at the end of the time step, $^{t+\Delta t}\mathbf{e}^C$, is

$$^{t+\Delta t}\mathbf{e}^C = {^t\mathbf{e}^C} + \Delta\mathbf{e}^C \tag{5.4.13}$$

and then the effective creep strain $^{t+\Delta t}\bar e^C$ is

$$^{t+\Delta t}\bar e^C = (\frac{2}{3}\,{^{t+\Delta t}\mathbf{e}^C}\cdot{^{t+\Delta t}\mathbf{e}^C})^{1/2} \tag{5.4.14}$$

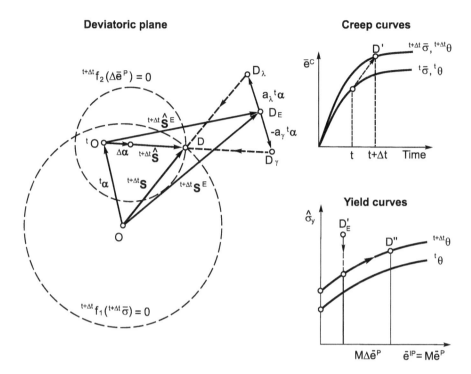

Fig. 5.4.1. Graphical representation of the stress integration for thermoplasticity and creep; deviatoric plane, creep and yield curves

In the calculation of $^{t+\Delta t}\bar{e}^{C}$ we have used the definition (5.2.18) of the effective creep strain rate $\dot{\bar{e}}^{C}$. If the O.R.N.L. hardening rule for cyclic loading is employed, then instead of $^{t+\Delta t}\mathbf{e}^{C}$ the modified creep strains (5.2.25) and the modified effective creep strain must be calculated. Note that the effective creep strain $\bar{e}^{C}$ can increase or decrease with time, unlike the effective plastic strain $\bar{e}^{P}$ which can only increase (see (4.4.25)).

The above numerical procedure can be employed for *thermoplastic deformations without creep*, or when there are *creep deformations only*, as special cases. In the case of *thermoplasticity* we use (5.4.12) with $a_{\gamma} = 0$ as the governing equation (which has the form (4.4.29)). Figure 5.4.2 shows graphically the solution in this case. We are searching for the solution along the yield curve corresponding to temperature $^{t+\Delta t}\theta$ and along the line defined by the unit normal $^{t+\Delta t}\mathbf{n} = {}^{t+\Delta t}\hat{\mathbf{S}}^{E}/||^{t+\Delta t}\hat{\mathbf{S}}^{E}||$.

In the case of *creep only*, we skip step 3 in Table 5.4.1, and use $a_{\lambda} = 0$ in (5.4.9) and (5.4.11). Note that in employing the strain hardening approach it is necessary to determine the pseudo-time τ_{p} for each trial value $^{t+\Delta t}\bar{\sigma}$. The pseudo-time can be calculated using (5.2.3) and (5.2.7) to obtain

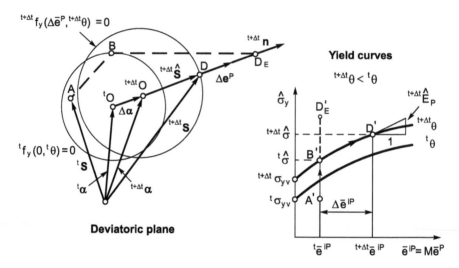

Fig. 5.4.2. Stress states at time t and $t + \Delta t$ (thermoplasticity, mixed hardening), deviatoric plane and yield curves

$$t\bar{e}^C + f_1(^{t+\Delta t}\bar{\sigma}) \, f_3(^{t+\Delta t}\theta) \left[\Delta t \dot{f}_2(\tau_p) - f_2(\tau_p) \right] = 0 \qquad (5.4.15)$$

and solving for τ_p. Instead of $^{t+\Delta t}\bar{\sigma}$ and $^{t+\Delta t}\theta$, the weighted values for the current time step may be employed (Kojic and Bathe 1987a,b). Also, the modified effective creep strain $^t\bar{e}^H$ and the O.R.N.L. hardening rule may be used. Once τ_p has been determined, $^{t+\Delta t}\gamma$ can be calculated (see (5.2.22))

$$^{t+\Delta t}\gamma = \frac{3}{2} \frac{^{t+\Delta t}\dot{\bar{e}}^C}{^{t+\Delta t}\bar{\sigma}} \qquad (5.4.16)$$

where $^{t+\Delta t}\dot{\bar{e}}^C$ corresponds to time τ_p. In the case of the power creep law, $^{t+\Delta t}\gamma$ can be determined analytically. Using $\bar{\sigma} = \,^{t+\Delta t}\bar{\sigma}$ in time step Δt, we integrate the differential equation

$$d\bar{e}^C = a_0^{1/a_2} \,^{t+\Delta t}\bar{\sigma}^{a_1/a_2} (\bar{e}^C)^{(1-1/a_2)} dt$$

which follows from (5.2.11). Hence, we have

$$^{t+\Delta t}\bar{e}^C = \left[a_0^{1/a_2} \Delta t \,^{t+\Delta t}\bar{\sigma}^{a_1/a_2} + \left(^t\bar{e}^C \right)^{1/a_2} \right]^{a_2} \qquad (5.4.17)$$

We then use $^{t+\Delta t}\bar{e}^C$ in (5.4.16) to obtain

$$\boxed{{}^{t+\Delta t}\gamma = \frac{3}{2}\frac{{}^{t+\Delta t}\bar{e}^{C} - {}^{t}\bar{e}^{C}}{\Delta t\;{}^{t+\Delta t}\bar{\sigma}}}$$
(5.4.18)

An extension of the above computational procedure to thermoplasticity and creep of orthotropic materials is given in Kojic et al. (1995c).

5.4.2 Shell (Plane Stress) Conditions

We follow the approach presented for plastic deformations in Section 4.4.2. The mean stress has the form (4.4.41), i.e.,

$$\boxed{{}^{t+\Delta t}\sigma_{m} = {}^{t+\Delta t}\bar{C}_{m}\left({}^{t+\Delta t}e^{E}_{xx} + {}^{t+\Delta t}e^{E}_{yy}\right)}$$
(5.4.19)

where the elastic strains are

$$
\begin{aligned}
{}^{t+\Delta t}e^{E}_{xx} &= {}^{t+\Delta t}e_{xx} - {}^{t+\Delta t}e^{P}_{xx} - {}^{t+\Delta t}e^{C}_{xx} - {}^{t+\Delta t}e^{TH}\\
{}^{t+\Delta t}e^{E}_{yy} &= {}^{t+\Delta t}e_{yy} - {}^{t+\Delta t}e^{P}_{yy} - {}^{t+\Delta t}e^{C}_{yy} - {}^{t+\Delta t}e^{TH}
\end{aligned}
$$
(5.4.20)

From (4.4.43), (4.4.18), (5.4.3) and (5.4.20) we obtain the following expressions for the deviatoric stresses ${}^{t+\Delta t}S_{xx}$ and ${}^{t+\Delta t}S_{yy}$:

$$
\begin{aligned}
{}^{t+\Delta t}S_{xx} &= {}^{t+\Delta t}S^{E}_{xx} - \Delta\lambda\left(\bar{C}'^{E}_{11}\,{}^{t+\Delta t}\hat{S}_{xx} + \bar{C}'^{E}_{12}\,{}^{t+\Delta t}\hat{S}_{yy}\right)\\
&\quad -\Delta t\,{}^{t+\Delta t}\gamma\left(\bar{C}'^{E}_{11}\,{}^{t+\Delta t}S_{xx} + \bar{C}'^{E}_{12}\,{}^{t+\Delta t}S_{yy}\right)\\
{}^{t+\Delta t}S_{yy} &= {}^{t+\Delta t}S^{E}_{yy} - \Delta\lambda\left(\bar{C}'^{E}_{21}\,{}^{t+\Delta t}\hat{S}_{xx} + \bar{C}'^{E}_{22}\,{}^{t+\Delta t}\hat{S}_{yy}\right)\\
&\quad -\Delta t\,{}^{t+\Delta t}\gamma\left(\bar{C}'^{E}_{21}\,{}^{t+\Delta t}S_{xx} + \bar{C}'^{E}_{22}\,{}^{t+\Delta t}S_{yy}\right)
\end{aligned}
$$
(5.4.21)

where

$$
\begin{aligned}
{}^{t+\Delta t}S^{E}_{xx} &= \bar{C}'^{E}_{11}\left({}^{t+\Delta t}e_{xx} - {}^{t}e^{P}_{xx} - {}^{t}e^{C}_{xx} - {}^{t+\Delta t}e^{TH}\right)\\
&\quad + \bar{C}'^{E}_{12}\left({}^{t+\Delta t}e_{yy} - {}^{t}e^{P}_{yy} - {}^{t}e^{C}_{yy} - {}^{t+\Delta t}e^{TH}\right)\\
{}^{t+\Delta t}S^{E}_{yy} &= \bar{C}'^{E}_{21}\left({}^{t+\Delta t}e_{xx} - {}^{t}e^{P}_{xx} - {}^{t}e^{C}_{xx} - {}^{t+\Delta t}e^{TH}\right)\\
&\quad + \bar{C}'^{E}_{22}\left({}^{t+\Delta t}e_{yy} - {}^{t}e^{P}_{yy} - {}^{t}e^{C}_{yy} - {}^{t+\Delta t}e^{TH}\right)
\end{aligned}
$$
(5.4.22)

are the elastic solutions for the deviatoric stresses. From the system of equations (5.4.21) and the relation (4.4.22) we first solve for the components of the stress radius ${}^{t+\Delta t}\hat{S}_{xx}$ and ${}^{t+\Delta t}\hat{S}_{yy}$,

$$
\begin{aligned}
{}^{t+\Delta t}\hat{S}_{xx} &= \left(\hat{a}_{11} \,{}^{t+\Delta t}\hat{S}_{xx}^{\gamma} - \hat{a}_{12} \,{}^{t+\Delta t}\hat{S}_{yy}^{\gamma} \right) / \hat{D} \\
{}^{t+\Delta t}\hat{S}_{yy} &= \left(-\hat{a}_{12} \,{}^{t+\Delta t}\hat{S}_{xx}^{\gamma} + \hat{a}_{11} \,{}^{t+\Delta t}\hat{S}_{yy}^{\gamma} \right) / \hat{D}
\end{aligned}
\tag{5.4.23}
$$

where

$$
\begin{aligned}
\hat{a}_{11} &= 1 + \Delta\lambda \left[\left(1 + \Delta t \,{}^{t+\Delta t}\gamma\,\hat{C} \right) \bar{C}_{11}'^{E} + \hat{C} \right] \\
\hat{a}_{12} &= \Delta\lambda \left(1 + \Delta t \,{}^{t+\Delta t}\gamma\,\hat{C} \right) \bar{C}_{12}'^{E} + \Delta t \,{}^{t+\Delta t}\gamma\,\bar{C}_{12}'^{E} \\
\hat{D} &= \hat{a}_{11}^{2} - \hat{a}_{12}^{2}
\end{aligned}
\tag{5.4.24}
$$

and

$$
\begin{aligned}
{}^{t+\Delta t}\hat{S}_{xx}^{\gamma} &= {}^{t+\Delta t}S_{xx}^{E} - \left(1 + \Delta t \,{}^{t+\Delta t}\gamma\,\bar{C}_{11}'^{E} \right) {}^{t}\alpha_{xx} - \Delta t \,{}^{t+\Delta t}\gamma\,\bar{C}_{12}'^{E}\,{}^{t}\alpha_{yy} \\
{}^{t+\Delta t}\hat{S}_{yy}^{\gamma} &= {}^{t+\Delta t}S_{yy}^{E} - \Delta t \,{}^{t+\Delta t}\gamma\,\bar{C}_{12}'^{E}\,{}^{t}\alpha_{xx} - \left(1 + \Delta t \,{}^{t+\Delta t}\gamma\,\bar{C}_{11}'^{E} \right) {}^{t}\alpha_{yy}
\end{aligned}
\tag{5.4.25}
$$

In this derivation we have taken into account the symmetry of the elastic matrix $\bar{\mathbf{C}}_{2}'^{E}$ given by (4.4.44), and also $\bar{C}_{11}'^{E} = \bar{C}_{22}'^{E}$. The coefficients $\bar{C}_{ij}'^{E}$ must be evaluated at the temperature ${}^{t+\Delta t}\theta$, while for the modulus $\hat{C}$ a weighted value with respect to temperature may be used. Note that the solutions for ${}^{t+\Delta t}\hat{S}_{xx}$ and ${}^{t+\Delta t}\hat{S}_{yy}$ reduce to (4.4.47) when creep is not present. The deviatoric stresses ${}^{t+\Delta t}S_{xx}$ and ${}^{t+\Delta t}S_{yy}$ can be expressed in terms of ${}^{t+\Delta t}\hat{S}_{xx}$ and ${}^{t+\Delta t}\hat{S}_{yy}$, either from (4.4.22) or (5.4.21). It is simpler to use (4.4.22), hence

$$
\begin{aligned}
{}^{t+\Delta t}S_{xx} &= {}^{t}\alpha_{xx} + \left(1 + \Delta\lambda\hat{C} \right) {}^{t+\Delta t}\hat{S}_{xx} \\
{}^{t+\Delta t}S_{yy} &= {}^{t}\alpha_{yy} + \left(1 + \Delta\lambda\hat{C} \right) {}^{t+\Delta t}\hat{S}_{yy}
\end{aligned}
\tag{5.4.26}
$$

The third components ${}^{t+\Delta t}\hat{S}_{zz}$ and ${}^{t+\Delta t}S_{zz}$ follow from the deviatoric character of $\hat{\mathbf{S}}$ and $\mathbf{S}$. The determination of the shear components in ${}^{t+\Delta t}\hat{\mathbf{S}}$ and ${}^{t+\Delta t}\mathbf{S}$ is accomplished in the same manner as for the general three-dimensional deformation because the thermoelastic constitutive relations are the same. Finally, instead of (5.4.11) and (5.4.12) we have the following equations:

$$
f_{1}\left({}^{t+\Delta t}\bar{\sigma}\right) = \frac{1}{2}\,{}^{t+\Delta t}\mathbf{S} \cdot {}^{t+\Delta t}\mathbf{S} - \frac{1}{3}\,{}^{t+\Delta t}\bar{\sigma}^{2} = 0
\tag{5.4.27}
$$

and

$$f_2 \left(\Delta \bar{e}^P \right) = \frac{1}{2} {}^{t+\Delta t}\hat{\mathbf{S}} \cdot {}^{t+\Delta t}\hat{\mathbf{S}} - \frac{1}{3} {}^{t+\Delta t}\hat{\sigma}_y^2 = 0 \qquad (5.4.28)$$

which are a result of the definition (3.2.39) of the effective stress and the yield condition (5.2.31).

The computational steps are as for the general three-dimensional deformations, Table 5.4.1: assume ${}^{t+\Delta t}\bar{\sigma}$, calculate ${}^{t+\Delta t}\gamma$ from (5.4.16), then calculate ${}^{t+\Delta t}\hat{\mathbf{S}}^\gamma$ from (5.4.25) and (5.4.10); iterate on $\Delta \bar{e}^P$ by using ${}^{t+\Delta t}\hat{\mathbf{S}}$ from (5.4.23) and (5.4.7) until (5.4.28) is satisfied; determine ${}^{t+\Delta t}\mathbf{S}$ from (4.4.22) and substitute into (5.4.27); repeat the calculation with a new value of ${}^{t+\Delta t}\bar{\sigma}$ until (5.4.27) is satisfied within a selected tolerance. If we compare these computations with those performed when considering fully three-dimensional conditions, the number of numerical operations is now larger because we need to calculate all components of ${}^{t+\Delta t}\hat{\mathbf{S}}$ and ${}^{t+\Delta t}\mathbf{S}$, and cannot simply work with the scalars in equations (5.4.11) and (5.4.12).

The above solution procedure is applicable to the special cases of thermoplasticity without creep, and when there are only creep effects. In the case of thermoplasticity, we only iterate on $\Delta \bar{e}^P$ to solve (5.4.28), with ${}^{t+\Delta t}\gamma = 0$ in (5.4.23) and (5.4.7), and set ${}^t\mathbf{e}^C = 0$ in the expressions for ${}^{t+\Delta t}\mathbf{S}^E$ (the calculations reduce to those in Section 4.4.2).

If there is no plastic deformation in the time step, we iterate on ${}^{t+\Delta t}\bar{\sigma}$ to solve (5.4.27) and skip the calculation of ${}^{t+\Delta t}\hat{\mathbf{S}}$. Instead of the system of equations (5.4.21) and (4.4.22), we use $\Delta\lambda = 0$ in (5.4.21) and find solutions for ${}^{t+\Delta t}S_{xx}$ and ${}^{t+\Delta t}S_{yy}$ in the form analogous to (5.4.23). We also use $\Delta\lambda = 0$ in (5.4.6) and (5.4.9) to express the shear terms of ${}^{t+\Delta t}\mathbf{S}$. Using these components of ${}^{t+\Delta t}\mathbf{S}$ we check whether (5.4.27) is satisfied.

5.4.3 Elastic-Plastic-Creep Constitutive Matrix

The consistent tangent constitutive matrices ${}^{t+\Delta t}\mathbf{C}^{EC}$ and ${}^{t+\Delta t}\mathbf{C}^{EPC}$ for the elastic-creep and elastic-plastic-creep material models, respectively, have the form (4.4.53). We here summarize the basic steps for the calculation of ${}^{t+\Delta t}\mathbf{C}^{EPC}$, following the general procedure of Section 4.2 and that for plasticity in Section 4.4.3. Since the plastic and creep deformations are isochoric, the form of the constitutive matrix given in Table 4.4.2 is applicable. Therefore, we need to determine the derivatives ${}^{t+\Delta t}\bar{C}'_{ij} \equiv {}^{t+\Delta t}S_{i,j} = \partial {}^{t+\Delta t}S_i / \partial {}^{t+\Delta t}e''_j$. As in Section 4.4.3, we consider general three-dimensional deformation.

It follows from (5.4.6) that

$$ {}^{t+\Delta t}S_{i,j} = \frac{1}{1 + a_\gamma + a_\lambda} \left[{}^{t+\Delta t}S^\lambda_{i,j} - {}^{t+\Delta t}S_i \left(a_{\gamma,j} + a_{\lambda,j} \right) \right] \qquad (5.4.29)$$

The derivatives on the right hand side can be obtained from (5.4.8) and (5.4.9) as

$$a_{\gamma,j} = 2\,^{t+\Delta t}G\,\Delta t\,^{t+\Delta t}\gamma'\,^{t+\Delta t}\bar{\sigma}_{,j}$$

$$a_{\lambda,j} = \frac{1}{1+\Delta\lambda\hat{C}}\left[2\,^{t+\Delta t}G\ell_P - a_\lambda\left(\ell_P\hat{C} + \Delta\lambda\hat{C}'\right)\right]\Delta\bar{e}^P_{,j}$$

$$^{t+\Delta t}S^\lambda_{i,j} = 2\,^{t+\Delta t}G\,\delta_{ij} + {}^t\alpha_i\,a_{\lambda,j} \tag{5.4.30}$$

where $^{t+\Delta t}\gamma' = \partial\,^{t+\Delta t}\gamma/\partial^{t+\Delta t}\bar{\sigma}$ can be calculated from (5.4.16) and the creep law, $\hat{C}' = \partial\hat{C}/\partial^{t+\Delta t}\bar{e}^P$ from the yield curve, as in Section 4.4.3 ; and ℓ_P is the coefficient defined by (4.4.65). Therefore, the derivatives in (5.4.29) can be computed if we determine the derivatives $^{t+\Delta t}\bar{\sigma}_{,j} = \partial^{t+\Delta t}\bar{\sigma}/\partial^{t+\Delta t}e''_j$ and $\Delta\bar{e}^P_{,j} = \partial(\Delta\bar{e}^P)/\partial^{t+\Delta t}e''_j$. By differentiation of the equations (5.4.11) and (5.4.12) we obtain the system of equations

$$\frac{\partial f_1}{\partial^{t+\Delta t}\bar{\sigma}}\,^{t+\Delta t}\bar{\sigma}_{,j} + \frac{\partial f_1}{\partial(\Delta\bar{e}^P)}\,\Delta\bar{e}^P_{,j} + \frac{\partial f_1}{\partial^{t+\Delta t}e''_j} = 0$$

$$\frac{\partial f_2}{\partial^{t+\Delta t}\bar{\sigma}}\,^{t+\Delta t}\bar{\sigma}_{,j} + \frac{\partial f_2}{\partial(\Delta\bar{e}^P)}\,\Delta\bar{e}^P_{,j} + \frac{\partial f_2}{\partial^{t+\Delta t}e''_j} = 0 \tag{5.4.31}$$

from which $^{t+\Delta t}\bar{\sigma}_{,j}$ and $\Delta\bar{e}^P_{,j}$ can be calculated. Note that, in accordance with (4.4.69), we have

$$\frac{\partial f_1}{\partial^{t+\Delta t}e''_j} = \frac{3\,^{t+\Delta t}G}{^{t+\Delta t}\bar{\sigma}_\lambda(1+a_\gamma+a_\lambda)}\,^{t+\Delta t}S'^\lambda_j$$

$$\frac{\partial f_2}{\partial^{t+\Delta t}e''_j} = \frac{3\,^{t+\Delta t}G}{(^{t+\Delta t}\hat{\sigma}_\gamma)^2}\,^{t+\Delta t}\hat{S}'^\gamma_j \tag{5.4.32}$$

where $^{t+\Delta t}S'^\lambda_j$ and $^{t+\Delta t}\hat{S}'^\gamma_j$ are equal to $^{t+\Delta t}S^\lambda_j$ and $^{t+\Delta t}S^\gamma_j$ for $j = 1, 2, 3$, and to $2^{t+\Delta t}S^\lambda_j$ and $2^{t+\Delta t}S^\gamma_j$ for $j = 1, 2, 3$.

The other computational details are analogous to those presented in Section 4.4.3. Also, it is straightforward to reduce the above derivations to thermoplasticity with no creep, or to conditions of creep only.

5.4.4 Examples

Example 5.4.1. Thermoelastic-Plastic and Creep Deformation of a Beam. Derive the basic relations for the stress integration considering the thermoelastic-plastic and creep deformations of a beam.

The physical conditions for the beam deformation are described in Example 4.5.2. Hence, the constitutive relations for the nonzero shear terms (x, y)

and (x, z) remain as for the general three-dimensional deformation in Section 5.4.1, and we next analyze the axial normal components of stress, creep and plastic strain. We start with the first relation of (b) of Example 4.5.2, which now becomes

$$^{t+\Delta t}S_{xx} = {}^{t+\Delta t}S_{xx}^E - {}^{t+\Delta t}C_{11}'^E(\Delta e_{xx}^P + \Delta e_{xx}^C) \tag{a}$$

where

$$^{t+\Delta t}S_{xx}^E = {}^{t+\Delta t}C_{11}'^E \left({}^{t+\Delta t}e_{xx} - {}^t e_{xx}^P - {}^t e_{xx}^C - {}^{t+\Delta t}e^{TH} \right) \tag{b}$$

and

$$^{t+\Delta t}C_{11}'^E = 2/3\,{}^{t+\Delta t}E \tag{c}$$

Next we express Δe_{xx}^P and Δe_{xx}^C by using (4.4.18) and (5.4.3), and write (a) in the form

$$^{t+\Delta t}S_{xx} = \left({}^{t+\Delta t}S_{xx}^E - \Delta\lambda\,{}^{t+\Delta t}C_{11}'^E\,{}^{t+\Delta t}\hat{S}_{xx} \right)/(1 + b_\gamma) \tag{d}$$

where

$$b_\gamma = \Delta t\,{}^{t+\Delta t}\gamma\,{}^{t+\Delta t}C_{11}'^E \tag{e}$$

We further solve (4.4.22) and (d) for ${}^{t+\Delta t}S_{xx}$ and ${}^{t+\Delta t}\hat{S}_{xx}$ as

$$^{t+\Delta t}S_{xx} = {}^{t+\Delta t}S_{xx}^\lambda/(1 + b_\gamma + b_\lambda) \tag{f}$$

$$^{t+\Delta t}\hat{S}_{xx} = \frac{{}^{t+\Delta t}\hat{S}_{xx}^\gamma}{1 + b_\gamma + \left[{}^{t+\Delta t}C_{11}'^E + (1 + b_\gamma)\,\hat{C} \right]\Delta\lambda} \tag{g}$$

where

$$^{t+\Delta t}S_{xx}^\lambda = {}^{t+\Delta t}S_{xx}^E + b_\lambda\,{}^t\alpha_{xx} \qquad {}^{t+\Delta t}\hat{S}_{xx}^\gamma = {}^{t+\Delta t}S_{xx}^E - (1 + b_\gamma)\,{}^t\alpha_{xx} \tag{h}$$

and

$$b_\lambda = \Delta\lambda\,{}^{t+\Delta t}C_{11}'^E / (1 + \Delta\lambda\hat{C}) \tag{i}$$

The equation for the calculation of $\Delta\bar{e}^P$ has the form (5.4.28), where we use the nonzero shear components ${}^{t+\Delta t}\hat{S}_{xy}$ and ${}^{t+\Delta t}\hat{S}_{xz}$ from (5.4.7), the normal component ${}^{t+\Delta t}\hat{S}_{xx}$ from (g), and

$$^{t+\Delta t}\hat{S}_{yy} = {}^{t+\Delta t}\hat{S}_{zz} = -\frac{1}{2}\,{}^{t+\Delta t}\hat{S}_{xx} \tag{j}$$

We employ the expression (4.4.24) to calculate $\Delta\lambda$ for each trial value of $\Delta\bar{e}^P$. Finally, the governing equation $f_1({}^{t+\Delta t}\bar\sigma) = 0$ is (5.4.27), with ${}^{t+\Delta t}S_{xy}$ and ${}^{t+\Delta t}S_{xz}$ given in (5.4.6), ${}^{t+\Delta t}S_{xx}$ in (f), and

$$^{t+\Delta t}S_{yy} = {}^{t+\Delta t}S_{zz} = -\frac{1}{2}{}^{t+\Delta t}S_{xx} \qquad (k)$$

The computational steps are as described in Table 5.4.1. Also, the special cases: creep only, or thermoplasticity without creep, presented for the shell conditions in Section 5.4.2, can be implemented (see also Kojic and Bathe 1987b). The comments regarding the constitutive matrix $^{t+\Delta t}\mathbf{C}^{EPC}$ are as for $^{t+\Delta t}\mathbf{C}^{EP}$ in Example 4.5.2.

Example 5.4.2. Thermoelastic-Plastic and Creep Deformation of a Pipe. Derive the basic relations for the stress integration considering the thermoelastic-plastic and creep deformation of a pipe under general loading and internal pressure.

The physical conditions for the deformation of a pipe with internal pressure are described in detail in Example 4.5.2. Hence, we start with (j) of Example 4.5.2, which is now

$$^{t+\Delta t}\sigma_{aa} = {}^{t+\Delta t}E({}^{t+\Delta t}e_{aa} - {}^{t+\Delta t}e^{TH} - {}^{t+\Delta t}e^P_{aa} - {}^{t+\Delta t}e^C_{aa}) + {}^{t+\Delta t}\nu\,{}^{t+\Delta t}\sigma_{cc} \qquad (a)$$

The normal deviatoric stress components are (see (k) of Example 4.5.2)

$$^{t+\Delta t}S_{aa} = {}^{t+\Delta t}S^E_{aa} - {}^{t+\Delta t}C'^E_{aa}(\Delta e^P_{aa} + \Delta e^C_{aa}) \qquad (b)$$

$$^{t+\Delta t}S_{cc} = {}^{t+\Delta t}S^E_{cc} + {}^{t+\Delta t}C'^E_{ac}(\Delta e^P_{aa} + \Delta e^C_{aa}) \qquad (c)$$

$$^{t+\Delta t}S_{bb} = -{}^{t+\Delta t}S_{aa} - {}^{t+\Delta t}S_{cc} \qquad (d)$$

where

$$^{t+\Delta t}S^E_{aa} = {}^{t+\Delta t}C'^E_{aa}\,{}^{t+\Delta t}e''_{aa} - \frac{1 - 2\,{}^{t+\Delta t}\nu}{3}\,{}^{t+\Delta t}\sigma_{cc} \qquad (e)$$

$$^{t+\Delta t}S^E_{cc} = -{}^{t+\Delta t}C'^E_{ac}\,{}^{t+\Delta t}e''_{aa} + \frac{2 - {}^{t+\Delta t}\nu}{3}\,{}^{t+\Delta t}\sigma_{cc} \qquad (f)$$

and

$$^{t+\Delta t}C'^E_{aa} = \frac{2}{3}\,{}^{t+\Delta t}E \qquad {}^{t+\Delta t}C'^E_{ac} = \frac{1}{3}\,{}^{t+\Delta t}E \qquad (g)$$

$$^{t+\Delta t}e''_{aa} = {}^{t+\Delta t}e_{aa} - {}^{t+\Delta t}e^{TH} - {}^{t}e^P_{aa} - {}^{t}e^C_{aa} \qquad (h)$$

We next use the expressions (4.4.18) and (5.4.3) for Δe^P_{aa} and Δe^C_{aa} in (b), and employ the relation (4.4.22). From (b) and (4.4.22) follow the solutions for $^{t+\Delta t}S_{aa}$ and $^{t+\Delta t}\hat{S}_{aa}$ as

$$^{t+\Delta t}S_{aa} = {}^{t+\Delta t}S^\lambda_{aa}/(1 + b_\gamma + b_\lambda) \qquad (i)$$

$$t + \Delta t \hat{S}_{aa} = \frac{t + \Delta t \hat{S}^{\gamma}_{aa}}{1 + b_{\gamma} + [t + \Delta t C'^{E}_{aa} + (1 + b_{\gamma})\hat{C})]\Delta\lambda} \tag{j}$$

where

$$t + \Delta t S^{\lambda}_{aa} = t + \Delta t S^{E}_{aa} + b_{\lambda}{}^{t}\alpha_{aa} \qquad t + \Delta t \hat{S}^{\gamma}_{aa} = t + \Delta t S^{E}_{aa} - (1 + b_{\gamma})^{t}\alpha_{aa} \tag{k}$$

and

$$b_{\lambda} = \Delta\lambda \, t + \Delta t C'^{E}_{aa} / (1 + \Delta\lambda\hat{C}) \tag{l}$$

while b_{γ} is defined in (e) of Example 5.4.1. The solutions for $t + \Delta t S_{cc}$ and $t + \Delta t \hat{S}_{cc}$ are

$$t + \Delta t S_{cc} = -\frac{1}{2} t + \Delta t S_{aa} + \frac{1}{2} t + \Delta t \sigma_{cc} \tag{m}$$

$$t + \Delta t \hat{S}_{cc} = (t + \Delta t S_{cc} - {}^{t}\alpha_{cc}) / (1 + \Delta\lambda\hat{C}) \tag{n}$$

The expressions for the nonzero shear terms $t + \Delta t S_{ab}$, $t + \Delta t S_{ac}$ of the stress, and $t + \Delta t \hat{S}_{ab}$, $t + \Delta t \hat{S}_{ac}$ of the yield surface radius, are given in (5.4.6) and (5.4.7), respectively.

The computational steps are in principle as described in Table 5.4.1, and are analogous to those for shell conditions in Section 5.4.2. Also, the calculation of the tangent constitutive matrix $t + \Delta t \mathbf{C}^{EPC}$ is analogous to that described for plasticity in Example 4.5.2.

The application of the above derived relations to an engineering problem is given in Example 5.4.7.

Example 5.4.3. Thermoplastic Deformation in Proportional Cyclic Loading. Assume that a material point of a von Mises material with mixed hardening characteristics is subjected to thermal loading and to cyclic proportional mechanical loading.

  a) Derive the general relations for the calculation of the increments of the effective plastic strain and the plastic strains.
  b) Use the derived relations to analyze a thin-walled tube subjected to cyclic tension and torsion.

This example represents a generalization of the two-step loading solution in Example 4.5.7 to multi-step conditions and thermoplasticity.

a) We consider the general proportional cyclic loading of a material, so that the deviatoric stress ${}^{t}\mathbf{S}$ at any load level can be expressed as (see (a) of Example 4.5.7)

$${}^{t}\mathbf{S} = f(t) {}^{0}\mathbf{S} \tag{a}$$

where ${}^{0}\mathbf{S}$ is the stress deviator at the start of first yielding. Therefore, the components of the unit normal vector $\mathbf{n}$ are

$$n_{ij} = {}^0S_{ij}/\|{}^0\mathbf{S}\| \tag{b}$$

and remain unchanged during loading. In a cycle we distinguish the positive and negative loading periods (in the positive and negative $\mathbf{n}$-directions, respectively), as shown in Fig. E.5.4-3a. Using (d) of Example 4.5.7 we obtain

$$^{t+\Delta t}\bar\sigma = s\,s_\alpha\,{}^t\bar\alpha + {}^{t+\Delta t}\hat\sigma_y + ({}^{t+\Delta t}E_P - M\,{}^{t+\Delta t}\hat E_P)\Delta\bar e^P \tag{c}$$

where

$$\begin{aligned} s &= 1 \quad \text{for the positive loading in the time step} \\ s &= -1 \quad \text{for the negative loading in the time step} \end{aligned} \tag{d}$$

and

$$\begin{aligned} s_\alpha &= 1 \quad \text{for } {}^t\boldsymbol\alpha \cdot \mathbf{n} > 0 \\ s_\alpha &= -1 \quad \text{for } {}^t\boldsymbol\alpha \cdot \mathbf{n} < 0 \end{aligned} \tag{e}$$

The equation (c) is applicable when plastic deformation takes place in the current time (load) step. In general, this equation is nonlinear when the yield curve $\sigma_y(\bar e^P)$ is nonlinear. If the Ramberg-Osgood formula (3.2.7) is employed, we have that the governing equation to be solved for $\Delta\bar e^P$ is

$$\begin{aligned} f(\Delta\bar e^P) = {}^{t+\Delta t}\bar\sigma &- s\,s_\alpha\,{}^t\bar\alpha - {}^{t+\Delta t}\sigma_{yv} - {}^{t+\Delta t}C_y\left[M({}^t\bar e^P + \Delta\bar e^P)\right]^{{}^{t+\Delta t}n} \\ &- {}^{t+\Delta t}n\,{}^{t+\Delta t}C_y(1 - M\,{}^{t+\Delta t}n)({}^t\bar e^P + \Delta\bar e^P)^{{}^{t+\Delta t}n-1}\Delta\bar e^P \\ = 0 & \end{aligned} \tag{f}$$

where we note that the material parameters σ_{yv}, C_y and n are temperature-dependent (and thus given with the superscript $t + \Delta t$). We give the expressions for other variables in the plasticity calculations:

$$\Delta e^P = s\sqrt{\frac{3}{2}}\,\Delta\bar e^P\,\mathbf{n} \tag{g}$$

$$\Delta\alpha = s\sqrt{\frac{2}{3}}\,{}^{t+\Delta t}n\,{}^{t+\Delta t}C_y(1 - M\,{}^{t+\Delta t}n)({}^t\bar e^P + \Delta\bar e^P)^{{}^{t+\Delta t}n-1}\Delta\bar e^P\mathbf{n} \tag{h}$$

$$\Delta\bar\alpha = {}^{t+\Delta t}n\,{}^{t+\Delta t}C_y(1 - M\,{}^{t+\Delta t}n)({}^t\bar e^P + \Delta\bar e^P)^{{}^{t+\Delta t}n-1}\Delta\bar e^P \tag{i}$$

where we have used (3.2.46), (3.2.73), and (3.2.80).

b) The thin-walled tube, schematically shown in Fig. E.5.4-3b, is subjected to the cyclic axial force F and torsional moment M_t,

$$F = f(t)\,F_0 \qquad M_t = f(t)\,M_0 \tag{j}$$

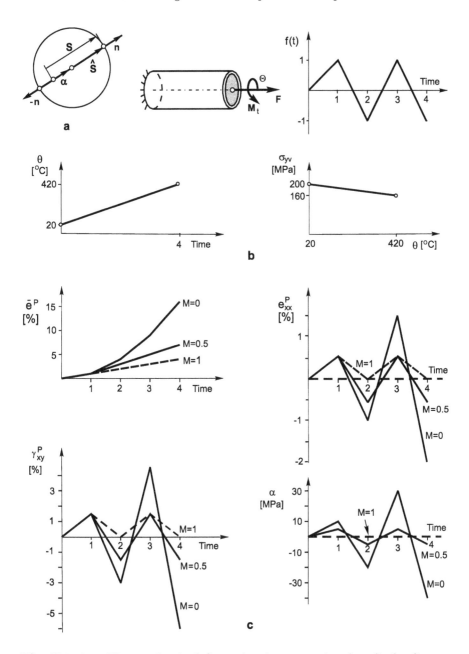

Fig. E.5.4-3. Thermoplastic deformation in proportional cyclic loading. **a** Yield surface and direction of loading **n**; **b** Thin-walled tube subjected to cyclic axial force F and torsional moment M_t, and thermal loading: time-function, temperature change, dependence $\sigma_{yv}(\theta)$; **c** Solutions for three values of M for: effective plastic strain $\bar{e}^P(t)$, plastic strains $e_{xx}^P(t)$, $\gamma_{xy}^P(t)$, and the yield surface position $\alpha = s_\alpha \bar{\alpha}(t)$

where $f(t)$ is the time-function shown in the figure, while F_0 and M_0 are the force and moment causing the stresses

$$\sigma_0 = \tau_0 = 100 \quad \text{MPa} \tag{k}$$

The tube is subjected to thermal loading with the temperature change given in the figure. The material hardening is linear,

$$\sigma_y = \sigma_{yv} + 1000\,\bar{e}^P \quad \text{MPa} \tag{l}$$

and the yield stress σ_{yv} dependence on temperature is as shown in the figure, but the hardening is independent of temperature.

The components of the normal $\mathbf{n}$ (see (b) and (k)) are:

$$n_{xx} = \frac{1}{2}\sqrt{\frac{2}{3}} \quad n_{yy} = n_{zz} = -\frac{1}{2}n_{xx} \quad n_{xy} = \frac{1}{2}\sqrt{\frac{3}{2}} \tag{m}$$

where x is the axial axis, while y and z lie in the cross-section, in the wall circumferential and normal directions. Since the hardening is linear, we solve (f) for $\Delta\bar{e}^P$,

$$\Delta\bar{e}^P = \frac{\bar{\sigma} - s\,s_\alpha\,{}^t\bar{\alpha} - {}^{t+\Delta t}\sigma_{yv}}{{}^{t+\Delta t}E_P} - M\,{}^t\bar{e}^P \tag{n}$$

where

$$\bar{\sigma} = 2\sigma_0 = 200 \tag{o}$$

We have calculated the effective plastic strain $\bar{e}^P$, plastic strains e_{xx}^P and γ_{xy}^P, and $\alpha = s_\alpha\,\bar{\alpha}$, by employing four steps. The results are given in Table E.5.4-3 and in Fig. E.5.4-3c, for $M = 0$ (kinematic hardening), $M = 0.5$, and $M = 1$ (isotropic hardening). The most severe plastic flow corresponds to kinematic hardening, and the smallest plastic flow corresponds to isotropic hardening. Note that the same results are obtained if more than one step per loading period is employed (see Section 4.4.4 and Example 4.5.7).

Example 5.4.4. Creep Deformation in Proportional Cyclic Loading. Consider a material point of a metal in creep subjected to proportional cyclic loading, and derive the general relations for the calculation of the creep strains. Use the derived relations for uniaxial cyclic loading, with creep defined by the exponential creep law.

The derivations presented below are analogous to those in Example 5.4.3. In general, if the material is subjected to proportional loading (see (3.2.29)), we have

$$\mathbf{S} = f(t)^0\mathbf{S} \tag{a}$$

where $^0\mathbf{S}$ is a reference deviatoric stress, and $f(t)$ is the time-function. According to (5.2.17) the creep strain rate $\dot{\mathbf{e}}^C$, represented as a six-component vector, is in the direction of the normal vector $\mathbf{n}$,

Table E.5.4-3. Solutions for thermoplastic deformation of material in proportional cyclic loading

Variable	M=0 step				M=0.5 step				M=1 step			
	1	2	3	4	1	2	3	4	1	2	3	4
$\bar{e}^P$ [%]	1.	4.	9.	16.	1.	3.	5.	7.	1.	2.	3.	4.
e_{xx}^P [%]	0.5	-1.	1.5	-2.	0.5	-0.5	0.5	-0.5	0.5	0.	0.5	0.
γ_{xy}^P [%]	1.5	-3.	4.5	-6.	1.5	-1.5	1.5	-1.5	1.5	0.	1.5	0.
α [Mpa]	10.	-20.	30.	-40.	5.	-5.	5.	-5.	0.	0.	0.	0.

$$\mathbf{n} = {}^0\mathbf{S} / \| {}^0\mathbf{S} \| \tag{b}$$

We can use the uniaxial creep law as in Section 5.2.1, with the use of Table
5.2.1. For a loading period t_n, t_{n+1} in which the loading does not change sign,
the increment of creep strain is

$$^{n+1}\mathbf{e}^C - {}^n\mathbf{e}^C = s \sqrt{\frac{3}{2}} \left({}^{n+1}\bar{e}^C - {}^n\bar{e}^C \right) \mathbf{n} \tag{c}$$

where $\bar{e}^C$ is the effective creep strain, while $s = 1$ for the loading in the
direction of $\mathbf{n}$, and $s = -1$ for the reverse loading. The relations shown in
Fig. 5.2.10 are applicable, with the origins $\mathbf{e}^+$ and $\mathbf{e}^-$ lying on the $\mathbf{n}$-axis. In
the incremental analysis we have that (5.4.3) becomes

$$\Delta\mathbf{e}^C = s \sqrt{\frac{3}{2}} \Delta\bar{e}^C \mathbf{n} \tag{d}$$

where

$$\Delta\bar{e}^C = \Delta t \; {}^{t+\Delta t}\dot{\bar{e}}^C \tag{e}$$

The effective creep strain rate ${}^{t+\Delta t}\dot{\bar{e}}^C$ corresponds to the pseudo-time τ_P
calculated from (5.4.15) and to the effective stress ${}^{t+\Delta t}\bar{\sigma}$. In the case of creep
only, the governing equation (5.4.11) reduces to

$$f_1({}^{t+\Delta t}\bar{\sigma}) = \frac{{}^{t+\Delta t}\bar{\sigma}^E}{1 + 2\Delta t \; {}^{t+\Delta t}G \; {}^{t+\Delta t}\gamma} - {}^{t+\Delta t}\bar{\sigma} = 0 \tag{f}$$

where the function ${}^{t+\Delta t}\gamma$ is defined by (5.2.23). With the solution of (f) we
obtain (see (5.4.6))

$$^{t+\Delta t}\mathbf{S} = s \sqrt{\frac{2}{3}} \; {}^{t+\Delta t}\bar{\sigma} \mathbf{n} \tag{g}$$

We next apply the derived relations to a case of creep in variable loading. The history of loading and the exponential creep law data are given in Fig. E.5.4-4 (see Snyder and Bathe 1980). The analytical solution for the creep strain is determined using the creep law and the origins at $t = 0, 100, 200, 300$ for the corresponding loading period. Numerical (incremental) solutions are calculated with the time step $\Delta t = 10$ and $\Delta t = 100$. The $\Delta t = 10$ solution is already very close to the analytical solution.

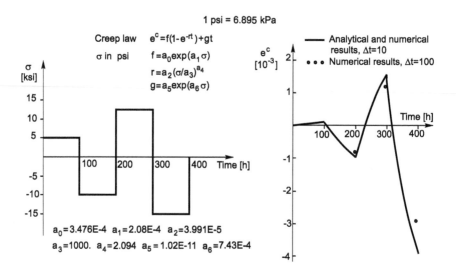

Fig. E.5.4-4. Uniaxial creep in variable loading. **a** Stress history and material data; **b** Creep strain as a function of time

Example 5.4.5. Creep of Cantilever. The cantilever shown in Fig. E.5.4-5 is loaded by a constant bending moment. Determine the stationary stress distribution for the creep deformation. The material and geometric data are given in the figure. Use the Newton-Cotes integration formula (see Bathe 1996) in the calculation of the integrals.

The only nonzero stress is σ_{xx}, hence we employ the constitutive relation

$$^{t+\Delta t}\sigma_{xx} = E\left(^{t+\Delta t}e_{xx} - {}^{t+\Delta t}e_{xx}^{C}\right) \tag{a}$$

Further, using (5.4.3), (5.4.17) and (5.4.18), with $a_2 = 1$, we obtain using (a)

$$f\left(^{t+\Delta t}\sigma_{xx}\right) = {}^{t+\Delta t}\sigma_{xx} + E\,\Delta t\,a_o\,{}^{t+\Delta t}\sigma_{xx}^{a_1} - E\left(^{t+\Delta t}e_{xx} - {}^{t}e_{xx}^{C}\right) = 0 \tag{b}$$

This equation replaces the governing equation $f_1\left(^{t+\Delta t}\bar{\sigma}\right) = 0$ given by (5.4.11). We have assumed that there is no plastic deformation of the material.

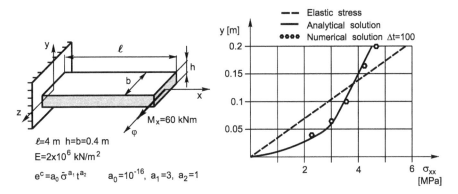

Fig. E.5.4-5. Creep of cantilever under constant moment

The moment in the beam is

$$2b \int_0^{h/2} (^t\sigma_{xx} + \Delta\sigma_{xx})\, y\, dy = M_x \tag{c}$$

Since the bending moment is constant along the cantilever axis x, the stress and strain distributions are the same for any cross-section, and the angle of the cross-section rotation increases linearly with x. With the beam assumption for strain we can write

$$^{t+\Delta t}e_{xx} = \frac{y}{\ell}\, ^{t+\Delta t}\varphi \tag{d}$$

where φ is the rotation of cross-section at the cantilever end. With use of the constitutive relation

$$\Delta\sigma_{xx} = {}^{t+\Delta t}C_{xx}^{EC}\,\Delta e_{xx} = \frac{1}{\ell}\,{}^{t+\Delta t}C_{xx}^{EC}\, y\,\Delta\varphi \tag{e}$$

we obtain from (c) and (d) (see Table 4.1.1)

$$^{t+\Delta t}K_{xx}^{(i-1)}\,\Delta\varphi^{(i)} = M_x - M_\sigma^{(i-1)} \tag{f}$$

where

$$^{t+\Delta t}K_{xx}^{(i-1)} = \frac{2b}{\ell} \int_0^{h/2} {}^{t+\Delta t}C_{xx}^{EC\,(i-1)}\, y^2\, dy \tag{g}$$

is the beam stiffness in bending,

$$M_\sigma^{(i-1)} = 2b \int_0^{h/2} y\,{}^{t+\Delta t}\sigma_{xx}^{(i-1)}\, dy \tag{h}$$

is the bending moment due to stresses $\sigma_{xx}^{(i-1)}$, and $\Delta\varphi^{(i)}$ is the increment corresponding to the iteration "i".

The constitutive coefficient $^{t+\Delta t}C_{xx}^{EC}$ can be obtained by differentiation of (b) with respect to $^{t+\Delta t}e_{xx}$, as

$$^{t+\Delta t}C_{xx}^{EC} = E \,/\, ^{t+\Delta t}f_{,\sigma} \tag{i}$$

where

$$^{t+\Delta t}f_{,\sigma} = \frac{\partial f}{\partial\,^{t+\Delta t}\sigma_{xx}} = 1 + a_0 a_1 E\,\Delta t\,\,^{t+\Delta t}\sigma_{xx}^{a_1-1} \tag{j}$$

We also use this derivative in solving (b). Namely, according to the Newton iteration scheme (2.1.14), the k-th trial $^{t+\Delta t}\sigma_{xx}^{(k)}$ is

$$^{t+\Delta t}\sigma_{xx}^{(k)} = \,^{t+\Delta t}\sigma_{xx}^{(k-1)} - f^{(k-1)} \,/\, f_{,\sigma}^{(k-1)} \tag{k}$$

The computational procedure is summarized in Table E.5.4-5a. All integrals are calculated using *six equal intervals* along the $y-$axis, i.e.,

$$J = \int_0^{h/2} \phi(y)dy = \frac{h}{2}\sum_{k=0}^{6} C_k^{(6)}\Phi(y_k) \tag{l}$$

where $C_k^{(6)}$ are the Newton-Cotes constants, given in, e.g., Bathe (1996).

We have used 10 steps with equal time step $\Delta t = 100$. The stationary stress state is practically reached after 5 steps. The stress distribution for the stationary stress state is shown in Fig. E.5.4-5, together with the analytical solution (Penny and Marriott 1971). The same results are obtained numerically by using beam or shell finite elements (Kojic and Bathe 1987b). Table E.5.4-5b gives the numerical values of the stress and strains for the last step.

The number of equilibrium iterations in each step was 4 for an energy convergence tolerance of 1×10^{-10}. In this iterative procedure we have used the estimated creep strain increment $(\Delta e_{xx}^C)^{est} = \Delta t\,^{t+\Delta t}\dot{e}_{xx}^C$, and the corresponding stress increment $\Delta\sigma_{xx}^{est} = E\,(\Delta e_{xx}^C)^{est}$ for the first iteration, in order to decrease the number of iterations. We give in Table E.5.4-5c the unbalanced energies in the last step which show the quadratic convergence rate.

Example 5.4.6. Thermoplastic and Creep Deformation of a Plate. The plate, shown in Fig. E.5.4-6, is subjected to the given in-plane loads. The changes of loads with time and the material data are given in the same figure. Determine the strains using 4 steps of $\Delta t = 0.5$ and one step of $\Delta t = 100$. Consider that the stress/strain state is uniform within the plate.

The incremental equilibrium equations for a unit volume have the following form (see Tables 4.1.1 and E.5.4-5a)

Table E.5.4-5a. Computational steps for creep of cantilever

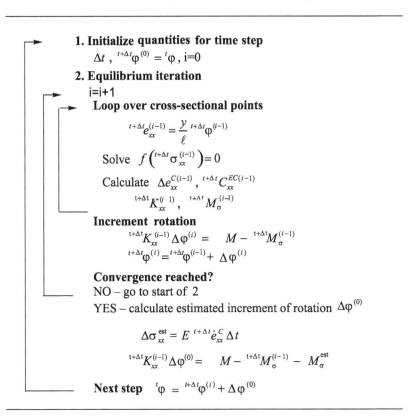

1. Initialize quantities for time step

$$\Delta t \,, \ ^{t+\Delta t}\varphi^{(0)} = {}^t\varphi \,, \ i{=}0$$

2. Equilibrium iteration

$i{=}i{+}1$

Loop over cross-sectional points

$$^{t+\Delta t}e_{xx}^{(i-1)} = \frac{y}{\ell} \, ^{t+\Delta t}\varphi^{(i-1)}$$

Solve $f\left(^{t+\Delta t}\sigma_{xx}^{(i-1)} \right) = 0$

Calculate $\Delta e_{xx}^{C(i-1)} \,, \ ^{t+\Delta t}C_{xx}^{EC(i-1)}$

$$^{t+\Delta t}K_{xx}^{(i-1)} \,, \ ^{t+\Delta t}M_{\sigma}^{(i-1)}$$

Increment rotation

$$^{t+\Delta t}K_{xx}^{(i-1)} \, \Delta\varphi^{(i)} = \ M - {}^{t+\Delta t}M_{\sigma}^{(i-1)}$$

$$^{t+\Delta t}\varphi^{(i)} = {}^{t+\Delta t}\varphi^{(i-1)} + \Delta\varphi^{(i)}$$

Convergence reached?

NO – go to start of 2

YES – calculate estimated increment of rotation $\Delta\varphi^{(0)}$

$$\Delta\sigma_{xx}^{\text{est}} = E \, ^{t+\Delta t}\dot{e}_{xx}^{C} \, \Delta t$$

$$^{t+\Delta t}K_{xx}^{(i-1)} \Delta\varphi^{(0)} = \ M - {}^{t+\Delta t}M_{\sigma}^{(i-1)} - M_{\sigma}^{\text{est}}$$

Next step $\ ^t\varphi = {}^{t+\Delta t}\varphi^{(i)} + \Delta\varphi^{(0)}$

Table E.5.4-5b Stationary stress state (time $= 1000$)

Coordinate y (m)	Stress σ_{xx} (MPa)	Strain e_{xx} (%)	Creep strain e_{xx}^{C} (%)
0.0	0.0	0.0	0.0
0.03333	2.3021	0.19204	0.07693
0.06667	3.0274	0.38407	0.23270
0.10000	3.4798	0.57611	0.40212
0.13333	3.8316	0.76814	0.57656
0.16667	4.1270	0.96018	0.75382
0.20000	4.3850	1.15224	0.93296

Table E.5.4-5c Unbalanced energies during iterations in the last step

Iteration	1	2	3	4
$\Delta E = (M - M_{\sigma})\Delta\varphi$	1.35E-1	6.07E-4	1.4E-8	7.0E-18

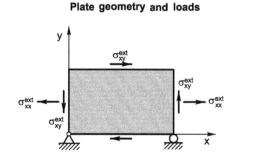

Plate geometry and loads

Material data

Plasticity data

$M = 0.6$

$\sigma_y = \sigma_{yv} + C_y (\bar{e}^P)^n$

$\sigma_{yv} = 10$

$C_y = 500$

$n = 1.2$

Elasticity constants

$E = 2 \times 10^5$ $\nu = 0.3$

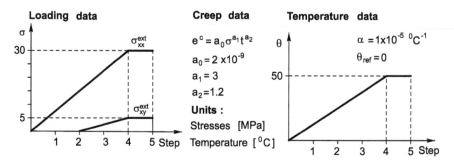

Loading data

Creep data

$e^C = a_0 \sigma^{a_1} t^{a_2}$

$a_0 = 2 \times 10^{-9}$

$a_1 = 3$

$a_2 = 1.2$

Units:

Stresses [MPa]

Temperature [°C]

Temperature data

$\alpha = 1 \times 10^{-5} \ ^\circ C^{-1}$

$\theta_{ref} = 0$

Fig. E.5.4-6. Creep and plastic deformation of a plate; material and loading data

$$^{t+\Delta t}C_{js}^{(i-1)} \Delta e_s^{(i)} = {}^{t+\Delta t}\sigma_j^{ext} - {}^{t+\Delta t}\sigma_j^{(i-1)} \qquad j, s = 1, 2, 3 \qquad \text{(a)}$$

where $e_1 = e_{xx}$, $e_2 = e_{yy}$, $e_3 = \gamma_{xy}$; $^{t+\Delta t}\sigma_j^{ext}$ and $^{t+\Delta t}\sigma_j^{(i-1)}$ are the given and calculated stresses, with $\sigma_1 = \sigma_{xx}$, $\sigma_2 = \sigma_{yy}$, $\sigma_3 = \sigma_{xy}$; and $^{t+\Delta t}C_{js}^{(i-1)}$ are the components of the constitutive matrix $^{t+\Delta t}\mathbf{C}^{EPC}$ (see Section 5.4.3).

We follow the computational procedure of Table 5.4.1 and Section 5.4.2. The system of equations (a) is solved by Gauss elimination, and the solutions of (5.4.27) for $^{t+\Delta t}\bar{\sigma}$ and (5.4.28) for $\Delta\bar{e}^P$ are obtained by using a bisection procedure.

The tangent constitutive matrix is obtained by using the relations of Section 5.4.3, with static condensation. To improve the convergence rate, we have employed a stress correction $\Delta\sigma^{est}$ for the initial equilibrium iteration, with the estimated creep strain increment $(\Delta e^C)^{est}$ and the thermal strain increment Δe^{TH},

$$\Delta\sigma^{est} = {}^t\mathbf{C}^{EPC} \left[(\Delta\hat{e}^C)^{est} + \Delta e^{TH}\right] \qquad \text{(b)}$$

where $(\Delta\hat{e}^C)^{est}$ is the increment of creep strain vector (see (A1.3)) corresponding to the stress ${}^t\sigma$ (see Example 5.4.5), and Δe^{TH} is due to the temperature change $\Delta\theta = {}^{t+\Delta t}\theta - {}^t\theta$.

Table E.5.4-6a lists some of the calculated results. We also give the constitutive matrix for the first three steps, at the start of the last iteration. Note that by employing static condensation on the matrices ${}^1C$ and ${}^2C$ to satisfy the condition $\sigma_{yy} = 0$, we can obtain the coefficient C_{xx}^{EC} for the uniaxial solution (see (i) of Example 5.4.5). The matrices are as follows:

$$
{}^1C = \begin{bmatrix} 2.1464 \times 10^5 & 6.6568 \times 10^4 & 0. \\ & 2.1775 \times 10^5 & 0. \\ \text{symmetric} & & 7.6063 \times 10^4 \end{bmatrix}
$$

$$
{}^2C = \begin{bmatrix} 5.1800 \times 10^2 & 4.7010 \times 10^2 & 0. \\ & 9.4073 \times 10^2 & 0. \\ \text{symmetric} & & 2.3551 \times 10^2 \end{bmatrix}
$$

$$
{}^3C = \begin{bmatrix} 5.9512 \times 10^2 & 5.0788 \times 10^2 & -5.2315 \times 10^1 \\ & 1.0164 \times 10^3 & 7.9743 \times 10^0 \\ \text{symmetric} & & 2.4800 \times 10^2 \end{bmatrix}
$$

Table E.5.4-6a. Results of stress integration for the plate elastic-plastic-creep deformation

Variable	Step				
	1	2	3	4	5
e_{xx}^C	$3.673\ 10^{-7}$	$3.572\ 10^{-6}$	$1.557\ 10^{-5}$	$4.682\ 10^{-5}$	$1.485\ 10^{-2}$
$e_{yy}^C = e_{zz}^C$	$-1.836\ 10^{-7}$	$-1.786\ 10^{-6}$	$-7.785\ 10^{-6}$	$-2.341\ 10^{-5}$	$-7.423\ 10^{-3}$
γ_{xy}^C	$0.$	$0.$	$-3.999\ 10^{-6}$	$1.963\ 10^{-5}$	$7.419\ 10^{-3}$
e_{xx}^P	$0.$	$2.003\ 10^{-2}$	$4.500\ 10^{-2}$	$6.808\ 10^{-2}$	$6.808\ 10^{-2}$
$e_{yy}^P = e_{zz}^P$	$0.$	$-1.001\ 10^{-2}$	$-2.250\ 10^{-3}$	$-3.404\ 10^{-2}$	$-3.404\ 10^{-2}$
γ_{xy}^P	$0.$	$0.$	9.37310^{-3}	$2.258\ 10^{-2}$	$2.258\ 10^{-2}$

The results presented in Table E.5.4-6b show that large strain increments may be used (e.g., steps 2 and 5) resulting still in good convergence (improved also due to the stress estimation terms).

Example 5.4.7. Creep and Plastic Deformations of Restrained Pipe. A pipe (Fig. E.5.4-7) restrained at both ends, is subjected to internal pressure. Determine the stresses, strains and the support reactions R_A and R_B, taking into account the creep and plastic deformations of the material. The material data and change of pressure with time are given in the figure.

Table E.5.4-6b. Unbalanced energies during equilibrium iterations

	$\Delta E = (\sigma^{ext} - \sigma^{(i-1)}) \cdot \Delta e^{(i)}$				
		S	t e p		
Iteration (i)	1	2	3	4	5
1	$2.732\ 10^{-8}$	$3.535\ 10^{-1}$	$1.498\ 10^{-4}$	$4.042\ 10^{-4}$	$3.123\ 10^{-18}$
2	$2.530\ 10^{-14}$	$6.426\ 10^{-1}$	$7.419\ 10^{-8}$	$3.578\ 10^{-7}$	
3	$2.282\ 10^{-28}$	$4.843\ 10^{-3}$	$3.865\ 10^{-11}$	$3.735\ 10^{-10}$	
4		$2.616\ 10^{-6}$	$2.126\ 10^{-14}$	$4.414\ 10^{-13}$	
5		$5.573\ 10^{-12}$	$1.211\ 10^{-16}$	$5.384\ 10^{-16}$	
6		$2.459\ 10^{-23}$			

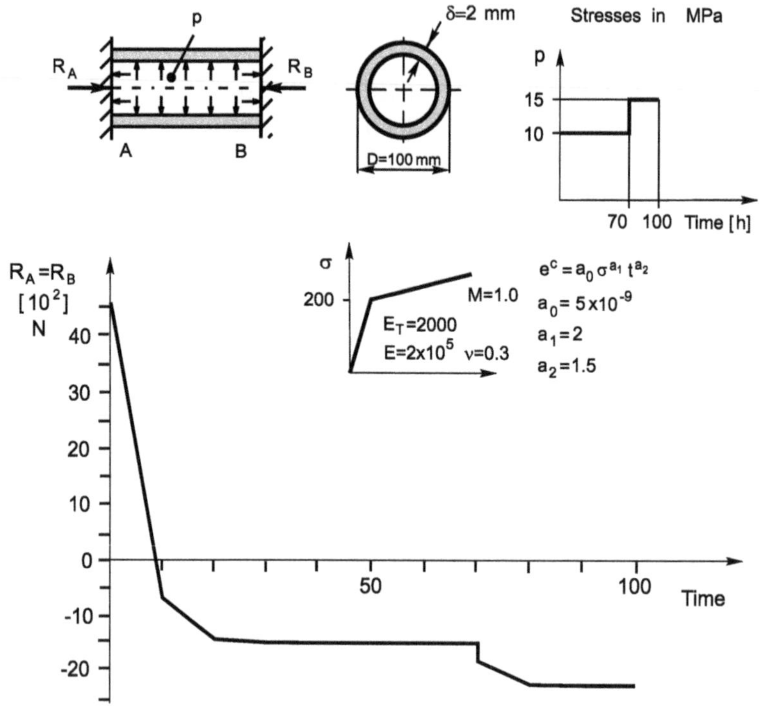

Fig. E.5.4-7. Creep and plastic deformations of restrained pipe. Change of support reactions with time

The physical conditions described in Example 4.5.4 are applicable here. We employ the basic relations of Example 5.4.2 and calculate the support reactions as in Example 4.5.4.

We have employed 12 steps. The first step ($\Delta t = 1. \times 10^{-4}$) is applied to reach the initial plastic state in the material. We use $\Delta t = 1. \times 10^{-4}$ at step 9 to take into account the jump in loading at time $t = 70$, and $\Delta t = 10$ in the other 10 steps. Plastic deformation occurs only in steps 1 and 9, while in the other steps the creep deformation progresses without plastic flow. Instead of (1) of Example 4.5.4 for the support reaction, we have now

$$R_A = \pi a_i \left(a_i - \nu D_m \right) p + \pi \delta D_m E \left(e_{aa}^P + e_{aa}^C \right) \tag{a}$$

Figure E.5.4-7 shows the change of the support reactions with time.

Note that this pipe problem can be solved using only one plane stress finite element restrained to deform in one direction, while subjected to tension by the given hoop stress (see (b) of Example 4.5.4) in the other direction.

We give some of the results for the last step. They are as follows (stresses in MPa and forces in N):

$$\sigma_{aa} = 180.00 \qquad\qquad \sigma_{cc} = 360.00$$
$$e_{aa}^C = -5.3071 \times 10^{-5} \qquad e_{cc}^C = 2.5059 \times 10^{-1}$$
$$e_{aa}^P = -3.0693 \times 10^{-4} \qquad e_{cc}^P = 4.8063 \times 10^{-2}$$
$$R_A = R_B = -2277.0$$

5.5 Stress Integration for Viscoplastic Models

In this section we first present some general relations used for the development of stress integration procedures in viscoplasticity, and then define the computational steps of the governing parameter method (Kojic 1996a). These basic equations are then used to formulate the stress integration algorithm for the viscoplastic model of an initially isotropic metal.

5.5.1 General Considerations

We first recall the basic conditions that are used in developing the computational procedure for a viscoplastic model. These conditions to satisfy are:

- The elastic constitutive relations for the end of the time step,
- The viscoplastic constitutive law for the time step, discretized in time using the Euler backward method, and
- The yield condition at the end of the time step for the conjugate stresses.

To fulfill these requirements we start with the elastic constitutive law for the end of the time step (see also (4.4.4) and (4.4.5))

$$^{t+\Delta t}\boldsymbol{\sigma} = \mathbf{C}^E (^{t+\Delta t}\hat{\mathbf{e}} - {}^t\hat{\mathbf{e}}^{VP} - \Delta\hat{\mathbf{e}}^{VP}) \tag{5.5.1}$$

where $^{t+\Delta t}\hat{\mathbf{e}}$, ${}^t\hat{\mathbf{e}}^{VP}$ and $\Delta\hat{\mathbf{e}}^{VP}$ are the strain vectors (see (A1.3)). For the tensor $\Delta\mathbf{e}^{VP}$ we use the viscoplastic constitutive law (5.3.19) and employ the Euler implicit scheme to obtain

$$\Delta e^{VP} = \frac{\Delta t}{\eta\,^{t+\Delta t}\eta_F} \left(^{t+\Delta t}\sigma - \,^{t+\Delta t}\sigma^*\right) \qquad (5.5.2)$$

Figure 5.5.1 gives a schematic representation of this relation. Note that the viscoplastic strain in general is not deviatoric; but, when the yield surface is defined in the deviatoric stress space, as in case of metal plasticity, the viscoplastic strain has the deviatoric character, see (5.3.22) and (5.3.23). Since the stress $^{t+\Delta t}\sigma$ and the conjugate stress $^{t+\Delta t}\sigma^*$ lie on the normal to the yield surface, the following relation, *independent* of (5.5.2), can be written:

$$\Delta e^{VP} = k\,\frac{\Delta t}{\eta}\,\frac{\partial\,^{t+\Delta t}f_y}{\partial\,^{t+\Delta t}\sigma^*} \qquad (5.5.3)$$

where k is a proportionality coefficient $(k > 0)$. We obtain from the last two equations for Δe^{VP} the relation suitable for development of the computational algorithm,

$$^{t+\Delta t}\sigma - \,^{t+\Delta t}\sigma^* = k\,\frac{\partial\,^{t+\Delta t}f_y}{\partial\,^{t+\Delta t}\sigma^*} \qquad (5.5.4)$$

Next, we need the constitutive relation for the internal variables β^* in the yield condition (5.3.20). In general, we have

$$\Delta\beta^* = \Delta\beta^* \left(^t\sigma, \,^t\sigma^*, \,^t\beta^*, \,^{t+\Delta t}\sigma, \,^{t+\Delta t}\sigma^*, \,^{t+\Delta t}e\right) \qquad (5.5.5)$$

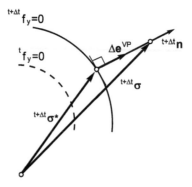

Fig. 5.5.1. Schematic representation of implicit integration of viscoplastic constitutive relations

Finally, the yield condition at the end of the time step must be satisfied, i.e.,

$$t+\Delta t f_y = f_y({}^{t+\Delta t}\boldsymbol{\sigma}^*, {}^{t+\Delta t}\boldsymbol{\beta}^*) = 0 \qquad (5.5.6)$$

Note that when the yield condition depends on the deviatoric stresses rather than the stresses, as given in (5.3.21), we will have in the expressions (5.5.2) to (5.5.6) the deviatoric stresses ${}^{t+\Delta t}\mathbf{S}$ instead of the stresses ${}^{t+\Delta t}\boldsymbol{\sigma}$, and the conjugate deviatoric stresses ${}^{t+\Delta t}\mathbf{S}^*$ instead of the conjugate stresses ${}^{t+\Delta t}\boldsymbol{\sigma}^*$. Then the increment of the viscoplastic strain $\Delta \mathbf{e}^{VP}$ has the deviatoric character.

5.5.2 The Governing Parameter Method for Stress Integration in Viscoplasticity

Following the ideas of the implicit stress integration in Section 4.2 we can now specify the computational steps for the stress calculation of viscoplastic material models. Namely, we assume that the unknowns at the end of the time step

$$t+\Delta t \boldsymbol{\sigma}, {}^{t+\Delta t}\boldsymbol{\sigma}^*, {}^{t+\Delta t}\boldsymbol{\beta}^*, {}^{t+\Delta t}\mathbf{e}^{VP} \qquad (5.5.7)$$

can be expressed in terms of one variable, the governing parameter p. Then we form the governing scalar equation

$$f(p) = 0 \qquad (5.5.8)$$

and solve for the unknown ${}^{t+\Delta t}p$. Using this solution we then find the unknowns (5.5.7). Also, the consistent tangent viscoplastic matrix ${}^{t+\Delta t}\mathbf{C}^{VP}$ can be determined by using the expression (4.2.4).

The computational steps for calculation of the unknowns (5.5.7) are given in Table 4.2.1.

5.5.3 Isotropic Metal

In this section we apply the general relations given above to an isotropic viscoplastic model based on the von Mises yield condition. We first consider general three-dimensional deformations and then shell structural conditions. Finally, we derive the consistent tangent elastic-viscoplastic matrix for the three-dimensional conditions from which the viscoplastic matrix for other conditions can be derived (see Section 4.4.3).

Three-Dimensional Deformations. The yield condition of an isotropic metal with mixed hardening is given in (3.2.57), hence at the end of the time step we have for the conjugate stresses

$$
{}^{t+\Delta t}f_y = \frac{1}{2}\,{}^{t+\Delta t}\hat{\mathbf{S}}^* \cdot {}^{t+\Delta t}\hat{\mathbf{S}}^* - \frac{1}{3}\,{}^{t+\Delta t}\hat{\sigma}_y^{*2} = 0
\qquad (5.5.9)
$$

where ${}^{t+\Delta t}\hat{\mathbf{S}}^*$ is the radius of the yield surface shown in Fig. 5.5.2, and ${}^{t+\Delta t}\hat{\sigma}_y^*$ is the yield stress defined by the yield curve. For the mixed hardening assumption,

$$
{}^{t+\Delta t}\hat{\sigma}_y^* = \hat{\sigma}_y\left(M\,{}^{t+\Delta t}\bar{e}^{VP}\right)
\qquad (5.5.10)
$$

where the increment of the effective viscoplastic strain is defined in accordance with (3.2.46),

$$
\Delta\bar{e}^{VP} = \left(\frac{2}{3}\Delta\mathbf{e}^{VP}\cdot\Delta\mathbf{e}^{VP}\right)^{1/2}
\qquad (5.5.11)
$$

For the back stress $\boldsymbol{\alpha}$ we employ the constitutive relation (4.4.20), therefore we can write

$$
\Delta\boldsymbol{\alpha} = \hat{C}\,\Delta\mathbf{e}^{VP}
\qquad (5.5.12)
$$

where $\hat{C}$ is given in (4.4.21).

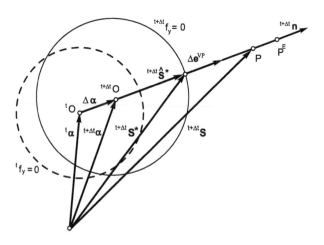

Fig. 5.5.2. Stresses and yield surfaces in deviatoric plane at the start and end of time step. Isotropic metal viscoplasticity

From the yield condition (5.5.9) and the viscoplastic constitutive law (5.3.22) follows the deviatoric character of viscoplastic strains. Therefore, as in Sections 4.4.1 and 5.4.1, we determine the mean stress $^{t+\Delta t}\sigma_m$ independently of the viscoplastic strains

$$^{t+\Delta t}\sigma_m = K\,^{t+\Delta t}e_V$$

(5.5.13)

where K is the bulk modulus, and $^{t+\Delta t}e_V$ is the volumetric strain. The deviatoric stress $^{t+\Delta t}\mathbf{S}$ can be expressed as (see (4.4.11))

$$^{t+\Delta t}\mathbf{S} = {}^{t+\Delta t}\mathbf{S}^E - 2G\,\Delta\mathbf{e}^{VP}$$

(5.5.14)

where

$$^{t+\Delta t}\mathbf{S}^E = 2G\,^{t+\Delta t}\mathbf{e}''$$

(5.5.15)

is the elastic solution, and

$$^{t+\Delta t}\mathbf{e}'' = {}^{t+\Delta t}\mathbf{e}' - {}^{t}\mathbf{e}^{VP}$$

(5.5.16)

Here $^{t+\Delta t}\mathbf{e}'$ is the deviatoric strain as in (4.4.13), and G is the shear modulus.

We next employ the viscoplastic constitutive relations (5.3.23) (see also (5.5.2)) from which

$$\Delta\mathbf{e}^{VP} = \frac{\Delta t}{\eta\,^{t+\Delta t}\eta_F}\left(^{t+\Delta t}\mathbf{S} - {}^{t+\Delta t}\mathbf{S}^*\right)$$

(5.5.17)

Using the geometric relations shown in Fig. 5.5.2 and using (5.5.12) we obtain (see also (4.4.22))

$$^{t+\Delta t}\mathbf{S}^* = {}^{t}\boldsymbol{\alpha} + \hat{C}\,\Delta\mathbf{e}^{VP} + {}^{t+\Delta t}\hat{\mathbf{S}}^*$$

(5.5.18)

From (5.5.14) to (5.5.18) we obtain the following equation

$$^{t+\Delta t}\hat{\mathbf{S}}^* + \left(^{t+\Delta t}a_{VP} + \hat{C}\right)\Delta\mathbf{e}^{VP} - {}^{t+\Delta t}\hat{\mathbf{S}}^E = 0$$

(5.5.19)

where

$$^{t+\Delta t}a_{VP} = 2G + \frac{\eta}{\Delta t}\,^{t+\Delta t}\eta_F \tag{5.5.20}$$

$$^{t+\Delta t}\hat{\mathbf{S}}^E = \,^{t+\Delta t}\mathbf{S}^E - \,^t\alpha \tag{5.5.21}$$

Here $^{t+\Delta t}\hat{\mathbf{S}}^E$ is the elastic solution for the stress radius, represented in the figure by the stress vector $^tOP^E$. By inspecting the terms in (5.5.19) we conclude that they all are vectors along the normal

$$^{t+\Delta t}\mathbf{n}^E = \,^{t+\Delta t}\hat{\mathbf{S}}^E / \|^{t+\Delta t}\hat{\mathbf{S}}^E\| \tag{5.5.22}$$

where $\|^{t+\Delta t}\hat{\mathbf{S}}^E\| = (^{t+\Delta t}\hat{\mathbf{S}}^E \cdot \,^{t+\Delta t}\hat{\mathbf{S}}^E)^{1/2}$. This "colinearity" of the vectors can also be seen in the figure. Hence, the scalar equation

$$\hat{f} = \,^{t+\Delta t}\hat{\sigma}_y^* + \frac{3}{2}\left(^{t+\Delta t}a_{VP} + \hat{C}\right)\Delta\bar{e}^{VP} - \sqrt{\frac{3}{2}}\,\|^{t+\Delta t}\hat{\mathbf{S}}^E\| = 0$$

$$\tag{5.5.23}$$

directly follows from (5.5.19). We have here used (3.2.60) and (5.5.11).

From the above derivations we see that the stress integration is reduced to solving the nonlinear equation (5.5.23). The solution process is geometrically represented as a search for the stress point along the line $^tOP^E$ in Fig. 5.5.2. In accordance with Table 4.2.1, the governing parameter is $\Delta\bar{e}^{VP}$ and the computational steps are as given in Table 5.5.1. The computational procedure is analogous to that given in Table 4.4.1 for isotropic metal plasticity.

Table 5.5.1. Computational steps for stress calculation - isotropic metal (three-dimensional deformation)

---

1. **Assume** $\Delta\bar{e}^{VP}$
2. **Determine** $^{t+\Delta t}\hat{\sigma}_y^*$ from the yield curve (5.5.10)
3. **Iterate** until (5.5.23) is satisfied within a selected tolerance

---

Shell (Plane Stress) Deformation. In addition to the conditions stated at the beginning of Section 5.5.1, the following condition must be satisfied during the deformation:

$$\sigma_{zz} = 0 \tag{5.5.24}$$

where z is the axis "normal" to the shell.

In strain-driven problems, the normal $^{t+\Delta t}\mathbf{n}^E$ given by (5.5.22) cannot be determined because the strain through the shell thickness cannot be calculated from the displacements (see also Sections 4.4.2 and 5.4.2). Therefore we employ the relation (5.5.3) which, with the yield condition (5.5.9), leads to

$$\Delta \mathbf{e}^{VP} = k \frac{\Delta t}{\eta} \, {}^{t+\Delta t} \hat{\mathbf{S}}^* \tag{5.5.25}$$

The constitutive relation (5.5.14) is now (with use of the one-index notation for stresses and strains, see, e.g., (3.3.4) and (3.3.10))

$$\boxed{{}^{t+\Delta t}\mathbf{S} = {}^{t+\Delta t}\mathbf{S}^E - \bar{\mathbf{C}}'^E \Delta \mathbf{e}^{VP}} \tag{5.5.26}$$

The components in this relation correspond to the normal in-plane stress and strain components (see (4.4.43)), and to all shear components. Hence, we have five scalar equations, two for the normal and three for the shear components. The normal deviatoric stress component corresponding to the shell normal is given by the third equation in (4.4.43). The elastic deviator ${}^{t+\Delta t}\mathbf{S}^E$ is obtained as

$$ {}^{t+\Delta t}\mathbf{S}^E = \bar{\mathbf{C}}'^E \left({}^{t+\Delta t}\mathbf{e} - {}^{t}\mathbf{e}^{VP} \right) \tag{5.5.27}$$

The nonzero terms in the first two rows of the 5×5 matrix $\bar{\mathbf{C}}'^E$ are $\bar{C}'^E_{ij}$, $i, j = 1, 2$, (for the normal stresses and strains) and are given in (4.4.44). The constitutive relations for the shear terms are as for the three-dimensional case, i.e., (5.5.14), therefore the nonzero terms in the last three rows are $\bar{C}'^E_{ii} = 2G$, $i = 3, 4, 5$.

With the use of (5.5.17), (5.5.18), (5.5.25) and (5.5.26), we obtain the stress radius ${}^{t+\Delta t}\hat{\mathbf{S}}^*$ components as

$$
\begin{aligned}
{}^{t+\Delta t}\hat{S}^*_{xx} &= \left(D^{(k)}_{xx} \, {}^{t+\Delta t}\hat{S}^E_{xx} - D^{(k)}_{xy} \, {}^{t+\Delta t}\hat{S}^E_{yy} \right) / D^{(k)} \\
{}^{t+\Delta t}\hat{S}^*_{yy} &= \left(-D^{(k)}_{xy} \, {}^{t+\Delta t}\hat{S}^E_{xx} + D^{(k)}_{xx} \, {}^{t+\Delta t}\hat{S}^E_{yy} \right) / D^{(k)} \\
{}^{t+\Delta t}\hat{S}^*_{zz} &= - {}^{t+\Delta t}\hat{S}^*_{xx} - {}^{t+\Delta t}\hat{S}^*_{yy} \\
{}^{t+\Delta t}\hat{S}^*_i &= {}^{t+\Delta t}\hat{S}^E_i / D^{(k)}_{sh} \qquad i = 3, 4, 5
\end{aligned}
\tag{5.5.28}
$$

where

$$
\begin{aligned}
D^{(k)}_{xx} &= 1 + k \, {}^{t+\Delta t}\eta_F + k \frac{\Delta t}{\eta} \left(\bar{C}'^E_{11} + \hat{C} \right) \\
D^{(k)}_{xy} &= k \frac{\Delta t}{\eta} \bar{C}'^E_{12} \\
D^{(k)} &= \left(D^{(k)}_{xx} \right)^2 - \left(D^{(k)}_{xy} \right)^2
\end{aligned}
\tag{5.5.29}
$$

and

$$D_{sh}^{(k)} = 1 + k\,^{t+\Delta t}\eta_F + k\frac{\Delta t}{\eta}\left(2G + \hat{C}\right) \tag{5.5.30}$$

Here, the x and y axes lie in the shell tangential plane and the z axis for the shell conditions is applicable (see Appendix A1), while the indices $i = 3, 4, 5$ correspond to shear stresses and strains in the x-y, y-z and x-z planes, respectively. Also, $^{t+\Delta t}\hat{\mathbf{S}}^E$ is given in (5.5.21), with $^{t+\Delta t}\mathbf{S}^E$ defined by (5.5.27). In addition to the solution (5.5.28) for the stress radius, we also employ the scalar equation which follows from (5.5.25), (5.5.9), and (5.5.11),

$$\boxed{k = \frac{3}{2}\frac{\eta}{\Delta t}\frac{\Delta\bar{e}^{VP}}{^{t+\Delta t}\hat{\sigma}_y^*}} \tag{5.5.31}$$

The stress radius $^{t+\Delta t}\hat{\mathbf{S}}^*$ must be determined so that the yield condition (5.5.9) is satisfied.

Finally, we can summarize the computational steps as given in Table 5.5.2. Note that the expressions for the shell conditions are applicable to plane stress conditions by not including the transversal shear terms.

Table 5.5.2. Computational steps for stress calculation - isotropic metal (shell conditions)

---

1. **Assume** $\Delta\bar{e}^{VP}$
2. **Determine** $^{t+\Delta t}\hat{\sigma}_y^*$ from the yield curve (5.5.10)
3. **Calculate** k from (5.5.31)
4. **Calculate** $^{t+\Delta t}\hat{\mathbf{S}}^*$ from (5.5.28)
5. **Iterate on** $\Delta\bar{e}^{VP}$ until (5.5.9) is satisfied within a selected tolerance

---

Elastic-Viscoplastic Matrix. The coefficients $^{t+\Delta t}C_{ij}^{VP}$ of the consistent tangent viscoplastic matrix can be written as

$$\boxed{^{t+\Delta t}C_{ij}^{VP} = \frac{\partial}{\partial\,^{t+\Delta t}e_j}\left(^{t+\Delta t}\sigma_{(m)i} + {}^{t+\Delta t}S_i\right)} \tag{5.5.32}$$

where we have used the one-subscript notation of Section 4.4.3. General three-dimensional deformation is considered only and the tangent moduli for the shell (plane stress) conditions can be obtained by static condensation.

Using (5.5.13) the nonzero derivatives of the mean stress $^{t+\Delta t}\sigma_m$ are

$$C_{ij}^m = \frac{\partial^{t+\Delta t}\sigma_{(m)i}}{\partial^{t+\Delta t}e_j} = \frac{1}{3}c_m \qquad i,j = 1,2,3 \tag{5.5.33}$$

where δ_{ij} is the Kronecker-delta symbol, and $c_m = 3K$. The derivatives of the deviatoric stresses can be obtained from (5.5.14) and (5.5.15)

$$\mathbf{C}' = \left[\frac{\partial^{t+\Delta t}S_i}{\partial^{t+\Delta t}e_j}\right] = \begin{bmatrix} \bar{\mathbf{C}}'^{(1)}\,\mathbf{A} & \frac{1}{2}\bar{\mathbf{C}}'^{(2)} \\[2mm] \bar{\mathbf{C}}'^{(3)}\,\mathbf{A} & \frac{1}{2}\bar{\mathbf{C}}'^{(4)} \end{bmatrix} \tag{5.5.34}$$

where $\bar{\mathbf{C}}'^{(s)}$, $s = 1,2,3,4$ are the (3×3) submatrices of the matrix $\bar{\mathbf{C}}'$,

$$\bar{\mathbf{C}}' = \left[\frac{\partial^{t+\Delta t}S_i}{\partial^{t+\Delta t}e_j''}\right] = \begin{bmatrix} \bar{\mathbf{C}}'^{(1)} & \bar{\mathbf{C}}'^{(2)} \\ \bar{\mathbf{C}}'^{(3)} & \bar{\mathbf{C}}'^{(4)} \end{bmatrix} \tag{5.5.35}$$

The matrix $\mathbf{A}$ is given by (3.2.70) and $\mathbf{I}_3$ is the (3×3) identity matrix. Therefore, the determination of $^{t+\Delta t}\mathbf{C}^{VP}$ reduces to the calculation of the derivatives $\bar{C}'_{ij}$.

From (5.5.14) we obtain

$$\bar{C}'_{ij} = 2G\left(\delta_{ij} - \Delta e_{i,j}^{VP}\right) \tag{5.5.36}$$

where $\Delta e_{i,j}^{VP} = \partial(\Delta e_i^{VP})/\partial^{t+\Delta t}e_j''$. Using (5.5.11) we can write

$$\Delta e_i^{VP} = \sqrt{\frac{3}{2}}\,\Delta\bar{e}^{VP}\,{}^{t+\Delta t}n_i \tag{5.5.37}$$

and then

$$\Delta e_{i,j}^{VP} = \sqrt{\frac{3}{2}}\left({}^{t+\Delta t}n_i\,\Delta\bar{e}_{,j}^{VP} + \Delta\bar{e}^{VP}\,{}^{t+\Delta t}n_{i,j}\right) \tag{5.5.38}$$

The derivatives of the normal $^{t+\Delta t}n_i$ can be obtained from (5.5.22) and the definition (5.5.15) of $^{t+\Delta t}\mathbf{S}^E$,

$$^{t+\Delta t}n_{i,j} = \frac{2G}{||\,^{t+\Delta t}\hat{\mathbf{S}}E\,||}\left(\delta_{ij} - {}^{t+\Delta t}n_i\,{}^{t+\Delta t}\hat{n}_j\right) \tag{5.5.39}$$

where $^{t+\Delta t}\hat{\mathbf{n}}$ contains twice the shear terms of $^{t+\Delta t}\mathbf{n}$. By differentiation of (5.5.23) with respect to $^{t+\Delta t}e_j''$ we obtain the derivatives $\Delta\bar{e}_{,j}^{VP}$ as

$$\Delta\bar{e}^{VP}_{,j} = \sqrt{\frac{3}{2}}\, A_P^{-1}\, {}^{t+\Delta t}\hat{n}_j - {}^{t+\Delta t}b_j \tag{5.5.40}$$

where

$$A_P = \left[M\,{}^{t+\Delta t}\hat{E}_P + \frac{3}{2}\left({}^{t+\Delta t}a_{VP} + \hat{C}\right) + (1-M)\,\Delta\bar{e}^{VP}\,{}^{t+\Delta t}E'_P + \right.$$

$$\left. \frac{3}{2}\frac{\Delta\bar{e}^{VP}}{\Delta t}\,\eta\,{}^{t+\Delta t}\eta'_F \right] / (2G) \tag{5.5.41}$$

and

$$^{t+\Delta t}b_j = \frac{3}{4}A_P^{-1}\frac{\eta\Delta\bar{e}^{VP}}{G\Delta t}\,{}^{t+\Delta t}\eta_{F,j} \tag{5.5.42}$$

with ${}^{t+\Delta t}\eta'_F = \partial\,{}^{t+\Delta t}\eta_F / \partial\left(\Delta\bar{e}^{VP}\right)$ and ${}^{t+\Delta t}\eta_{F,j} = \partial\,{}^{t+\Delta t}\eta_F / \partial\,{}^{t+\Delta t}e''_j$. Also,

$$^{t+\Delta t}\hat{E}_P = \frac{\partial\,{}^{t+\Delta t}\hat{\sigma}_y}{\partial\,{}^{t+\Delta t}\bar{e}^{VP}}\bigg|_{(M\,{}^{t+\Delta t}\bar{e}^{VP})} \qquad {}^{t+\Delta t}E'_P = \frac{\partial\,{}^{t+\Delta t}E_P}{\partial\,{}^{t+\Delta t}\bar{e}^{VP}} \tag{5.5.43}$$

Finally, substituting (5.5.39) and (5.5.40) into (5.5.38), and then the resulting expression for $\Delta e^{VP}_{i,j}$ into (5.5.36), we obtain

$$\boxed{\bar{C}'_{ij} = 2G\left(B_P\,\delta_{ij} - D_P\,{}^{t+\Delta t}n_i\,{}^{t+\Delta t}\hat{n}_j + \sqrt{\frac{3}{2}}\,{}^{t+\Delta t}n_i\,{}^{t+\Delta t}b_j\right)}$$

$$\tag{5.5.44}$$

where the dimensionless coefficients B_P and D_P are

$$B_P = 1 - b_P$$
$$D_P = \frac{3}{2}A_P^{-1} - b_P \tag{5.5.45}$$

with

$$b_P = 2\sqrt{\frac{3}{2}}\frac{G\,\Delta\bar{e}^{VP}}{\|{}^{t+\Delta t}\hat{S}^E\|} \tag{5.5.46}$$

Then the coefficients ${}^{t+\Delta t}C^{VP}_{ij}$ follow from (5.5.32), with C^m_{ij} given by (5.5.33) and C'_{ij} by (5.5.34). Note that, with use of (5.5.44) and (3.2.70), it can be shown that the matrix ${}^{t+\Delta t}\mathbf{C}^{VP}$ is symmetric.

5.5.4 Examples

Example 5.5.1. Viscoplastic Deformation of Isotropic Beam. Derive the expressions for stress integration and the tangent constitutive matrix for the viscoplastic deformation of a beam. The axial axis of the beam is x, and y and z are the axes in the beam cross-section (see Fig. A1.2b in Appendix A1). The material is assumed to be an isotropic metal with mixed hardening.

The elastic matrix $\mathbf{C}'^E$, relating the deviatoric stresses and elastic strains, is given in (d) of Example 4.5.2. The relations (5.5.2), (5.5.3) and (5.5.1) are now

$$\Delta \mathbf{e}^{VP} = \frac{\Delta t}{\eta \, {}^{t+\Delta t}\eta_F}({}^{t+\Delta t}\mathbf{S} - {}^{t+\Delta t}\mathbf{S}^*) = k\frac{\Delta t}{\eta} {}^{t+\Delta t}\hat{\mathbf{S}}^* \tag{a}$$

$$ {}^{t+\Delta t}\mathbf{S} = {}^{t+\Delta t}\mathbf{S}^E - \mathbf{C}'^E \Delta \mathbf{e}^{VP} \tag{b}$$

where all stress and strain tensors contain one normal and two shear components. With the use of the relation (5.5.18), we solve from (a) and (b) for the stress radius components

$$ {}^{t+\Delta t}\hat{S}^*_{xx} = {}^{t+\Delta t}\hat{S}^E_{xx}/(1 + b_k) \qquad {}^{t+\Delta t}\hat{S}^*_i = \frac{{}^{t+\Delta t}\hat{S}^E_i}{1 + a_k} \quad i = 2,3 \tag{c}$$

where the coefficients b_k and a_k are

$$b_k = \left[{}^{t+\Delta t}\eta_F + \frac{\Delta t}{\eta}\left(\frac{2}{3}E + \hat{C}\right)\right]k$$

$$a_k = \left[{}^{t+\Delta t}\eta_F + \frac{\Delta t}{\eta}\left(2G + \hat{C}\right)\right]k \tag{d}$$

and the indices $i = 2,3$ correspond to transversal shear (xy) and (xz). The yield condition (5.5.9) is now

$$ {}^{t+\Delta t}f_y = \frac{3}{4} {}^{t+\Delta t}\hat{S}^{*2}_{xx} + {}^{t+\Delta t}\hat{S}^{*2}_{xy} + {}^{t+\Delta t}\hat{S}^{*2}_{xz} - \frac{1}{3} {}^{t+\Delta t}\hat{\sigma}^{*2}_y = 0 \tag{e}$$

The computational steps are as for the shell conditions (Table 5.5.2), except that the components ${}^{t+\Delta t}\hat{S}^*_i$ are calculated from (c) and the yield condition is now given in (e).

We next give briefly the calculation of the tangent matrix components by using the above relations for the stress integration. Following the procedure of Section 5.5.3, we have now

$$ {}^{t+\Delta t}C^{VP}_{ij} = C^E_{ii} - C^E_{ii} \Delta e^{VP}_{i,j} \qquad \text{no sum on } i \tag{f}$$

where $C_{11}^E = E$, $C_{22}^E = C_{33}^E = G$. The derivatives $\Delta e_{i,j}^{VP}$ (the indices $j = 1, 2, 3$ correspond to strains $^{t+\Delta t}e_{xx}$, $^{t+\Delta t}\gamma_{xy}$ and $^{t+\Delta t}\gamma_{xz}$) can be determined from (a) as

$$\Delta e_{i,j}^{VP} = \frac{\Delta t}{\eta}\left(^{t+\Delta t}\hat{S}_i^*\, k_{,j} + k\, ^{t+\Delta t}\hat{S}_{i,j}^*\right) \tag{g}$$

and the $^{t+\Delta t}\hat{S}_{i,j}^*$ follow from (c),

$$^{t+\Delta t}\hat{S}_{xx,j}^* = \left(\frac{2}{3}E\,\delta_{1j} - {}^{t+\Delta t}\hat{S}_{xx}^*\, b_{k,j}\right)\big/(1 + b_k) \quad j = 1, 2, 3$$

$$^{t+\Delta t}\hat{S}_{i,j}^* = \frac{1}{1 + a_k}\left(2G\,\delta_{ij} - {}^{t+\Delta t}\hat{S}_i^*\, a_{k,j}\right) \quad i = 2, 3; j = 1, 2, 3 \tag{h}$$

To calculate $b_{k,j}$ and $a_{k,j}$ we use the derivatives of the viscosity function $^{t+\Delta t}\eta_F' = \partial^{t+\Delta t}\eta_F\,/\partial\left(\Delta\bar{e}^{VP}\right)$ and $^{t+\Delta t}\eta_{F,j} = \partial^{t+\Delta t}\eta_F\,/\partial^{t+\Delta t}e_j$, and also $^{t+\Delta t}E_P'$ and $k_{,j}$ that can be determined from (5.5.43) and (5.5.31). Hence, the derivatives $\hat{S}_{xx,j}^*$ and $^{t+\Delta t}\hat{S}_{i,j}^*$ transform into

$$^{t+\Delta t}\hat{S}_{xx,j}^* = b_{11}\delta_{1j} - b_{12}\,{}^{t+\Delta t}\hat{S}_{xx}^*\,\Delta\bar{e}_{,j}^{VP} - b_{13}\,{}^{t+\Delta t}\hat{S}_{xx}^*\,{}^{t+\Delta t}\eta_{F,j} \quad j = 1, 2, 3$$

$$^{t+\Delta t}\hat{S}_{i,j}^* = a_{11}\delta_{ij} - a_{12}\,{}^{t+\Delta t}\hat{S}_i^*\,\Delta\bar{e}_{,j}^{VP} - a_{13}\,{}^{t+\Delta t}\hat{S}_i^*\,{}^{t+\Delta t}\eta_{F,j}$$

$$i = 2, 3; j = 1, 2, 3 \tag{i}$$

where b_{ij} and a_{ij} are the known coefficients. The derivatives $\Delta\bar{e}_{,j}^{VP}$ can be determined by differentiating (e) with respect to $^{t+\Delta t}e_j$ and using the expressions (i). Finally, with $\Delta\bar{e}_{,j}^{VP}$ calculated, all derivatives in (f) to (i) can be obtained.

Example 5.5.2. Viscoplastic Deformations Under Constant Stress and in Case of Stress Relaxation (Uniaxial Loading). Consider a uniaxial stress state like in Fig. 5.3.1.

a) Calculate the total axial strain when the stress is constant at 400 MPa.
b) Calculate the stress when the total axial strain is held constant at 0.05.

In each case compare the numerical results obtained using the general algorithm of Section 5.5.3 with the analytical solutions given in Section 5.3.1. The material data are given in Fig. E.5.5-2.

We include this simple example merely to illustrate the time stepping needed for typical one-dimensional solutions considered in Section 5.3.1.

The viscoplastic law for the one-dimensional model is given in (5.3.4) and for the 3-D model is given in (5.3.23). In order to compare the analytical solutions based on the viscoplastic constitutive relation (5.3.4) with the numerical solutions obtained using the algorithm of Section 5.5.3, we write (5.3.23) for the uniaxial loading as

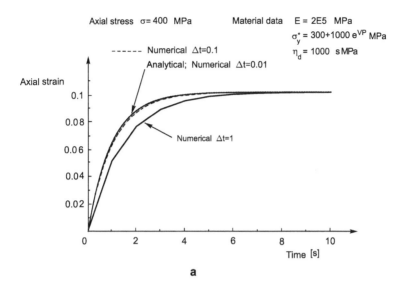

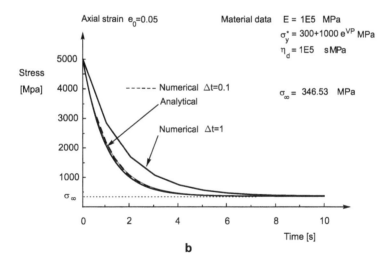

Fig. E.5.5-2. Viscoplastic deformations under uniaxial loading conditions. **a** Increase of axial strain during loading by a constant stress; **b** Stress decrease (stress relaxation) while the axial strain is held constant

$$\dot{e}^{VP} = \frac{2}{3}\frac{\sigma - \sigma^*}{\eta_F} \qquad (a)$$

where $\dot{e}^{VP}$, σ and σ^* are the viscoplastic strain rate, stress and the conjugate stress on the yield curve, respectively. Comparing (a) and (5.3.4) we obtain that the following equation must be satisfied

$$\eta \eta_F = \frac{2}{3}\eta_d \qquad (b)$$

and the viscosity function is

$$\eta_F = \frac{2}{3}\frac{\eta_d}{\eta} \qquad (c)$$

Since in this problem the coefficient η_d is taken to be constant, the viscosity function is constant,

$$\eta_F = const. \qquad (d)$$

satisfying (b), with the numerical values given in Fig. E.5.5-2.

The numerical solutions are obtained using the procedures for the general 3-D deformation and for the shell conditions, with several time steps.

a) The analytical solution is calculated from (5.3.8). Figure E.5.5-2a shows the increase of the axial strain with time due to viscoplastic flow. It can be seen that the increase of the axial strain diminishes as the conjugate stress σ^* approaches the given stress $\sigma = 400$.

b) Figure E.5.5-2b shows the relaxation curves obtained using two magnitudes of time steps, while the analytical curve is calculated using (5.3.10). The stress σ decreases with time and approaches the stress σ_∞ given in (5.3.11),

$$\sigma_\infty = 346.53 \text{ MPa} \qquad (e)$$

We see that, in each case for the time span of 10 sec, about 100 time steps are sufficient to obtain an accurate response prediction.

Example 5.5.3. Viscoplastic Flow of Material – Plane Strain Conditions. Consider the viscoplastic flow of a material under constant loading and plane strain conditions. The material is subjected to the constant normal and shear stresses shown in Fig. E.5.5-3a. The elastic material constants and the dependence of the yield stress σ_y^* on the effective viscoplastic strain $\bar{e}^{VP}$, assuming the isotropic hardening behavior, are as follows:

$$E = 2 \times 10^5 \quad \text{MPa} \qquad \nu = 0.3$$

$$\sigma_y^* = 305.8 + 405.4\,(\bar{e}^{VP})^{0.415} \quad \text{MPa}$$

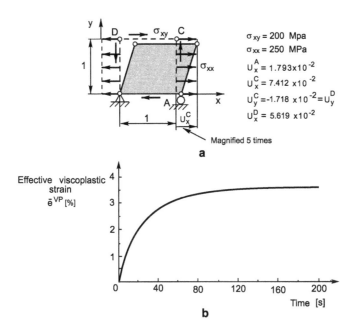

$\sigma_{xy} = 200$ Mpa
$\sigma_{xx} = 250$ MPa
$U_x^A = 1.793 \times 10^{-2}$
$U_x^C = 7.412 \times 10^{-2}$
$U_y^C = -1.718 \times 10^{-2} = U_y^D$
$U_x^D = 5.619 \times 10^{-2}$

Magnified 5 times

a

b

Fig. E.5.5-3. Plane strain finite element with viscoplastic material model. **a** Final deformed shape; **b** Increase of the effective viscoplastic strain with time

We use the viscoplastic law (5.3.15), with the function ϕ given in (5.3.16), which for the metal reduces to the constitutive relation (5.3.23). The viscosity function is

$$\eta_F = \frac{2}{3} (\bar{\sigma}^*)^N (\bar{\sigma} - \bar{\sigma}^*)^{(1-N)}$$

and the viscosity coefficient is $\eta = 1/\gamma$. The following values are employed in this example:

$$\eta = 100 \text{ s} \qquad N = 1$$

This is a very simple problem but serves to show a typical viscoplastic material response. We have used 400 steps of size $\Delta t = 0.5\,\text{s}$.

The final displacements of the element nodes are given in Fig. E.5.5-3a. Figure E.5.5-3b gives the increase of the effective viscoplastic strain as a function of time. It can be seen from the figure that the rate of increase of effective viscoplastic strain diminishes with time as the stress point approaches the yield surface.

6. Plasticity of Geological Materials

In this chapter we describe some of the most common plasticity material models used to represent the behavior of geomaterials and present computational procedures for stress integration of these models. The computational algorithms represent an application of the governing parameter method of Section 4.2.

In Section 6.1 we first define some fundamental notions of soil mechanics and present a general view of soil behavior under loading. Then we give a short historical review of soil plasticity. In Sections 6.2, 6.3 and 6.4 we consider the cap model, the Cam-clay model, and a general soil plasticity model, respectively. For these models we develop the computational procedures for implicit stress integration and derive expressions for the consistent tangent elastic-plastic matrices. With suitably solved examples we illustrate the characteristics of the material models and of the computational algorithms.

As it is usual in geomechanics, we use the *compressive stresses and strains as positive*. Hence, *the normal stresses and strains that in the other chapters are positive, are all negative in this chapter, and vice versa.*

6.1 Introduction to the Mechanical Response of Geomaterials

The mechanical response of geological materials is complex due to the heterogeneous material structure, and due to variation of both the mechanical properties of these materials and the physical conditions in the real in situ environment. The complexity of a geological material response is particularly pronounced when the material undergoes permanent (large) deformations. The formulation of material models and their applications require laboratory investigations and in situ measurements.

One of the basic characteristics of the material behavior considered in the previous chapters was the isochoric character of the inelastic strains. Only in the case of the Gurson model (Example 4.5.11) did we have volumetric plastic strain develop during plastic flow. In the cases of isochoric inelastic deformations considered the mean stress does not affect the inelastic deformations. This characteristic pertains mainly to metals. On the other hand, the elastic-plastic models for geological materials, such as soils, rocks and concrete, are more complex, with yield criteria dependent on the mean stress, and with a permanent (plastic) volumetric strain. Many yield criteria have been proposed for geological materials and we will present some of them in the subsequent sections.

In order to describe the main mechanical characteristics of geological materials we first introduce some basic notions generally used in geomechanics.

6.1.1 Basic Notions in Geomechanics

We begin with the fact that geological materials are porous, composed of *solid material* and *pores*. Pores can be filled with a gas (or a mixture of gases) and/or a liquid (or a mixture of liquids). If the whole pore volume is filled by a liquid, the state is called *saturated*. The geological materials are *inhomogeneous* in nature, as schematically shown in Fig. 6.1.1. However, in large scale considerations, common in engineering practice (and assumed here), the geological materials are represented as *continuous media*.

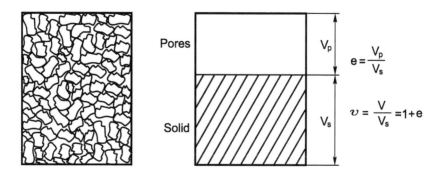

Fig. 6.1.1. Volume element V of a geological material

One of the main characteristics, relevant for the material response under a mechanical action, is the *void ratio e*, defined as

$$e = \frac{V_p}{V_s} \qquad (6.1.1)$$

where, as shown in Fig. 6.1.1, V_p and V_s are, respectively, the volumes occupied by pores and the solid material within a volume V. Another quantity of the same character, used as an inherent property of the material, is the *specific volume v*,

$$v = \frac{V}{V_s} = 1 + e \qquad (6.1.2)$$

Under a mechanical action, a geological material changes its shape and its volume. The change in the material shape is either due to the change of shape of the solid matrix (for rocks), or due to rearrangements of the solid particles (for soils). On the other hand, the volume change of the porous continuum can be attributed to the volume change of the solid material and the change of volume occupied by the pores. It is usually considered in engineering practice that the solid material by itself is incompressible (the solid material has a very high bulk modulus K_s when compared with the bulk modulus K of the porous continuum), and the volume change is only due to the change of volume of pores. Then the following relations between the increment of volumetric strain e_V of the continuum and the increments of void ratio e and specific volume v are applicable

$$de_V = -\frac{dV}{V} = -\frac{d(1+e)V_s}{V} = -\frac{de}{1+e} = -\frac{dv}{v} \qquad (6.1.3)$$

where the negative sign is due to the convention that the *compressive volumetric strain is positive*[1] . Note that the compressibility of a fluid does not enter (6.1.3).

A very common condition in nature, and important in engineering practice, corresponds to the porous material being saturated by a liquid. The load applied to this two-phase medium is carried by the solid matrix and the fluid. The stresses in the solid material are called the *effective stresses*[2] σ_{ij}^* and they are related to the total stresses σ_{ij} (forces per unit total area of the mixture) and the fluid pressure p_F as

$$\sigma_{ij}^* = \sigma_{ij} - p_F \delta_{ij} \qquad (6.1.4)$$

This relation expresses the *effective stress principle* introduced by Terzaghi (1936). Hence, the total normal stress is equal to the sum of the normal effective stress and the fluid pressure, while the shear components of the total and the effective stresses are equal. The principle is considered to be

---

[1] We note that these relations, and others to follow, seem to imply a large strain analysis (see Chapter 7); however, these relations are commonly used in kinematic small strain analysis (although the calculated strains can indeed be rather large) and we shall proceed in the same manner in this chapter.

[2] Note that the effective stresses defined here have no relation to the effective stress used in metal plasticity theory, see (3.2.39).

one of the most fundamental principles in soil mechanics. The principle is geometrically illustrated in Fig. 6.1.2 by use of the Mohr circles (note that compressive stresses are positive). We use the pole P of the Mohr circle[3] to

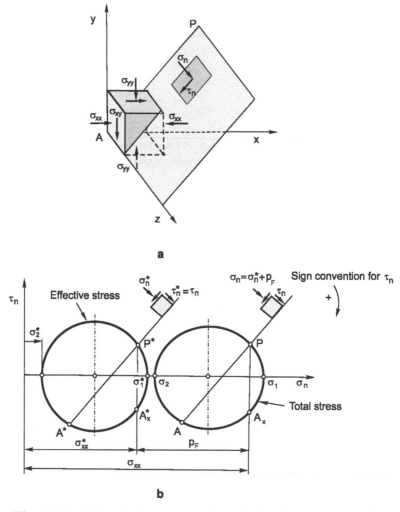

Fig. 6.1.2. Mohr circle representation of effective stress principle. **a** Physical space; **b** Stress space

---

[3] The pole P of the Mohr circle is obtained as follows. Consider the Mohr circle in Fig. 6.1.2b representing stress states on all planes containing the z-axis (Fig. 6.1.2a). Assume that we know the normal and shear stresses on a plane, for example σ_{xx} and σ_{xy} acting on the y, z plane. The stress state on this plane is represented by the stress point A_x on the circle. We draw a vertical straight

geometrically illustrate that the relation (6.1.4) is applicable for any plane, e.g., for the plane AP shown in the figure (because AP and A^*P^* in Fig. 6.1.2b are parallel and the τ_n coordinates of P and P^* are the same).

The mechanical characteristics of the soil skeleton can be considered unaffected by the fluid pressure, therefore the *constitutive relations*, in the elastic as well as in the plastic domain, *correlate the effective stresses and total strains*. Since the subject of this book is the stress integration, *we will use the effective stresses as the stresses* in the model descriptions and stress integrations of Sections 6.2 to 6.4.

The effective stress principle is applicable in the analyses of porous media with a static or flowing fluid, as well as in the cases of a compressible solid material and a compressible fluid. The fluid flow (seepage) is frequently assumed to be governed by *Darcy's law* in which the fluid velocity is proportional to the fluid pressure gradient. The proportionality coefficient is called the permeability coefficient and depends on the properties of the porous medium and the fluid. However, in each case the numerical procedures for the stress integration of the elastic-plastic constitutive relations correspond to conditions of the solid material subjected to the effective stresses (Lewis and Schrefler 1987).

We next introduce the notions of *drained* and *undrained* conditions for a saturated soil, both important in geotechnical engineering. The *drained conditions* at a material point correspond to the case when a change of the *total stress* does not produce a change of fluid pressure at that point. These physical conditions are illustrated in Fig. 6.1.3a. A change in the total normal stress σ_n is followed by the simultaneous changes of the effective stress σ_n^* and the volumetric strain. This situation is encountered when there is a slow change in the load or boundary conditions and the fluid is free to flow during this time period.

The *undrained conditions* are illustrated in Fig. 6.1.3b. Under these conditions a change of the total normal stress σ_n occurs in a short period of time, and it is assumed that the fluid does not have time to flow (or change the flow characteristics) in this short time period. Of course, the fluid may also be constrained due to an impermeability of the boundaries (in the case of a stationary fluid). In this short time period the change of σ_n is accompanied by an increase of the fluid pressure, while the volume (assuming an incompressible fluid) and the effective stresses remain constant. The load change is carried by the fluid. If the boundaries are permeable, the fluid pressure

---

line (which actually corresponds to the y, z plane on which the σ_{xx} and σ_{xy} act) through the stress point A_x, and at the intersection with the Mohr circle obtain the pole P. The pole has the following property. The intersection of a straight line through P with the Mohr circle gives the normal and shear stresses σ_n and τ_n on the plane defined by that line (e.g., the point A gives the stresses corresponding to the plane defined by the direction AP in Fig. 6.1.2b and the z-axis, see Fig. 6.1.2a), with the sign convention for the shear stress as shown in the figure. The pole of Mohr's circle has been used for graphical methods in plasticity (see, e.g., Prager 1959; De Jong 1959; Kojic and Cheatham 1974a,b).

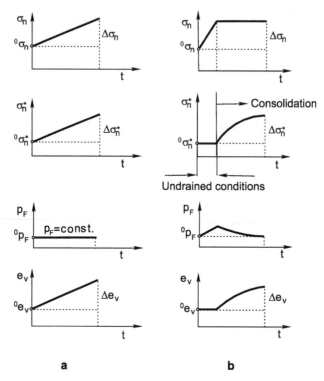

Fig. 6.1.3. Change of normal effective stress, fluid pressure and volumetric strain due to a change of normal total stress. **a** Drained conditions; **b** Undrained and consolidation conditions

subsequently decays slowly until the initial fluid pressure is reached again. This slow time process is known as *consolidation* of the material and it occurs within the time interval called the consolidation time. During the consolidation time both the volume of the solid-fluid mixture and the effective stresses change, as schematically represented in the figure.[4]

6.1.2 Mechanical Properties of Geological Materials

We consider some mechanical properties of geological materials typical for the behavior of these materials. The material models considered in the subsequent sections pertain mainly to soils and therefore we focus on the mechanical characteristics of soils.

In general, the response of a soil subjected to a *given loading/straining sequence* depends on the *mechanical properties of the material including the current state of the material*. The current state depends on the history of the

---

[4] We may note that clays behave more like in undrained conditions, whereas sands behave more like in drained conditions, because of the grain sizes in each case.

material deformation prior to the considered load application. The current state can be considered as a *disturbed state* with respect to a defined initial (reference) state (Desai 2001). For example, if a sample of a soil is subjected to high hydrostatic compression, the material will become dense, while in light compression it will be loose (less dense). If then the material is subjected to loading, the response of the dense and loose samples will be different. We will specify later (see hydrostatic compression) more precisely the conditions that define the state of the material.

The mechanical properties are extracted from experimental observations and serve as a basis for the formulation of material models, such as those given in the subsequent sections. In order to identify the material properties we consider particular loading/straining conditions generally accepted and applied in standard tests in laboratory (or in situ) measurements. A detailed description of the testing procedures is given in a number of books (see, e.g., Desai and Siriwardane 1984; Wood 1990; Atkinson 1993; Das 1997). We will illustrate the material behavior by showing some typical results obtained in laboratory investigations.

Uniaxial Straining. A soil is quite often subjected to one-dimensional straining in in-situ conditions. The material response under such conditions is measured in the so-called *oedometer test*, schematically shown in Fig. 6.1.4a. A cylindrical material sample is subjected to compressive loading σ_a while the lateral deformations are suppressed. This test is also used for determining the consolidation characteristics when the material is filled with a fluid (water). In this case the top and bottom surfaces of the oedometer apparatus are permeable, providing the conditions for one-dimensional flow. Figure 6.1.4b shows a typical dependence of the axial effective stress σ_a^* on the specific

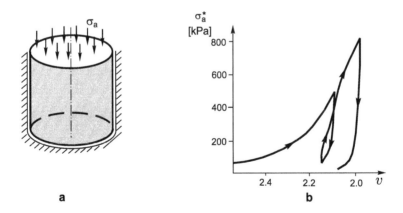

Fig. 6.1.4. Uniaxial straining (oedometer test). **a** Material sample; **b** Axial stress − specific volume dependence for speswhite kaolin, according to Al-Tabbaa (1987)

volume v. The axial stress increase causes a decrease of the specific volume (and volume) of the material. The curve has a nonlinear character, with a hysteresis response in the unloading-reloading cycles.

Shearing of the Material. A schematic representation of shearing of a soil sample is shown in Fig. 6.1.5a. The material is subjected to a constant normal stress, or effective normal stress σ_n^*, with increasing shear deformations. Testing of the material is performed in a *shear test apparatus* (Desai and Siriwardane 1984; Atkinson 1993). Figure 6.1.5b gives typical curves for soils when subjected to shear. We have two types of material response. One corresponds to a loose and the other to a dense state of the material. In the case of a loose state we have that the shear stress continuously increases with shear strain, until a critical shear stress τ_{ncrit} is reached. Then shearing of the material continues with no change of shear stress, that is, we say that the material behaves as a perfectly plastic solid. On the other hand, in the case of a dense state of the material, the shear stress first increases, reaches a maximum, and then decreases, approaching the same critical stress τ_{ncrit}.

A very important property of soils is the so-called *dilatancy* – the property that the volume of the material changes in shearing. A simple explanation of dilatancy is that the shearing of material causes rearrangements of the solid particles in the volume, leading to a volume change. Here we have a different material behavior for the loose and dense states, as shown in Figs. 6.1.5c,d. In the case of the loose state the volume continuously decreases

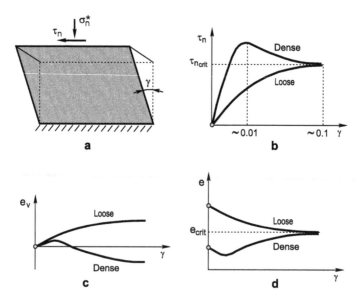

Fig. 6.1.5. Schematic shear of a soil. **a** Material element under shear; **b** Dependence of the shear stress τ_n on the shear strain γ; **c** Volumetric strain $e_V - \gamma$ dependence; **d** Void ratio $e - \gamma$ dependence

(the volumetric strain increases and the void ratio decreases). The dense material first displays a volume decrease and then a volume increase. In both, the loose and dense states of the material, the *ultimate void ratio* is e_{crit}. The ultimate void ratio and the critical shear stress are reached at a large shear strain γ, of order 10% (Fig. 6.1.5b). Note that the initial values of the void ratio ^0e are different and the signs of the total change $(e_{crit} - {}^0e)$ are different for the two states. The change of the volumetric strain is related to the change of the void ratio by (6.1.3) and the ultimate volumetric strains are different for the two states.

Hydrostatic Compression. Loading of the material by pressure in all directions is called hydrostatic compression, or isotropic compression. We first show some experimental results and then give a schematic representation of the material response in loading, unloading and reloading conditions.

Figures 6.1.6a,b give experimental results for two types of sand in the mean effective stress σ_m^* — specific volume v plane, and in the volumetric strain — pressure (mean stress) plane, for loose and dense states. It can be seen that the responses of the loose and dense materials are significantly different. The same character of response in hydrostatic compression is displayed by clay material, as can be seen from the experimental results given in Fig. 6.1.6c. This figure also shows that the relation $v - \ln \sigma_m^*$ is linear.

Schematic representations of the soil behavior based upon the experimental observations are given in Figs. 6.1.6d,e. Figure 6.1.6d shows a nonlinear $\sigma_m^* - e_V$ relation, with a hysteresis material response under unloading and reloading conditions. The *isotropic compression line (icl)* (also called isotropic virgin compression line or isotropic normal consolidation line) in Fig. 6.1.6e represents a material characteristic analogous to the yield curve for metals (see Section 3.2). If the material is unloaded from a point B on the icl-line, the volume and the specific volume will increase (the material swells), following the path $B - C$ shown in the figure. Hydrostatic reloading to the mean stress at the point B will be along the path $C - B$. The hysteresis is usually neglected so that the unloading-reloading occurs along the same line. The isotropic compression line and the unloading-reloading lines are approximately straight in the $\ln \sigma_m^* - v$ representation, as shown in Fig. 6.1.6e.

All states represented by points below the isotropic compression line correspond to an *overconsolidated* material. The overconsolidation ratio R_p defined as

$$R_p = \frac{(\sigma_m^*)_B}{(\sigma_m^*)_C} \tag{6.1.5}$$

represents a measure of the overconsolidation. The point C is a point below the isotropic compression line, and B is the corresponding point on the isotropic compression line as described above.

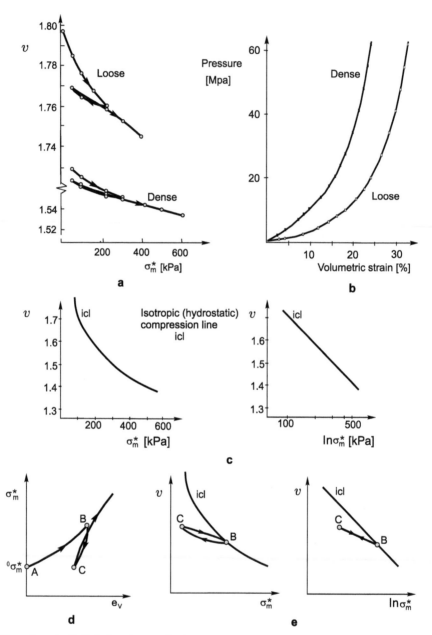

Fig. 6.1.6. Soil behavior in hydrostatic compression. **Experimental results: a** Fuji river sand, according to Tatsuoka (1972), cited in Wood (1990); **b** Chattahoochee River sand (dry), according to Vesic and Clough (1968); **c** Weald clay, according to Roscoe et al. (1958). **Schematic representation: d** Dependence mean effective stress σ_m^* – volumetric strain e_V; **e** Dependence specific volume $v - \sigma_m^*$ and $v - \ln \sigma_m^*$

General 3-D Loading. In order to establish material models for the material response in more general loading conditions, it is necessary to perform a series of adequate experiments. A rather general test for material response in 3-D loading conditions can be performed in a triaxial test device (Desai and Siriwardane 1984) which allows application of independent three normal stresses in three orthogonal directions, on six faces of a cubical specimen of a material. However, a special case of 3-D test commonly used in practice is the *triaxial compression test* which we describe in some detail. In this test a cylindrical material sample, shown in Fig 6.1.7**a**, is loaded by the axial stress σ_a and the radial stress σ_r, under drained or undrained conditions. As a consequence of the given normal stresses, shear stresses in the radial planes are generated. A special case of this test is the *conventional triaxial compression test*, in which case the radial stress is kept constant while the axial stress is increased.

We next introduce some mechanical quantities specific for the triaxial compression test, which can be related to stress and strain invariants necessary for general 3-D analyses. These specific quantities are the so-called "deviatoric" stress [5] q,

$$q = \sigma_a^* - \sigma_r^* \tag{6.1.6}$$

and the increment of "triaxial shear strain" de_q,

$$de_q = \frac{2}{3}(de_a - de_r) \tag{6.1.7}$$

The specific work dW for the material subjected to the triaxial compression test can be expressed in the form

$$dW = \sigma_a^* de_a + 2\sigma_r^* de_r = \sigma_m^* de_V + q\, de_q \tag{6.1.8}$$

Also, q and de_q represent the effective stress $\bar{\sigma}$ defined in (3.2.39), and the deviatoric effective strain increment $d\bar{e}$ (using (3.2.46) for the deviatoric strain increments)

$$q = \bar{\sigma} = \sqrt{3J_{2D}} \tag{6.1.9}$$

$$de_q = d\bar{e} \tag{6.1.10}$$

where J_{2D} is the second invariant of the deviatoric stresses defined in (3.2.13).

Typical results obtained in the test are shown in Figs. 6.1.7 to 6.1.9. It can be seen that sands and clays have hardening characteristics: the deviatoric stress q, the axial stress σ_a^* and the mean stress σ_m^* increase with the strains (Figs. 6.1.7 **b,c**; 6.1.8; 6.1.9**a,b**). The experimental results given in Fig. 6.1.8 show that at a certain stage of loading the material reaches a critical state

---

[5] Note that q is not a deviatoric stress as defined in (3.2.14)

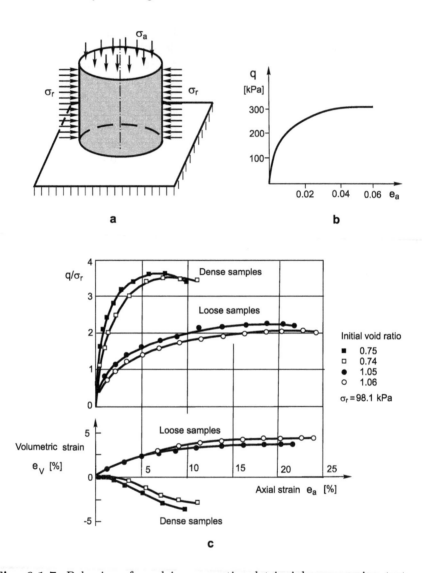

Fig. 6.1.7. Behavior of sand in conventional triaxial compression test. **a** Material sample ; **b** Dependence "deviatoric" stress q — axial strain e_a for Hostun sand, according to Kohata et al. (1997) **c** Dependence of the ratio q/σ_r and volumetric strain e_V on e_a for Chattahoochee River sand, drained conditions (according to Vesic and Clough 1968)

with the ultimate values q_{crit} and $(e_V)_{crit}$. Also, the material response is different for the loose and dense states. In the case of a loose state the deviatoric stress q and the volumetric strain increase continuously during the loading to point D; while in the case of the dense state the deviatoric stress increases,

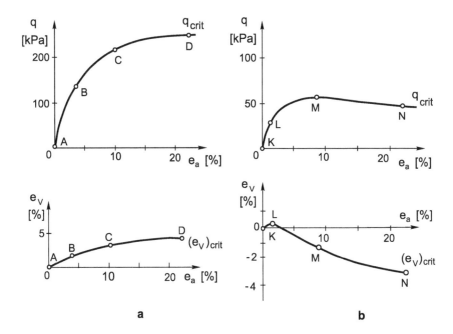

Fig. 6.1.8. Experimental results of Weald clay under conventional triaxial compression, drained and undrained conditions, according to Bishop and Henkel (1957); **a** The deviatoric stress $q-$ axial strain e_a, and the volumetric strain $e_V - e_a$ dependence for loose material; **b** The $q - e_a$ and $e_V - e_a$ dependence for dense material

reaches a maximum at a point M and then decreases to the critical value at point N, and the volumetric strain first increases (volume decrease) to a point L and then decreases to a critical value at point N. Sand also has the same character of the relation $e_V - e_a$ in loose and dense states, as can be seen in Fig. 6.1.7c.

Data shown in the $\sigma_m^* - q$ plane (Fig. 6.1.9c) demonstrate that yielding of the material depends on the preconsolidation pressure σ_{vc}. This pressure increases with the depth of the soil considered in the analysis. The yield curves drawn from the experimental results clearly indicate the hardening character in yielding under general loading conditions. It can be seen from Fig. 6.1.9d that to each yield curve corresponds a curve in the $\sigma_m^* - v$ plane.

Critical State. It can be concluded from the above description of the material behavior that there exists a *critical state* of the material which can be reached by various loading processes (see Figs. 6.1.5, 6.1.7 and 6.1.8). If loading continues after the critical state has been reached, the material behaves as perfectly plastic. The existence of the critical state in soils is con-

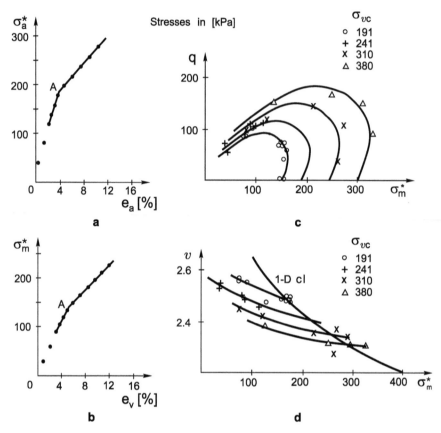

Fig. 6.1.9. Triaxial test results on undisturbed Winnipeg clay according to Graham et al. (1983). **a** Axial stress σ_a^* – axial strain e_a dependence; **b** Mean stress σ_m^* – volumetric strain e_V dependence; **c** Yield curves in $\sigma_m^* - q$ plane for various preconsolidation pressures σ_{vc}; **d** Results represented in the compression plane $\sigma_m^* - v$ (1-D cl is 1-D the compression line obtained in the uniaxial straining test)

sidered as one of the *crucial features of soil behavior* (Wood 1990; Atkinson 1993).

The critical states of the material obtained experimentally for various loading conditions can be represented by the *critical state line (csl)*. Figures 6.1.10a,b show the experimental results for the Weald clay in the $\sigma_m^* - q$, $\sigma_m^* - v$ and $\ln \sigma_m^* - v$ planes. It can be seen that to a given mean stress correspond the critical deviatoric stress q_{crit} and the value of the specific volume v_{crit} (and the void ratio e_{crit}), for the critical state. Note that the critical states lie on the same critical state line for the loose and dense states of the material.

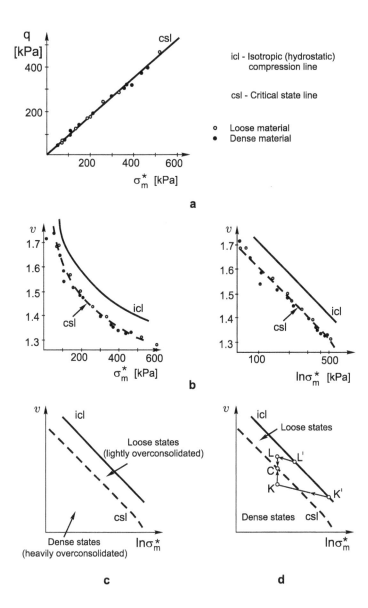

Fig. 6.1.10. Critical state of soil. Critical state line of Weald clay according to experiments by Roscoe et al. (1958): **a** In effective mean stress σ_m^*– deviatoric stress q plane; **b** In σ_m^*– specific volume v and in $\ln \sigma_m^* - v$ planes. Schematic: **c** Loose and dense state regimes in $\ln \sigma_m^* - v$ plane; **d** Shear of dense and loose material samples leading to same critical state

The states of the material above the critical state line correspond to loose or *lightly overconsolidated* states, while the points below the *csl*-line correspond to dense or *heavily overconsolidated* states, Figs. 6.1.10b,c,d . The material response is different for the loose and dense states, as already pointed out and shown in Figs. 6.1.5; 6.1.6a,b; 6.1.7c; and 6.1.8. To exemplify this difference we consider two material samples, one in the loose state and the other in the dense state, and assume that they are subjected to shear under the same constant mean stress, Fig 6.1.10d. The initial states of the samples are represented by the points L and K, which may be reached by different values of hydrostatic compression (to the points L' and K'), and then unloading. The volume of the loose material decreases, while the volume of the dense material increases, until the same specific volume is reached at the point C on the critical state line.

Plasticity Constitutive Relations. Figure 6.1.11 shows experimental data that illustrate the applicability of the associated flow rule (3.2.67b). The increments of plastic strain Δe^P are approximately normal to the yield surface of the undisturbed Winnipeg clay (Graham et al. 1983). However, it was found experimentally that for some soils, especially soils of sand-type, there can be a significant deviation from the normality rule. Then, the *nonassociated flow rule* should be used, with the potential Q instead of the yield function f_y, and instead of (3.2.67b) we have

$$de^P = d\lambda \frac{\partial Q}{\partial \sigma} \qquad (6.1.11)$$

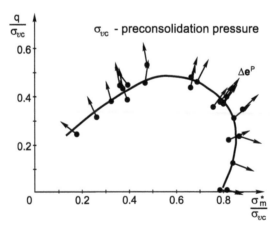

Fig. 6.1.11. Increments of plastic strain Δe^P obtained by triaxial test on undisturbed Winnipeg clay, according to Graham et al. (1983)

Various forms of the function Q have been proposed, see Desai and Siriwardane (1980b); Desai (1989).

6.1.3 Short Historical Review of Soil Plasticity

In Section 3.1 we outlined the history of metal plasticity. We mentioned that a seminal contribution to metal plasticity was made by Tresca who introduced the yield criterion for metals in 1864. However, the first yield criterion for soils was already formulated by Coulomb in 1773. This yield criterion has been used with the Mohr circle representation and is also called the Mohr-Coulomb yield criterion.

Important experimental investigations of soil behavior and theoretical considerations in the first half of the last century provided the basis for the formulation of soil models and solving engineering problems, e.g., Casagrande (1936); Rendulic (1936); Terzaghi (1936); Hvorslev (1937) (translated ref. (1969)).

Significant progress in soil plasticity was made by Drucker and Prager (Drucker and Prager 1952) who extended the Coulomb yield criterion to use of a smooth yield surface in three-dimensional soil mechanics problems. The yield criterion for soils formulated by Drucker and Prager has been used in the analysis of many soil plasticity problems. However, the application of the Drucker-Prager model with the constitutive relations based on the normality rule (3.2.71) leads to a prediction of excessive volume dilatation.

Drucker, Gibson and Henkel (Drucker et al. 1957) introduced the concept of work-hardening plasticity for soils. They defined yield surfaces consisting of a portion corresponding to perfectly plastic response, like an ultimate failure yield surface, and a cap with work-hardening characteristics. The void ratio, or the plastic volumetric strain, was used as the work-hardening parameter. Widely used models are the Drucker-Prager cap model (Drucker et al. 1957) and the generalized cap model (DiMaggio and Sandler 1971; Sandler et al. 1976).

Another approach proposed to model soil response more accurately is based upon introducing non-associated plasticity models, with a plastic potential different from the yield function (e.g., Davis 1968).

The introduction of work-hardening to soil plasticity led to experimental investigations and the formulation of other soil plasticity models. Generally accepted models with work-hardening characteristics are those relying on the critical state concept, developed at Cambridge University in the U.K. (Roscoe et al. 1958, 1963a). The Cam-clay and so-called modified Cam-clay models (Roscoe et al. 1963b; Burland 1965; Roscoe and Burland 1968) are models of this group.

Other models are the so-called nested models (Mroz 1967; Iwan 1967) with a set of nested yield surfaces which come into contact during plastic deformation of the material. Conceptually similar models are the bounding surface models (Dafalias and Popov 1975, 1977; Krieg 1975; Mroz et al. 1978, 1979; Dafalias 1979).

A general form of yield condition, expressed as a polynomial of stress invariants, was examined by Desai and coworkers (Desai 1980a; Desai and Siriwardane 1984).

Descriptions of various additional geomechanical models, which also include time effects (viscoplastic or creep behavior) are given in a number of books, e.g., Christian and Desai (1977); Keedwell (1988); Pietruszczak and Pande (1989); Wood (1990); Atkinson (1993); Cristescu and Goida (1994); Terzaghi et al. (1996); Desai (2001). A very important material in geomechanics and structural design is concrete for which various material models have been introduced (e.g., Kotsovos and Pavlovic 1995).

6.2 Cap Models

In this section we first give a description of two cap models; the model based on the Drucker-Prager line with a plane cap and the so-called generalized cap model. We develop the numerical algorithms for the stress integrations and the calculation of the consistent tangent elastic-plastic matrices in Sections 6.2.2 and 6.2.3. Finally, in Section 6.2.4 we give some numerical solutions.

The cap models are applicable to soils (claylike and sands) and to rocks (e.g., Desai and Siriwardane 1984; Chen and Baladi 1985; Chen and Mizuno 1990).

6.2.1 Description of Cap Models

It was found experimentally that soils with cohesion can undergo plastic deformations not only when subjected to shear loading but also when subjected to hydrostatic compression. In addition, it was observed that in hydrostatic compression the plastic deformations workharden the material. The experiments show that soils may have a permanent increase in volume or the volumetric strain may be positive (decrease in volume), see Section 6.1.2. The plasticity models of the Mohr-Coulomb or Drucker-Prager type, proposed for soils with cohesion (see Section 6.1.3) predict plastic flow of the material under shear, but cannot predict the plastic flow under hydrostatic compression. Also, the application of these models leads to an excessive plastic volumetric expansion, inconsistent with experimental findings. In order to model the plastic deformations in hydrostatic loading and in general obtain a better correspondence between the model predictions and experimental observations, the cap models were introduced.

The yield surfaces of cap models consist of a fixed yield surface and a cap which changes its size during plastic deformations. We describe two cap models that will be used in Section 6.2.2; a model with the Drucker-Prager fixed surface and a plane cap, and a so-called generalized cap model, with a general shape of the fixed surface and a spherical cap.

Drucker-Prager Model with Plane Cap. The simplest model with a hardening cap is shown in Fig. 6.2.1a. It consists of the fixed Drucker-Prager line defined by

$$f_{DP} = -\alpha I_1 + \sqrt{J_{2D}} - k = 0 \qquad (6.2.1)$$

and the plane cap

$$f_C = I_1 - X = 0 \qquad (6.2.2)$$

where α and k are material constants, and I_1 and J_{2D} are the first invariant of stress and the second invariant of the stress deviator (see (3.2.11) and (3.2.13)). The position of the cap, X, depends on the volumetric plastic strain e_V^P,

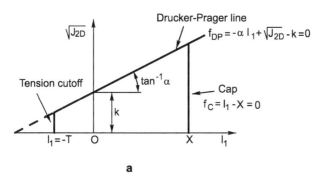

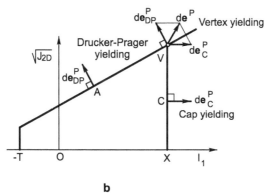

Fig. 6.2.1. Drucker-Prager model with plane cap and tension cutoff. **a** Yield surface; **b** Geometric interpretation of the three yielding regimes

$$X = X(e_V^P)$$

$$(6.2.3)$$

One possible form of the hardening law (6.2.3) is (DiMaggio and Sandler 1971; Sandler et al. 1976),

$$e_V^P = W\left[1 - \exp(-DX)\right]$$

$$(6.2.4)$$

where W and D are material constants. Since we consider the compressive stresses to be positive, W and D are positive. Note that the volumetric plastic strain rate with respect to the hydrostatic compressive stress decreases as plastic deformation progresses, i.e.,

$$\frac{\partial e_V^P}{\partial X} = WD \exp(-D\dot{X}) = WD \left(1 - \frac{e_V^P}{W}\right)$$

$$(6.2.5)$$

It follows from (6.2.4) and (6.2.5) that W represents the ultimate compressive volumetric strain

$$W = (e_V^P)_{\max}$$

$$(6.2.6)$$

while the constant D characterizes the volumetric strain rate with respect to the compressive hydrostatic stress,

$$D = \frac{\partial e_V^P}{\partial X}/(W - e_V^P)$$

$$(6.2.7)$$

The physical meaning of the material constants suggests their experimental determination. A detailed description of the experimental procedures for determination of the material constants is given, for example, in Desai and Siriwardane (1984).

The following form of the hardening law can be obtained from (6.2.4),

$$X = -\frac{1}{D}\ln(1 - \frac{e_V^P}{W}) + {}^0X$$

$$(6.2.8)$$

where ${}^0X$ represents the initial position of the cap. Figure 6.2.2 gives a graphical representation of the dependence $X(e_V^P)$. We see that as e_V^P approaches W, the maximum plastic volumetric strain, the hardening of the material becomes zero, that is, a perfectly plastic material response is reached.

In order to model the material failure due to tensile stresses, a *tension cutoff* condition can be employed. A simple concept is to set the stresses to

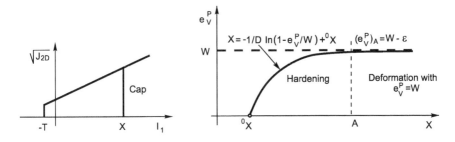

Fig. 6.2.2. Position of the cap X as a function of the volumetric plastic strain e_V^P

a maximum hydrostatic tension when the first invariant I_1 reaches a given value T (Sandler et al. 1976). Then the stress state is defined by

$$\sigma_{xx} = \sigma_{yy} = \sigma_{zz} = -\frac{1}{3}T$$
$$\sigma_{ij} = 0 \qquad i \neq j \tag{6.2.9}$$

and remains unchanged in subsequent deformations. Figure 6.2.1 shows the model with the tension cutoff conditions.

As shown in Fig. 6.2.1b, there are three yielding regimes:

a) Drucker-Prager yielding
b) cap yielding, and
c) vertex yielding

The *Drucker-Prager yielding* corresponds to stress states on the Drucker-Prager line, like the stress state A shown in Fig. 6.2.1b. The increment of plastic strain de_{DP}^P follows from the normality rule (3.2.67b) and the yield condition (6.2.1),

$$de_{DP}^P = -\alpha \, d\lambda_{DP} \, \mathbf{I} + \frac{d\lambda_{DP}}{2\sqrt{J_{2D}}} \, \mathbf{S} \tag{6.2.10}$$

where $d\lambda_{DP}$ is the proportionality factor and $\mathbf{I}$ is the identity matrix.

The *cap yielding* corresponds to the stress states on the cap, like the stress state C in Fig. 6.2.1b, and plastic deformation leads to a pure volumetric plastic compression. We obtain the increment of plastic strain de_C^P from the flow rule (3.2.67b) and the yield condition (6.2.2),

$$de_C^P = d\lambda_C \, \mathbf{I} \tag{6.2.11}$$

Hence, the increment of plastic volumetric strain is

$$de_V^P = 3\,d\lambda_C \qquad (6.2.12)$$

where $d\lambda_C$ is the proportionality factor corresponding to the cap yielding.

When the stress state is at the intersection V of the Drucker-Prager line and the cap, we have *vertex yielding*. To obtain the increment of plastic strain de^P we employ the relation (3.2.92), with the yield conditions (6.2.1) and (6.2.2). Therefore, we obtain

$$de^P = de_{DP}^P + de_C^P = d\lambda_{DP}\left(-\alpha\,\mathbf{I} + \frac{\mathbf{S}}{2\sqrt{J_{2D}}}\right) + d\lambda_C\,\mathbf{I} \qquad (6.2.13)$$

Figure 6.2.1b gives a geometric representation of this relation. It can be seen from (6.2.13) that an increase or decrease of the plastic volumetric strain may occur, depending on the direction of the stress increment at point V.

The proportionality coefficients $d\lambda_{DP}$ and $d\lambda_C$ depend on the stress states in the three plasticity regimes. The coefficients can be obtained from the condition that the stress point remains on the corresponding yield surfaces (see Section 6.2.2).

Generalized Cap Model. The generalized cap model described below is based on experimental investigations by DiMaggio and Sandler (1971), and Sandler et al. (1976). The elastic region in this model is bounded by two surfaces in the stress space (Fig. 6.2.3):

1) the failure surface $f_1 = 0$, and
2) the hardening cap $f_C = 0$.

In the plane $(I_1, \sqrt{J_{2D}})$ the failure surface is defined as

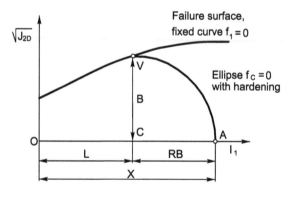

Fig. 6.2.3. Generalized cap model

$$f_1 = \sqrt{J_{2D}} - k + A \exp(-B_1 I_1) - \alpha I_1 = 0 \qquad (6.2.14)$$

where A, B_1, k and α are material constants. Note that the term $(-\alpha I_1)$ is included in the original formulation so that we have the Drucker-Prager line (6.2.1) when $A = 0$ (Desai and Siriwardane 1984; Kojic et al. 1995e).

The cap is defined by an ellipse whose equation is

$$f_C = (I_1 - L)^2 + R^2(J_{2D} - B^2) = 0 \qquad (6.2.15)$$

where L defines the center of the ellipse, B one semiaxis, and R the ratio between the semiaxes

$$R = (X - L)/B \qquad (6.2.16)$$

Here X is the intersection of the cap with the I_1 axis. It was found experimentally that the ratio R can be related to L as (DiMaggio and Sandler 1971)

$$R = \frac{R_0}{1 + R_1} [1 + R_1 \exp(-R_2 L)] + R_3 \left[\exp\left(-R_4(R_5 - L)^2\right)\right] \qquad (6.2.17)$$

where $R_0, R_1, \ldots, R_5$ are material constants.

The material exhibits work-hardening behavior with a change of the cap size. The hardening parameter is the volumetric plastic strain e_V^P, and the hardening law is given by

$$e_V^P = W \left[1 - \exp(-D\,X) - \beta\,D\,X\,\exp(-D_1 X - D_2 X^2)\right] \\ + W_1\,X^2\,\exp(-D_3 X) \qquad (6.2.18)$$

where $W, W_1, \beta, D, D_1, D_2$ and D_3 are additional material constants. Note that the hardening law has the character shown in Fig. 6.2.2, and that the expression (6.2.4) is a special case of (6.2.18).

6.2.2 Stress Integration Procedure

In this section we develop the computational procedure of implicit stress integration for the generalized cap model (Kojic et al. 1995e). The procedure is applicable to the Drucker-Prager cap model as a special case. We consider general three-dimensional deformations and use the derivations of Section 4.2.

For a time (load) step Δt we first determine the elastic solutions $^{t+\Delta t}\mathbf{S}^E$ and $^{t+\Delta t}\sigma_m^E$ for the deviatoric and mean stresses corresponding to given strains $^{t+\Delta t}\mathbf{e}$ and the known plastic strain $^t\mathbf{e}^P$ (see (A1.19) and (A1.12)),

$$^{t+\Delta t}\mathbf{S}^E = 2G\ ^{t+\Delta t}\mathbf{e}''$$

(6.2.19)

$$^{t+\Delta t}\sigma_m^E = c_m\ ^{t+\Delta t}e_m'' \qquad \text{no sum on } m$$

(6.2.20)

where

$$^{t+\Delta t}\mathbf{e}'' = {}^{t+\Delta t}\mathbf{e}' - {}^t\mathbf{e}'^P$$

(6.2.21)

$$^{t+\Delta t}e_m'' = {}^{t+\Delta t}e_m - {}^te_m^P$$

(6.2.22)

Here $^t\mathbf{e}'^P$ is the deviatoric plastic strain tensor and $^te_m^P$ is the mean plastic strain.

The bulk modulus K and the shear modulus G may depend on the current stress/strain state. Then, instead of (6.2.19) and (6.2.20) we have

$$^{t+\Delta t}\mathbf{S}^E = {}^t\mathbf{S} + 2\ ^tG\Delta\mathbf{e}'$$

(6.2.23)

$$^{t+\Delta t}\sigma_m^E = {}^t\sigma_m + {}^tc_m\,\Delta e_m$$

(6.2.24)

where $\Delta\mathbf{e}'$ and Δe_m are the increments of the deviatoric strain and mean strain in the time step; and ^tG and ^tc_m are the moduli corresponding to the start of the time step. Note that weighted values may be employed in the time step, based on ^tG and ^tc_m as well as $^{t+\Delta t}G$ and $^{t+\Delta t}c_m$. In order to simplify the derivations we adopt ^tG and ^tc_m, as it is usual in geomechanics (e.g., Borja and Lee 1990).

It is common for geomaterials to take the bulk modulus and the shear modulus dependent on the stress/strain state (these dependencies are determined experimentally), and use these in the incremental "elastic" constitutive matrix. One approach is to use a variable bulk modulus and a constant Poisson ratio. Then the shear modulus is variable and follows from the relations (A1.13) and (A1.9). Various other approaches are discussed in, e.g., Naylor and Pande (1981); Desai and Siriwardane (1984); Britto and Gunn (1987); Chen and Mizuno (1990); Wood (1990).

The next step is to check for yielding in the time (load) step. We substitute the elastic deviatoric and mean stresses $^{t+\Delta t}\mathbf{S}^E$ and $^{t+\Delta t}\sigma_m^E$ into the expressions (6.2.14) and (6.2.15), using ^tL, ^tR, ^tB from the previous step and obtain $^{t+\Delta t}f_1^E$ and $^{t+\Delta t}f_C^E$. When $^{t+\Delta t}f_1^E \leq 0$ and $^{t+\Delta t}f_C^E \leq 0$, the deformation in the current step is elastic. Otherwise, we have elasto-plastic deformations in the current load step.

Yielding on the Failure Surface. We consider that yielding on the failure surface $f_1 = 0$ occurs when $^{t+\Delta t}f_1^E > 0$ and $(3\ ^{t+\Delta t}\sigma_m^E) < \ ^tL$ (see Fig. 6.2.4). Then the increments of the mean and deviatoric plastic strains Δe_m^P and $\Delta \mathbf{e}'^P$ are

$$\Delta e_m^P = -\Delta\lambda \left[B_1\, A \exp\left(-B_1\ ^{t+\Delta t}I_1 \right) + \alpha \right] \tag{6.2.25}$$

$$\Delta \mathbf{e}'^P = \frac{\Delta\lambda}{2}\ ^{t+\Delta t}J_{2D}^{-1/2}\ ^{t+\Delta t}\mathbf{S} \tag{6.2.26}$$

where we have employed the flow rule (4.2.10) and $^{t+\Delta t}f_1$ given by (6.2.14). It follows from (6.2.25) that $\Delta e_m^P < 0$ (*volume expansion*). This character of Δe_m^P can be seen from Fig. 6.2.4 (the projection of $\Delta \mathbf{e}^P$ on the I_1 axis is negative). The mean stress $^{t+\Delta t}\sigma_m$ and deviatoric stresses $^{t+\Delta t}\mathbf{S}$ can be obtained using (4.2.15) and the last two equations, as

$$^{t+\Delta t}\sigma_m = \ ^{t+\Delta t}\sigma_m^E - \ ^tc_m\,\Delta e_m^P \qquad \text{no sum on } m \tag{6.2.27}$$

$$^{t+\Delta t}\mathbf{S} = \frac{^{t+\Delta t}\mathbf{S}^E}{1 + \Delta\lambda\,^tG\ ^{t+\Delta t}J_{2D}^{-1/2}} \tag{6.2.28}$$

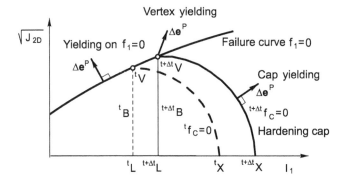

Fig. 6.2.4. Notation in stress integration using the generalized cap model

and hence

$$t+\Delta t\, J_{2D}^{1/2} = \left(t+\Delta t\, J_{2D}^{E}\right)^{1/2} - \Delta\lambda\, {}^{t}G \tag{6.2.29}$$

where $t+\Delta t\, J_{2D}^{E}$ corresponds to $t+\Delta t\mathbf{S}^{E}$. Finally, the yield condition (6.2.14) must be satisfied at the end of the time step,

$$\boxed{t+\Delta t\, f_1 = \; t+\Delta t\, J_{2D}^{1/2} - k + A\exp(-B_1\; t+\Delta t\, I_1) - \alpha\; t+\Delta t\, I_1 = 0}$$

$$(6.2.30)$$

Summarizing the above relations we see that in order to find the unknown quantities at the end of the time step, we need to solve the nonlinear equation (6.2.30) with respect to the governing parameter Δe_m^P (see Table 4.2.1). Namely, for a given Δe_m^P we calculate $t+\Delta t\sigma_m$ from (6.2.27), $\Delta\lambda$ from (6.2.25), and $t+\Delta t\, J_{2D}^{1/2}$ from (6.2.29). We repeat this sequence of calculation until (6.2.30) is satisfied within a selected tolerance. When the solution for the governing parameter Δe_m^P has been obtained, $t+\Delta t\mathbf{S}$ and $\Delta\mathbf{e}'^{P}$ follow from (6.2.28) and (6.2.26). The position $t+\Delta t X$ can be obtained from the hardening law (6.2.18). Table 6.2.1 lists the computational steps for these calculations.

Table 6.2.1. Computational steps for yielding on the failure surface

---

1. **Assume Δe_m^P**
2. **Calculate** $t+\Delta t\sigma_m$ from (6.2.27), $\Delta\lambda$ from (6.2.25),
   $t+\Delta t\, J_{2D}^{1/2}$ from (6.2.29)
3. **Calculate** $t+\Delta t\, f_1$ according to (6.2.30)
   If $\left|t+\Delta t\, f_1\right| \leq \varepsilon$ go to step 4; otherwise go to step 1 with a new trial value of Δe_m^P
4. **Final calculations:** $t+\Delta t\mathbf{S}$ from (6.2.28)
   $\Delta\mathbf{e}'^{P}$ from (6.2.26)

Note that we determined the mean stress such that the stress point in the $I_1 - \sqrt{J_{2D}}$ plane is on the curve $f_1 = 0$. Of course, the stress integration presented above is applicable to the Drucker-Prager model (with yielding on the Drucker-Prager line (6.2.1)).

Since $\Delta e_m^P < 0$ we have $t+\Delta t X < {}^t X$, therefore the plastic flow on $f_1 = 0$ leads to a decrease of the cap size and represents *strain softening*. This type of material behavior is illustrated in Example 6.2.2. There are some materials, such as rocks, for which a cap model with no strain softening behavior is adequate (Chen and Mizuno 1990). Then, during yielding on the failure surface the elastic domain in the stress space does not decrease, and the cap position remains unchanged.

Cap Yielding. We consider that cap yielding occurs in the time step when $^{t+\Delta t}f_C^E > 0$, $(3\,^{t+\Delta t}\sigma_m^E) > \,^tL$, and $^{t+\Delta t}f_1^E \leq 0$. The elastic stress point is below or on the failure surface $f_1 = 0$ and to the right of the line $I_1 = \,^tL$ (see Fig. 6.2.4). The computational steps are analogous to the above steps for yielding on the failure surface. From (4.2.10) and (6.2.15) we have

$$\Delta e_m^P = 2\Delta\lambda_C \left(3\,^{t+\Delta t}\sigma_m - \,^{t+\Delta t}L\right) \qquad (6.2.31)$$

$$\Delta \mathbf{e}'^P = \Delta\lambda_C \,^{t+\Delta t}R^2 \,^{t+\Delta t}\mathbf{S} \qquad (6.2.32)$$

where $\Delta\lambda_C$ is a positive scalar. From (A1.19), (6.2.23) and (6.2.32) we solve for $^{t+\Delta t}\mathbf{S}$ as

$$^{t+\Delta t}\mathbf{S} = \,^{t+\Delta t}\mathbf{S}^E / \left(1 + 2\Delta\lambda_C \,^tG \,^{t+\Delta t}R^2\right) \qquad (6.2.33)$$

and further, in analogy with the derivation of (6.2.29), we obtain

$$^{t+\Delta t}J_{2D} = \,^{t+\Delta t}J_{2D}^E / \left(1 + 2\Delta\lambda_C \,^tG \,^{t+\Delta t}R^2\right)^2 \qquad (6.2.34)$$

Assuming a value for ΔL, we have $^{t+\Delta t}L = \,^tL + \Delta L$, and then from (6.2.17) we obtain $^{t+\Delta t}R$. Also, using Fig. 6.2.4 and the expression (6.2.14), we obtain

$$^{t+\Delta t}B = k - A\exp(-B_1 \,^{t+\Delta t}L) + \alpha\,^{t+\Delta t}L \qquad (6.2.35)$$

where we have substituted $I_1 = \,^{t+\Delta t}L$ in (6.2.14). It follows from (6.2.16) that

$$^{t+\Delta t}X = \,^{t+\Delta t}R\,^{t+\Delta t}B + \,^{t+\Delta t}L \qquad (6.2.36)$$

and then the volumetric plastic strain $^{t+\Delta t}e_V^P = 3\,^{t+\Delta t}e_m^P$ can be obtained from (6.2.18). Here the governing parameter is ΔL and the computational steps are as given in Table 6.2.2.

The above stress integration procedure for cap yielding is also applicable to the plane cap case; but the following changes must be used. Firstly, we have $^{t+\Delta t}L = \,^{t+\Delta t}X$ and consequently $^{t+\Delta t}R = 0$. It follows from (6.2.32)

Table 6.2.2. Computational steps for cap yielding

---

1. Assume a value for ΔL

2. Calculate $^{t+\Delta t}B$ from (6.2.35), $^{t+\Delta t}R$ from (6.2.17), $^{t+\Delta t}X$
from (6.2.36), $^{t+\Delta t}e_V^P$ from (6.2.18), $^{t+\Delta t}\sigma_m$ from (6.2.27),
$\Delta\lambda_C$ from (6.2.31), and $^{t+\Delta t}J_{2D}$ from (6.2.34)

3. Calculate $^{t+\Delta t}f_C$ according to (6.2.15)

If $|^{t+\Delta t}f_C| < \varepsilon$ go to step 4; otherwise go to step 1 with a new
trial value ΔL

4. Final calculations: $^{t+\Delta t}\mathbf{S}$ from (6.2.33)
$\Delta\mathbf{e}'^P$ from (6.2.32)

---

that $\Delta\mathbf{e}'^P = 0$. Secondly, the increment Δe_m^P can be determined from the
condition that the expression in the parentheses of (6.2.31) is equal to zero
and using (6.2.27) and the hardening law (6.2.18). Note that now $\Delta\lambda_C = \Delta e_m^P$
(see (6.2.12)).

Vertex Yielding. Yielding occurs at the vertex, that is, at the intersection between the two yield surfaces, when:

a) $3\,^{t+\Delta t}\sigma_m^E = \,^{t+\Delta t}I_1^E = \,^tL$ and $\,^{t+\Delta t}f_1^E > 0$, or

b) $\,^{t+\Delta t}I_1^E > \,^tL$ and $\,^{t+\Delta t}f_1^E > 0$

In case a) the elastic stress point is on the line $I_1 = \,^tL$ (within a selected
numerical tolerance), while in case b) the point is on the right from this line
and above the failure line (see Fig. 6.2.4).

The analysis given in Example 6.2.1 shows that case a) reduces to
cap yielding. Then we have $\,^{t+\Delta t}I_1 = \,^tI_1 = \,^tL$, $\,^{t+\Delta t}R = \,^tR$, $\Delta e_m^P = 0$, $\,^{t+\Delta t}J_{2D} = \,^tJ_{2D} = \,^tB^2$, and $\Delta\lambda_C$ can be determined from (6.2.34),

$$\Delta\lambda_C = \frac{1}{2\,^tG\,^tR^2}\left(\sqrt{^{t+\Delta t}J_{2D}^E\,/\,^tB} - 1\right) \qquad (6.2.37)$$

Then $\,^{t+\Delta t}\mathbf{S}$ follows from (6.2.33) and $\Delta\mathbf{e}'^P$ can be calculated from (6.2.32).
The yielding corresponds to a *perfectly plastic response* with the shear plastic
flow (no change of the volumetric plastic strain) at the fixed stress point in
the plane $(I_1, \sqrt{J_{2D}})$.

Case b) is considered in Example 6.2.1.

6.2.3 Elastic-Plastic Matrices

We follow the derivation of the tangent moduli using the general concepts of
Section 4.2 (see also Section 4.4.3). Hence, using the one-index notation, we
have

$$
\begin{aligned}
{}^{t+\Delta t}C_{ij}^{EP} &= \frac{\partial\, {}^{t+\Delta t}S_i}{\partial\, {}^{t+\Delta t}e_j} + \frac{\partial\, {}^{t+\Delta t}\sigma_m}{\partial\, {}^{t+\Delta t}e_j} \qquad i = 1,2,3;\ j = 1,2,\ldots,6 \\[2mm]
{}^{t+\Delta t}C_{ij}^{EP} &= \frac{\partial\, {}^{t+\Delta t}S_i}{\partial\, {}^{t+\Delta t}e_j} \qquad\qquad\qquad i = 4,5,6;\ j = 1,2,\ldots,6
\end{aligned}
$$

$$(6.2.38)$$

Since we have three regimes of plastic deformations, we have three sets of expressions for ${}^{t+\Delta t}C_{ij}^{EP}$.

Yielding on the Failure Surface. Using (6.2.28) we have

$$
{}^{t+\Delta t}S_{i,j} = \frac{1}{D_\lambda}\left({}^{t+\Delta t}S_{i,j}^{E} - {}^{t+\Delta t}S_i D_{\lambda,j}\right)
$$

$$(6.2.39)$$

where we use the notation $_{,j} \equiv \partial/\partial^{t+\Delta t}e_j$, and

$$
D_\lambda = 1 + \Delta\lambda\,{}^tG\,{}^{t+\Delta t}J_{2D}^{-1/2}
$$

$$(6.2.40)$$

It follows from (6.2.23) that

$$
\begin{aligned}
{}^{t+\Delta t}S_{i,j}^{E} &= 2\,{}^tG\,A_{ij}, \quad i,j = 1,2,3 \\[1mm]
{}^{t+\Delta t}S_{i,j}^{E} &= {}^tG\,\delta_{ij} \qquad i = 4,5,6;\ j = 1,2,\ldots,6
\end{aligned}
$$

$$(6.2.41)$$

where the matrix $\mathbf{A}$ is given by (3.2.70). Note that we use here $e_j = 2e'_j$, $j = 4,5,6$ (see 4.4.52). By differentiation of D_λ and with use of (6.2.29) we obtain

$$
D_{\lambda,j} = b_0\,\Delta\lambda_{,j} - b_j^{E}
$$

$$(6.2.42)$$

where

$$
\begin{aligned}
b_0 &= {}^tG\left({}^{t+\Delta t}J_{2D}^{-1/2} + \Delta\lambda\,{}^tG\,{}^{t+\Delta t}J_{2D}^{-1}\right) \\[1mm]
b_j^{E} &= \Delta\lambda\,{}^tG^2\,{}^{t+\Delta t}J_{2D}^{-1}\left({}^{t+\Delta t}J_{2D}^{E}\right)^{-1/2}{}^{t+\Delta t}S_j^{E}
\end{aligned}
$$

$$(6.2.43)$$

For the determination of b_j^{E} we have used the derivatives

$$
{}^{t+\Delta t}J_{2D,j}^{E} = 2\,{}^tG\,{}^{t+\Delta t}S_j^{E}
$$

$$(6.2.44)$$

which follow from (6.2.23), and the definition (3.2.13). Using (6.2.25), (6.2.27) and (6.2.24) it follows that

$$\Delta\lambda_{,j} = d_j - d_0 \Delta e^P_{m,j}$$

(6.2.45)

where

$$d_j = a_j/c_0$$
$$d_0 = (1 + a_0)/c_0$$

(6.2.46)

with

$$c_0 = B_1 A \exp\left(-B_1 \, {}^{t+\Delta t}I_1\right) + \alpha$$
$$a_0 = 3\,\Delta\lambda\, c_m\, A\, B_1^2 \exp\left(-B_1 \, {}^{t+\Delta t}I_1\right)$$
$$a_j = a_0/3 \qquad j = 1, 2, 3$$
$$a_j = 0 \qquad j = 4, 5, 6$$

(6.2.47)

Finally, by differentiation of (6.2.30) and using the above expressions for the derivatives with respect to ${}^{t+\Delta t}e_j$ we obtain $\Delta e^P_{m,j}$ as

$$\Delta e^P_{m,j} = -\frac{W_j}{{}^{t+\Delta t}f'_1}$$

(6.2.48)

where

$${}^{t+\Delta t}f'_1 = \frac{\partial\, {}^{t+\Delta t}f_1}{\partial\left(\Delta e^P_m\right)} = {}^tG\left[c_0 - 3\,{}^tc_m B_1 (c_0 - \alpha)\Delta e^P_m\right]/c_0^2 + 3\,{}^tc_m c_0$$

$$W_j = {}^{t+\Delta t}f_{1,j} = {}^tG\left[\left({}^{t+\Delta t}J^E_{2D}\right)^{-1/2} \, {}^{t+\Delta t}S^E_j - d_j\right] - c_0\,\bar{a}_j$$

(6.2.49)

with

$$\bar{a}_j = {}^tc_m \qquad j = 1, 2, 3$$
$$\bar{a}_j = 0 \qquad j = 4, 5, 6$$

(6.2.50)

When $\Delta e^P_{m,j}$ is determined from (6.2.48), we calculate $\Delta\lambda_{,j}$ from (6.2.45), and $D_{\lambda,j}$ from (6.2.42). Then, with use of (6.2.41), we determine ${}^{t+\Delta t}S_{i,j}$ from (6.2.39) and substitute into (6.2.38).

Then (6.2.27) and (6.2.24) give

$$^{t+\Delta t}\sigma_{m,j} = \left(\frac{1}{3} - \Delta e^P_{m,j}\right){}^t c_m \quad j = 1,2,3 \quad \text{no sum on m}$$

$$^{t+\Delta t}\sigma_{m,j} = -\,{}^t c_m\,\Delta e^P_{m,j} \qquad j = 4,5,6$$

(6.2.51)

Using (6.2.48) the values of $^{t+\Delta t}\sigma_{m,j}$ can be computed and then substituted into (6.2.38).

Cap Yielding. The derivatives $^{t+\Delta t}S_{i,j}$ have the form (6.2.39), but instead of D_λ we have D_C given as

$$D_C = 1 + 2\Delta\lambda_C\,{}^t G\,{}^{t+\Delta t}R^2$$

(6.2.52)

which follows from (6.2.33). Further,

$$D_{C,j} = 2\,{}^t G\,{}^{t+\Delta t}R^2\,\Delta\lambda_{C,j} + 4\Delta\lambda_C\,{}^t G\,{}^{t+\Delta t}R\,{}^{t+\Delta t}R_{,j}$$

(6.2.53)

The governing parameter is now ΔL, or $^{t+\Delta t}L$, and we need $^{t+\Delta t}L_{,j}$. The relations (6.2.18) and (6.2.31) give

$$\Delta e^P_{m,j} = \Delta e^P_{m,X}\,{}^{t+\Delta t}X'\,{}^{t+\Delta t}L_{,j}$$

(6.2.54)

$$\Delta\lambda_{C,j} = q_0\,{}^{t+\Delta t}L_{,j} - q_j$$

(6.2.55)

where

$$\Delta e^P_{m,X} = \frac{1}{3}\frac{\partial\,{}^{t+\Delta t}e^P_V}{\partial\,{}^{t+\Delta t}X}$$

(6.2.56)

$$^{t+\Delta t}X' = \frac{\partial\,{}^{t+\Delta t}X}{\partial\,{}^{t+\Delta t}L} = {}^{t+\Delta t}R'\,{}^{t+\Delta t}B + {}^{t+\Delta t}R\,{}^{t+\Delta t}B' + 1$$

(6.2.57)

$$q_0 = \frac{2\Delta\lambda_C + (1 + 6\,\Delta\lambda_C\,{}^t c_m)\,\Delta e^P_{m,X}\,{}^{t+\Delta t}X'}{2\,(\,{}^{t+\Delta t}I_1 - {}^{t+\Delta t}L)}$$

$$q_j = \frac{\Delta\lambda_C}{{}^{t+\Delta t}I_1 - {}^{t+\Delta t}L}\,\bar{a}_j$$

(6.2.58)

The derivatives $\Delta e^P_{m,X}$, $^{t+\Delta t}R' = \partial^{t+\Delta t}R/\partial^{t+\Delta t}L$ and $^{t+\Delta t}B' = \partial^{t+\Delta t}B/\partial^{t+\Delta t}L$ can be obtained from (6.2.18), (6.2.17) and (6.2.35), respectively. We next determine the derivatives $^{t+\Delta t}J_{2D,j}$ from (6.2.34) as

$$^{t+\Delta t}J_{2D,j} = -\bar{q}_0\,^{t+\Delta t}L_{,j} + \bar{q}_j \qquad (6.2.59)$$

where

$$\bar{q}_0 = \frac{4\,^tG\,^{t+\Delta t}R\,^{t+\Delta t}J_{2D}}{D_C}\left(q_0\,^{t+\Delta t}R + 2\Delta\lambda_C\,^{t+\Delta t}R'\right)$$

$$\bar{q}_j = \frac{2\,^tG}{D_C^2}\left(^{t+\Delta t}S^E_j + 2\,^{t+\Delta t}R^2\,^{t+\Delta t}J_{2D}D_C\,q_j\right) \qquad (6.2.60)$$

Finally, differentiating (6.2.15) with respect to $^{t+\Delta t}e_j$, and employing the relations $^{t+\Delta t}B_{,j} = {}^{t+\Delta t}B'\,^{t+\Delta t}L_{,j}$, $^{t+\Delta t}R_{,j} = {}^{t+\Delta t}R'\,^{t+\Delta t}L_{,j}$ and also (6.2.27), (6.2.54) and (6.2.59), we obtain

$$^{t+\Delta t}L_{,j} = -\frac{p_j}{f'_C} \qquad (6.2.61)$$

where

$$p_j = 2\left(^{t+\Delta t}I_1 - {}^{t+\Delta t}L\right)\bar{a}_j + {}^{t+\Delta t}R^2\,\bar{q}_j$$
$$f'_C = -2\left(^{t+\Delta t}I_1 - {}^{t+\Delta t}L\right)\left(3\,^tc_m\,\Delta e^P_{m,X}\,^{t+\Delta t}X' + 1\right)$$
$$- {}^{t+\Delta t}R^2\left(\bar{q}_0 + 2\,^{t+\Delta t}B\,^{t+\Delta t}B'\right)$$
$$+ 2\,^{t+\Delta t}R\left(^{t+\Delta t}J_{2D} - {}^{t+\Delta t}B^2\right)^{t+\Delta t}R' \qquad (6.2.62)$$

With the known values $^{t+\Delta t}L_{,j}$ we can calculate all other derivatives in (6.2.39).

The derivatives $^{t+\Delta t}\sigma_{m,j}$ follow from (6.2.51) where we use $\Delta e^P_{m,j}$ from (6.2.54) and $^{t+\Delta t}L_{,j}$ from (6.2.61).

With the derivatives $^{t+\Delta t}S_{i,j}$ and $^{t+\Delta t}\sigma_{m,j}$ determined, we have the elastic-plastic matrix according to (6.2.38).

Vertex yielding. With the assumption a), we have $\Delta e^P_m = 0$, $\Delta e^P_{m,j} = 0$, and $^{t+\Delta t}\sigma_{m,j}$ follows from (6.2.51). Also, we have $^{t+\Delta t}R = {}^tR$, $^{t+\Delta t}R_{,j} = 0$, and from (6.2.53) we obtain

$$D_{C,j} = 2\,^tG\,^tR^2\Delta\lambda_{C,j} \qquad (6.2.63)$$

The derivatives $\Delta\lambda_{C,j}$ follow from (6.2.37) as

$$\Delta\lambda_{C,j} = \frac{^{t+\Delta t}S_j^E}{2\,^tR^2\,^tB\,\sqrt{^{t+\Delta t}J_{2D}^E}} \qquad (6.2.64)$$

where (6.2.44) is used. With $\Delta\lambda_{C,j}$ determined, we calculate $D_{C,j}$ from (6.2.63), and $^{t+\Delta t}S_{i,j}$ from (6.2.39), where instead of D_λ and $D_{\lambda,j}$ we employ the last expressions for D_C and $D_{C,j}$.

The derivation of $^{t+\Delta t}\mathbf{C}^{EP}$ for the assumption b) is given in Example 6.2.1.

6.2.4 Examples

Example 6.2.1. Vertex Yielding with Contributions from Both Yield Surfaces. Derive the expressions for the stress integration of the generalized cap model for vertex yielding, taking into account the contributions to plastic flow of the failure surface and the cap.

As stated in Section 6.2.2, the case when (case b))

$$3^{t+\Delta t}\sigma_m^E > {}^tL, \quad {}^{t+\Delta t}f_1^E > 0 \qquad (a)$$

can be considered as vertex yielding, with contributions of both yield surfaces to plastic flow in the load step. We develop an implicit stress integration procedure for this case.

The basic relation for the increment of plastic strain $\Delta\mathbf{e}^P$ is now (3.2.92). With use of (6.2.26) and (6.2.32), we obtain

$$\Delta\mathbf{e}'^P = \left(\frac{\Delta\lambda_F}{2}\,^{t+\Delta t}J_{2D}^{-1/2} + \Delta\lambda_C\,R^2\right){}^{t+\Delta t}\mathbf{S} = \Delta\bar{\lambda}\,^{t+\Delta t}\mathbf{S} \qquad (b)$$

where $\Delta\lambda_F$ and $\Delta\lambda_C$ correspond to the failure surface and the cap, while $\Delta\bar{\lambda}$ is a positive scalar multiplying $^{t+\Delta t}\mathbf{S}$. Hence

$$^{t+\Delta t}\mathbf{S} = {}^{t+\Delta t}\mathbf{S}^E/(1 + 2\,^tG\Delta\bar{\lambda}) \qquad (c)$$

Using this equation we obtain

$$^{t+\Delta t}J_{2D}^{1/2} = \left({}^{t+\Delta t}J_{2D}^E\right)^{1/2}/(1 + 2\,^tG\Delta\bar{\lambda}) \qquad (d)$$

Since at the vertex V we have $^{t+\Delta t}J_{2D} = {}^{t+\Delta t}B$, the last equation can be solved with respect to $\Delta\bar{\lambda}$,

$$\Delta\bar{\lambda} = \left[\frac{\left(^{t+\Delta t}J^E_{2D}\right)^{1/2}}{^{t+\Delta t}B} - 1\right] / (2\,^tG) \tag{e}$$

Also, the expressions (6.2.35) and (6.2.36) are applicable, with the quantities that define the failure surface and the cap.

In addition we employ the relation

$$^{t+\Delta t}I_1 = \,^{t+\Delta t}L \tag{f}$$

to enforce the condition that the stress point at the end of the step lies at the intersection of the two yield surfaces. The next relation to be satisfied is the elastic constitutive law (6.2.27), which, with use of (f), gives

$$\Delta e^P_V = \left(^{t+\Delta t}I^E_1 - \,^{t+\Delta t}L\right) / \,^tc_m \tag{g}$$

where $^{t+\Delta t}I^E_1 = 3\,^{t+\Delta t}\sigma^E_m$.

Note that when $^{t+\Delta t}I^E_1 = \,^tL$ we have $\Delta e^P_V = 0$, with $\Delta \mathbf{e}'^P$ the same as in (6.2.32). Therefore we have the vertex yielding considered as case a) in Section 6.2.2.

Finally, the hardening law (6.2.18) must be satisfied, which, with use of (6.2.35), (6.2.36) and (g), reduces to

$$\begin{aligned}
f(\Delta L) = 0 = \,^te^P_V + \left(^{t+\Delta t}I^E_1 - \,^{t+\Delta t}L\right) / \,^tc_m &- W\left[1 - \exp\left(-D\,^{t+\Delta t}X\right)\right.\\
&\left.- \beta D\,^{t+\Delta t}X \exp\left(-D_1\,^{t+\Delta t}X - D_2\,^{t+\Delta t}X^2\right)\right]\\
&- W_1\,^{t+\Delta t}X^2 \exp\left(-D_3\,^{t+\Delta t}X\right)
\end{aligned} \tag{h}$$

Hence the governing parameter is, as in the case of cap yielding ΔL and using (f) $\Delta L = \Delta I_1 = 3\Delta\sigma_m$. In summary, we can state that all fundamental relations are satisfied:

- the elastic constitutive relations (6.2.23) and (6.2.24);
- the yield conditions (6.2.14) and (6.2.15) at the end of load step;
- the flow rule of the form (3.2.92).

Note that Δe^P_m is calculated from the elastic constitutive law (g) and the hardening law (h), rather than from the flow rule (3.2.92). As shown in Table E.6.2-1, after $\Delta L = \Delta I_1$ is determined, $\Delta\bar{\lambda}$ is calculated from (e), $^{t+\Delta t}\mathbf{S}$ from (c), and $\Delta\mathbf{e}'^P$ from (b). The above stress calculation can also be used when the vertex yielding occurs at the Drucker-Prager failure line ($A = 0$ should be used in (6.2.35)). The procedure can also be employed for the plane cap; then step 2 in Table E.6.2-1 should be skipped, and $^{t+\Delta t}B$ should be calculated once, from (6.2.35), after $^{t+\Delta t}L$ is determined.

The consistent tangent elastic-plastic matrix $^{t+\Delta t}\mathbf{C}^{EP}$ can be determined in accordance with Section 6.2.3. Following the details given for cap yielding, the denominator (6.2.52) is now

Table E.6.2-1. Computational steps for vertex yielding (contribution of both yield surfaces)

---

a) **Iterations on ΔL**

    1. Assume $\Delta I_1 = \Delta L$

    2. Calculate $^{t+\Delta t}R$ from (6.2.17), $^{t+\Delta t}B$ from (6.2.35), and $^{t+\Delta t}X$ from (6.2.36)

    3. Iterate until $f(\Delta L)$, given by (h), is satisfied

b) **Final calculations**

    Calculate: $\Delta\bar{\lambda}$ form (e), $^{t+\Delta t}\mathbf{S}$ from (c), and $\Delta\mathbf{e}'^P$ from (b)

$$D_C = 1 + 2\,{}^t G \Delta\bar{\lambda} \tag{i}$$

and instead of (6.2.53) we have

$$D_{C,j} = 2\,{}^t G \Delta\bar{\lambda}_{,j} \tag{j}$$

The derivatives $\Delta\bar{\lambda}_{,j}$ can be written in the form (6.2.55),

$$\Delta\bar{\lambda}_{,j} = q_0 L_{,j} - q_j \tag{k}$$

where

$$
\begin{aligned}
q_0 &= -\frac{\left({}^{t+\Delta t}J^E_{2D}\right)^{1/2}}{2\,{}^t G\,{}^{t+\Delta t}B^2}\left[AB_1\exp\left(-B_1\,{}^{t+\Delta t}L\right)+\alpha\right] \\
q_j &= {}^{t+\Delta t}S^E_j \Big/ \left[\left({}^{t+\Delta t}J^E_{2D}\right)^{1/2}{}^{t+\Delta t}B\right]
\end{aligned}
\tag{l}
$$

Finally, the derivatives $^{t+\Delta t}L_{,j}$ can be written in the form (6.2.61), i.e.,

$$^{t+\Delta t}L_{,j} = -p_j / f' \tag{m}$$

This relation follows from the differentiation of the governing equation (h) with respect to $^{t+\Delta t}e_j$. The coefficients p_j are

$$
\begin{aligned}
p_j &= 1 \qquad j = 1,2,3 \\
p_j &= 0 \qquad j = 4,5,6
\end{aligned}
\tag{n}
$$

while f' is (see (6.2.57))

$$f' = -\frac{1}{{}^t c_m} + \frac{\partial f}{\partial\,{}^{t+\Delta t}X}\left({}^{t+\Delta t}R'\,{}^{t+\Delta t}B + {}^{t+\Delta t}R\,{}^{t+\Delta t}B' + 1\right) \tag{o}$$

where $^{t+\Delta t}R' = \partial^{t+\Delta t}R/\partial^{t+\Delta t}L$ and $^{t+\Delta t}B' = \partial^{t+\Delta t}B/\partial^{t+\Delta t}L$ can be calculated from (6.2.17) and (6.2.35), respectively. Note that the derivatives $\Delta e_{m,j}^P$ follow from (g) and (m),

$$\Delta e_{m,j}^P = \frac{1}{3}\left[1 + 1/\left(^t c_m \, f'\right)\right] p_j \tag{p}$$

Example 6.2.2. Uniaxial Compression. The elastic-plastic material represented by the generalized cap model, with material constants given in Fig. E.6.2-2, is compressed in one direction, while the deformation is suppressed in the two other directions (oedometer test).

Determine the dependence of the axial stress σ_{xx} on the axial strain e_{xx} if the material is loaded to the stress level A shown in the figure and then unloaded until the zero axial stress is reached.

We implement the computational procedure of Section 6.2.2 for the stress calculation.

The solution in the loading to the level A is obtained using 20 equal steps. The plastic flow corresponds to cap yielding. The unloading is solved using 10 equal steps, the first step corresponds to elastic unloading and then plastic flow occurs on the failure surface $f_1 = 0$. The numerical results and experimental data according to DiMaggio and Sandler (1971) are shown in Fig. E.6.2-2. It can be seen that the model prediction of the material response

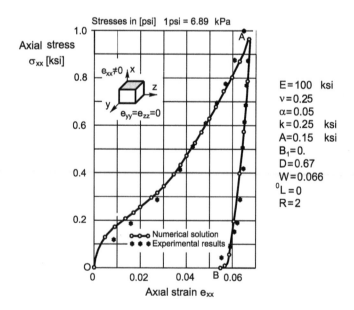

Fig. E.6.2-2. Uniaxial compression of soil (generalized cap model). Loading and unloading to the zero axial stress

agrees with the experimental observation, in both loading and unloading regimes.

Indeed, the model predicts a hysteretic behavior of the material under cyclic loading, see also Kojic et al. (1995e).

Example 6.2.3. Conventional Triaxial Compression Test. Calculate the material response assuming the cap model and triaxial loading according to the conventional triaxial compression test.

The test is modeled using one three-dimensional solid finite element. The nonzero material constants for the model are as follows (Desai and Siriwardane 1984):

$$E = 4000 \text{ psi} \quad \nu = 0.35$$

$$k = 5.6 \text{ psi} \quad A = 5.6 \text{ psi} \quad \alpha = 0.11 \quad B_1 = 0.062 \text{ (psi)}^{-1}$$

$$D = 0.05 \text{ (psi)}^{-1} \quad R = const. = 2.0 \quad W = 0.18$$

The material is first compressed by hydrostatic pressure $p_0 = 10$ psi; then the axial stress σ_{11} is linearly increased to the value $\sigma_{11} = 27$ psi, while the other two stresses are kept constant. The stress paths OA and AB in Fig. E.6.2-3a correspond to these two loading regimes. The hardening character of the material response is shown in Fig. E.6.2-3b. The same character of the material response is given in Desai and Siriwardane (1984).

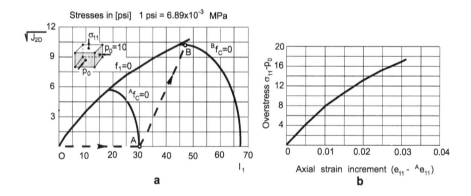

Fig. E.6.2-3. Modeling of a triaxial compression test using one 3-D finite element. **a** Loading path and caps corresponding to stress states A and B; **b** Dependence of the overstress on axial strain increment

6.3 Cam-Clay Model

We first define the Cam-clay material model in Section 6.3.1, and then develop the computational procedure for stress integration in Section 6.3.2. In Section 6.3.3 we derive the expressions for the consistent tangent elastic-plastic matrix. Finally, in Section 6.3.4 we present numerical examples that illustrate some characteristics of the material model and the computational algorithm.

The Cam-clay model is applicable to claylike soils and to sands (see, e.g., Desai and Siriwardane 1984; Chen and Mizuno 1990; Wood 1990).

6.3.1 Formulation of the Model

The main characteristics of the soil considered in this section are obtained experimentally and are shown in Figs. 6.1.6, and 6.1.8 to 6.1.10. It can be seen that the material can have (i) hardening response, (ii) softening response, and (iii) a perfectly plastic response. The responses (i) to (iii) correspond to the lightly overconsolidated (loose), heavily overconsolidated (dense) and critical states, respectively. The material model formulated in this section gives a calculated response aimed to agree with the experimental observations.

The model is known as the *Cam-clay* (Cambridge-clay) *model* or *modified Cam-clay model*, according to Roscoe and Schofield (1963b), and Roscoe and Burland (1968). We will use the formulation of the modified Cam-clay model and call it simply the Cam-clay model.

The mathematical form of the yield condition in the (σ_m, q) plane that follows the character of the experimental results shown in Fig. 6.1.9c is

$$f_y = q^2 - M^2 \sigma_m (p_0 - \sigma_m) = 0 \tag{6.3.1}$$

where M is a material parameter, and p_0 is the length of the horizontal axis of the ellipse, as shown in Fig. 6.3.1a. Only one half of the ellipse is applicable, since the material behavior is defined for $q \geq 0$. The critical state line, corresponding to the experimental results in Fig. 6.1.10a, has the slope M and passes through the midpoint V of the ellipse.

We next define the material hardening. The hardening is represented by the change of the ellipse size, observed experimentally and displayed in Fig. 6.1.9c. As shown in Fig. 6.3.1a, it is assumed that the ellipse changes its size, but passes through the origin. In accordance with Figs. 6.1.9c,d and 6.1.6c, there is a correspondence (mapping) between the points in the (σ_m, q), (σ_m, v) and the $(\ln \sigma_m, v)$ planes. The points A' and B' (and A'' and B''), as the images of the points A and B, lie on the isotropic compression line (*icl*) as shown in Figs. 6.3.1a,b,c, because the change of the stress AB represents a hydrostatic (isotropic) loading. The segment AB is at the same time the

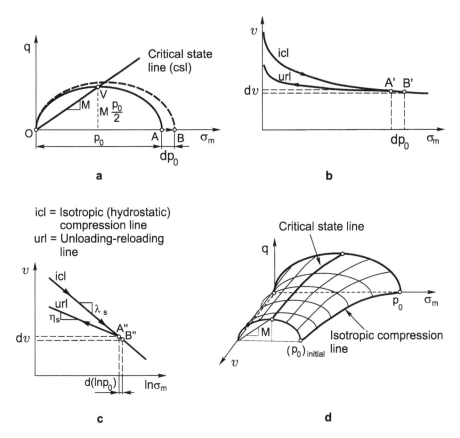

Fig. 6.3.1. Formulation of Cam-clay model. **a** Yield curves and critical state line in the mean stress σ_m - "deviatoric" stress q plane; **b,c** Isotropic compression and unloading-reloading lines in planes σ_m - specific volume v and $\ln \sigma_m - v$; **c** Yield surface in (σ_m, q, v) space

increment dp_0 of the ellipse axis. The isotropic compression line is taken to be straight in the $(\ln \sigma_m, v)$ plane, with the slope λ_s. The following relation holds:

$$dv = -\lambda_s \, d(\ln p_0) = -\lambda_s \frac{dp_0}{p_0} \qquad (6.3.2)$$

This relation defines the *hardening law* for the model assuming that we load along the *icl* line. The material constant for this law is λ_s, determined experimentally. Note that the *hardening parameter* is the specific volume v, or the void ratio e. The yield surface in the (σ_m, q, v) space is shown in Fig. 6.3.1d.

As shown in Figs. 6.3.1b,c, the unloading-reloading of the material is represented by the (url) line, which is also a straight line in the $(\ln \sigma_m, v)$ coordinate system. Hence, an "elastic" increment of the specific volume dv^E can be expressed by the slope of the (url) line in the $(\ln \sigma_m, v)$ coordinate system,

$$dv^E = -\eta_s \, d(\ln \sigma_m) = -\eta_s \frac{d\sigma_m}{\sigma_m} \tag{6.3.3}$$

where η_s is the material constant. If we take into account the relation (6.1.3), we find the following expression for the increment of elastic volumetric strain de_V^E:

$$de_V^E = \frac{\eta_s}{v} \frac{d\sigma_m}{\sigma_m} = \frac{\eta_s}{1+e} \frac{d\sigma_m}{\sigma_m} \tag{6.3.4}$$

Therefore, the bulk modulus K (see (A1.14)) is

$$K = \frac{1+e}{\eta_s} \sigma_m \tag{6.3.5}$$

which shows that the "elastic" bulk modulus depends on the void ratio e (or volumetric strain e_V) and on the mean stress σ_m. Further, using the relation (6.1.3) and the condition that in hydrostatic loading $p_0 = \sigma_m$ we obtain from (6.3.2) and (6.3.3)

$$de_V^P = de_V - de_V^E = \frac{k_s}{v} \frac{dp_0}{p_0} = \frac{k_s}{1+e} \frac{dp_0}{p_0} \tag{6.3.6}$$

where

$$k_s = \lambda_s - \eta_s \tag{6.3.7}$$

The hardening law (6.3.6) is more convenient for applications than (6.3.2) because it relates the increment of the ellipse size to the increment of volumetric plastic strain de_V^P.

Let us now consider the material response predicted when using the Cam-clay model. To compare the character of the response prediction with the experimental data we use the quantities of the triaxial compression test (see Section 6.1.2). Assume that we have a material sample for which the yield surface size is defined by the semiaxis $^1p_0$. Consider first the loading specified by the loading path $A_L B$ in Fig. 6.3.2.a. From the point A_L to the point B at the yield surface the material deformation is elastic, with a linear increase of the "deviatoric" stress q with the "shear strain" e_q, and a linear decrease of the specific volume v with e_q. The loading path BB' corresponds to plastic deformation, with an increase of the yield surface size, a nonlinear increase of the deviatoric stress, and a nonlinear decrease of the specific volume. The material work hardens until the point B' on the critical state line (csl) is reached. The material deformation at point B' corresponds to the critical state, described in Section 6.1.2, and the deformation continues with constant "deviatoric" stress $q_{crit}^{B'}$ and constant specific volume $v_{crit}^{B'}$. This material response represents the behavior of a lightly overconsolidated (loose) material (see Fig. 6.1.8a).

Consider next the same material (the same initial yield surface) subjected to the loading starting from the point A_H along the stress path $A_H D$ shown in Fig. 6.3.2b. The material deforms elastically until the point D on the yield surface is reached. Further deformation is elastic-plastic, with a decrease of the stresses q and σ_m, decrease of the yield surface size (the plastic volumetric strain decreases), and an increase of the specific volume (see graphs in Fig. 6.3.2b). The critical state is reached at the point D' and deformation con-

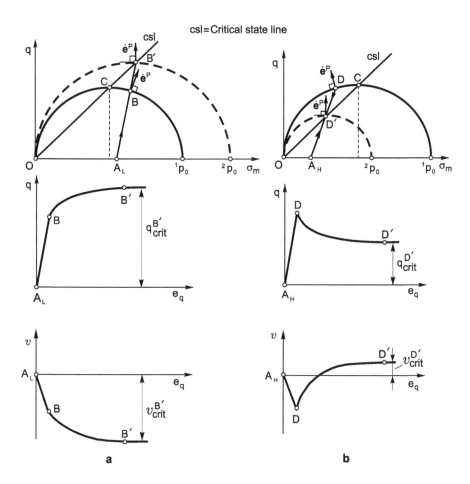

Fig. 6.3.2. Material response predicted using the Cam-clay model in planes: mean stress σ_m – deviatoric stress q, shear strain $e_q - q$, e_q – specific volume v. **a** Lightly overconsolidated material response; **b** Heavily overconsolidated material response

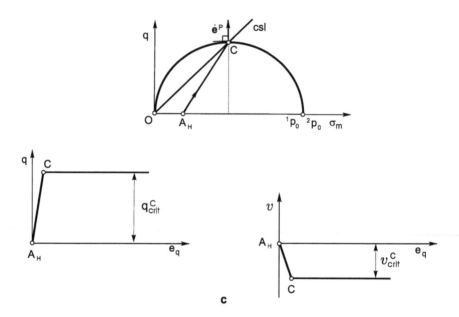

Fig. 6.3.2 (continued). c Response at the critical state

tinues with the critical "deviatoric" stress $q_{crit}^{D'}$ and critical specific volume $v_{crit}^{D'}$. Note that the critical values $q_{crit}^{D'}$ and $v_{crit}^{D'}$ are different from $q_{crit}^{B'}$ and $v_{crit}^{B'}$. The described elastic-plastic deformation corresponds to the behavior of a heavily overconsolidated (dense) material with the strain softening character. Note that by comparing Figs. 6.1.8b and 6.3.2b, the Cam-clay model presented here predicts only approximately the actual physical response.

Assume finally the loading path $A_H C$ of the same material shown in Fig. 6.3.2c. The material response is elastic until the critical state at the point C is reached. Then deformation continues with constant deviatoric stress q_{crit}^{C}, constant specific volume v_{crit}^{C}, with no change of the volumetric plastic strain and the yield surface size.

The above described material behavior has the same character as found experimentally (see Fig 6.1.8). A detailed analysis of the material response prediction using the Cam-clay model is given in Wood (1990).

Finally, we give the expression for the yield condition, based on the triaxial compression form (6.3.1), that corresponds to a general stress state (see (6.1.9)),

$$f_y = \sigma_m(\sigma_m - p_0) + \frac{3J_{2D}}{M^2} = 0 \qquad (6.3.8)$$

where J_{2D} is the second invariant of the stress deviator (see (3.2.13)). Note that J_{2D} represents a measure for the shear loading of the material.

In summary we note that the model is defined by the yield condition (6.3.8), the hardening law (6.3.6), and with only three material constants, M, λ_s and η_s and the initial conditions. The constants can be fitted to experimental data as shown in detail by Desai and Siriwardane (1984). The model can be used to predict the material response under complex loading conditions of soils in elastic-plastic deformations, resulting in hardening and softening, critical state and dilatancy. The model is suitable for general applications and it has been widely used to represent the behavior of claylike soils and sands.

6.3.2 Stress Integration Procedure

We develop a stress integration procedure for the modified Cam-clay material model in an analogous manner to that presented in Section 6.2.2 for the cap model. The governing relations of Sections 6.2.2 and 6.3.1 are followed and the formulas specific for the Cam-clay model are included (see also Kojic et al. 1994 and the references therein). A general three-dimensional deformation is considered, from which the solutions for the two-dimensional plane strain and axisymmetric problems can be directly obtained (see (4.4.36) in Section 4.4.1).

The material can exhibit hardening, softening or perfectly plastic behavior during plastic deformation (see Section 6.3.1) and we consider these three regimes separately. To find which deformation regime corresponds to the current time (load) step Δt, we first calculate the elastic deviatoric and mean stresses, $^{t+\Delta t}\mathbf{S}^E$ and $^{t+\Delta t}\sigma_m^E$, according to (6.2.19) and (6.2.20), using $^{t+\Delta t}\mathbf{e}''$ and $^{t+\Delta t}e_m''$ from (6.2.21) and (6.2.22). Note that the expressions (6.2.23) and (6.2.24) are also applicable. The use of constant volumetric and shear moduli c_m and G would be an approximation since the elastic *bulk modulus depends on the mean stress and volumetric strain*. Namely, from (6.3.4)

$$d^t e_V^E = \frac{\eta_s}{1 + {}^t e} \frac{d^t \sigma_m}{{}^t \sigma_m} \tag{6.3.9}$$

and

$$^t K = \frac{1}{3} {}^t c_m = \frac{1 + {}^t e}{\eta_s} {}^t \sigma_m \tag{6.3.10}$$

The calculation of the void ratio $^t e$ follows from (6.1.3),

$$^t e = (1 + {}^0 e) \exp(-{}^t e_V) - 1 \tag{6.3.11}$$

where $^0 e$ is the initial void ratio.

Taking Poisson's ratio as constant (see, e.g., Desai and Siriwardane 1984; Wood 1990; Atkinson 1993), we obtain the Young and shear moduli ^tE and ^tG as

$$^tE = 3(1 - 2\nu)\,^tK \qquad ^tG = \frac{^tE}{2(1 + \nu)} \tag{6.3.12}$$

We will further use ^tE and ^tG corresponding to the stress/strain state at the start of the load step, although weighted values that include $^{t+\Delta t}E$ and $^{t+\Delta t}G$ may be employed.

Substituting the elastic stresses $^{t+\Delta t}\sigma_m^E$ and $^{t+\Delta t}\mathbf{S}^E$ into (6.3.8), with $p_0 = {}^tp$, we obtain $^{t+\Delta t}f_y^E$. In the case $^{t+\Delta t}f_y^E \leq 0$ the deformation in the current step Δt is elastic. In the case $^{t+\Delta t}f_y^E > 0$ plastic deformation took place, and we decide about the deformation regime as follows:

$$^{t+\Delta t}\sigma_m^E > \frac{1}{2}\,^tp_0 \qquad \text{hardening}$$

$$^{t+\Delta t}\sigma_m^E < \frac{1}{2}\,^tp_0 \qquad \text{softening} \tag{6.3.13}$$

$$^{t+\Delta t}\sigma_m^E = \frac{1}{2}\,^tp_0 \qquad \text{perfect plasticity - critical state}$$

Figures 6.3.3a,b show schematically these conditions. Note that for the equality condition (perfect plasticity), a certain numerical tolerance is assumed. The stress integration procedures for these three conditions are presented next.

Hardening Regime. In this case the ellipse size increases in the time step Δt, as shown in Fig. 6.3.3a. The stress integration according to the general approach of Section 4.2, consists of the following calculations.

The increments of the mean and deviatoric plastic strains, Δe_m^P and $\Delta e'^P$, follow from the flow rule (3.2.67b) and the yield condition $^{t+\Delta t}f_y = 0$ (see (6.3.8))

$$\Delta e_m^P = \frac{\Delta\lambda}{3}\left(2\,^{t+\Delta t}\sigma_m - {}^{t+\Delta t}p_0\right) \tag{6.3.14}$$

$$\Delta e'^P = \frac{3\Delta\lambda}{M^2}\,^{t+\Delta t}\mathbf{S} \tag{6.3.15}$$

where $\Delta\lambda$ is a positive scalar. We solve for $^{t+\Delta t}\mathbf{S}$ using (6.3.15) and the elastic constitutive law (4.2.15)

$$^{t+\Delta t}\mathbf{S} = \frac{^{t+\Delta t}\mathbf{S}^E}{1 + 6\Delta\lambda\,^tG/M^2} \tag{6.3.16}$$

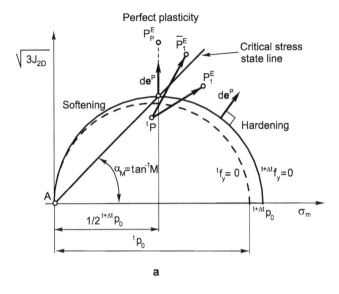

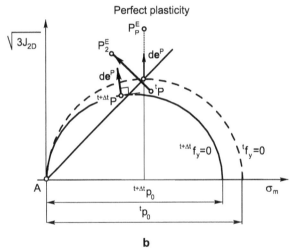

Fig. 6.3.3. Illustration regarding the stress integration with the Cam-clay model. **a** Hardening regime; **b** Softening regime

while $^{t+\Delta t}\sigma_m$ is given by (6.2.27). By scalar multiplication (see (A2.29)) on the both sides of (6.3.16) we obtain

$$^{t+\Delta t}J_{2D} = \frac{^{t+\Delta t}J_{2D}^E}{\left(1 + 6\Delta\lambda\,{}^tG/M^2\right)^2} \tag{6.3.17}$$

Integration of the hardening rule (6.3.6) in the time step Δt gives

$$t+\Delta t_{p_0} = {}^t p_0 \exp\left(\frac{\Delta e_m^P}{t+\Delta t_{b_V}}\right) \tag{6.3.18}$$

where

$$t+\Delta t_{b_V} = \frac{k_s}{3\left(1 + {}^{t+\Delta t}e\right)} \tag{6.3.19}$$

The void ratio ${}^{t+\Delta t}e$ can be obtained from the volumetric strain ${}^{t+\Delta t}e_V$ according to (6.3.11)

$$t+\Delta t_e = (1 + {}^0 e)\exp(-{}^{t+\Delta t}e_V) - 1 \tag{6.3.20}$$

Finally, the yield condition (6.3.8) at the end of the time step

$$t+\Delta t_{f_y} = {}^{t+\Delta t}\sigma_m\left({}^{t+\Delta t}\sigma_m - {}^{t+\Delta t}p_0\right) + 3\,{}^{t+\Delta t}J_{2D}/M^2 = 0 \tag{6.3.21}$$

must be satisfied. The governing parameter, in the sense of Section 4.2, is Δe_m^P and the governing equation is (6.3.21). Table 6.3.1 summarizes the computational steps.

Table 6.3.1. Computational steps for stress integration of the Cam-clay model (hardening regime)

---

1. **Assume Δe_m^P**
2. **Calculate ${}^{t+\Delta t}\sigma_m$** from (6.2.27) and ${}^{t+\Delta t}p_0$ from (6.3.18), $\Delta\lambda$ from (6.3.14) and ${}^{t+\Delta t}J_{2D}$ from (6.3.17)
3. **Check if $\left|{}^{t+\Delta t}f_y\right| < \varepsilon$** according to (6.3.21).
   If YES, go to step 4; otherwise, go to step 1
4. **Determine ${}^{t+\Delta t}\mathbf{S}$** from (6.3.16) and $\Delta e'^P$ from (6.3.15)

Softening Regime. In case the elastic solution is given by a point P_2^E in Fig. 6.3.3b, the volumetric plastic strain decreases (increment Δe_m^P represents the volume expansion) and leads to a decrease in the size of the yield surface. All the above derived expressions for the hardening regime are applicable. Since $\Delta e_m^P < 0$, we have ${}^{t+\Delta t}p_0 < {}^t p_0$ as shown in Fig. 6.3.3b.

Perfect Plasticity (Critical State). If the elastic solution ${}^{t+\Delta t}\sigma_m^E = {}^t p_0/2$, plastic flow occurs according to the perfect plasticity conditions and corresponds to the critical state material behavior, see Fig. 6.3.2. Therefore, the stress point must lie on the critical state line with

$$t+\Delta t \sigma_m = \frac{1}{2} {}^t p_0$$

$$\sqrt{t+\Delta t J_{2D}} = \frac{1}{2\sqrt{3}} M {}^t p_0 \tag{6.3.22}$$

In a graphical representation, we perform the calculations under the condition that the stress point moves from P_P^E (Fig. 6.3.3) along the vertical line to reach the yield surface ${}^t f_y = 0$. Note that the volumetric plastic strain increment is equal to zero due to the associated flow rule (3.2.67b). This also follows from (6.3.14) and ${}^{t+\Delta t} p_0 = {}^t p_0 = 2 {}^{t+\Delta t} \sigma_m^E$.

In order to determine the increments of the deviatoric plastic strains (6.3.15), we first calculate $\Delta\lambda$. Substituting

$$\begin{aligned} {}^{t+\Delta t} p_0 &= 2 {}^{t+\Delta t} \sigma_m^E \\ {}^{t+\Delta t} \sigma_m &= {}^{t+\Delta t} \sigma_m^E \end{aligned} \tag{6.3.23}$$

into (6.3.21) and using (6.3.17) we obtain an equation from which the solution for $\Delta\lambda$ follows

$$\Delta\lambda = \frac{M^2}{6 {}^t G} \left(\frac{\sqrt{3 {}^{t+\Delta t} J_{2D}^E}}{M {}^{t+\Delta t} \sigma_m^E} - 1 \right) \tag{6.3.24}$$

With $\Delta\lambda$ determined, we calculate ${}^{t+\Delta t}\mathbf{S}$ from (6.3.16) and $\Delta\mathbf{e}'^P$ from (6.3.15). Therefore, we have the shear plastic flow in the time step, with the change of the deviatoric plastic strains and deviatoric stresses.

6.3.3 Elastic-Plastic Matrices

We follow the derivations of Section 6.2.3. According to the above stress integration procedure we distinguish yielding on the ellipse (hardening and/ or softening) and on the critical state line (perfect plasticity).

Yielding on the Ellipse. The derivatives ${}^{t+\Delta t} S_{i,j} = \partial^{t+\Delta t} S_i / \partial^{t+\Delta t} e_j$ in the expressions (6.2.38) can be written in the form (6.2.39), where (see (6.3.16))

$$D_\lambda = 1 + 6\Delta\lambda {}^t G / M^2 \tag{6.3.25}$$

and

$$D_{\lambda,j} = \frac{6 {}^t G}{M^2} \Delta\lambda_{,j} \tag{6.3.26}$$

The derivatives $^{t+\Delta t}S^E_{i,j}$ are given by (6.2.41). Using (6.3.14), (6.3.18), (6.2.27) and (6.2.24), the derivatives $\Delta\lambda_{,j}$ can be written in the form (6.2.45), with the coefficients

$$d_j = -\frac{(\Delta\lambda)^2}{3}\left(\frac{2}{3} - \frac{^{t+\Delta t}p_0}{^t c_m \, ^{t+\Delta t}b_V}\right)\bar{a}_j$$

$$d_0 = -\frac{\Delta\lambda}{\Delta e^P_m}\left[1 + \frac{\Delta\lambda}{3}\left(2\,^t c_m + \,^{t+\Delta t}p'_0\right)\right] \tag{6.3.27}$$

where the $\bar{a}_j$ are given in (6.2.50). By $^{t+\Delta t}p'_0$ we have denoted

$$\boxed{^{t+\Delta t}p'_0 = \frac{\partial\,^{t+\Delta t}p_0}{\partial\left(\Delta e^P_m\right)} = \frac{^{t+\Delta t}p_0}{^{t+\Delta t}b_V}} \tag{6.3.28}$$

which follows from (6.3.18) to (6.3.20). Differentiating the governing equation (6.3.21) with respect to $^{t+\Delta t}e_j$, we obtain the derivatives $\Delta e^P_{m,j}$ in the form (6.2.48), where

$$W_j = \left[\frac{1}{3}\left(2\,^{t+\Delta t}\sigma_m - \,^{t+\Delta t}p_0\right) + \frac{^{t+\Delta t}\sigma_m \,^{t+\Delta t}p_0}{^t c_m \,^{t+\Delta t}b_V}\Delta e^P_m\right]\bar{a}_j$$
$$+ \frac{6\,^t G}{M^2 D^2_\lambda}\left(^{t+\Delta t}S^E_j - \frac{6\,^{t+\Delta t}J^E_{2D}}{M^2 D_\lambda}d_j\right) \tag{6.3.29}$$

and, instead of $^{t+\Delta t}f'_1$ we have $^{t+\Delta t}f'_y$,

$$^{t+\Delta t}f'_y = \frac{\partial\,^{t+\Delta t}f_y}{\partial\left(\Delta e^P_m\right)} = -\,^t c_m\left(2\,^{t+\Delta t}\sigma_m - \,^{t+\Delta t}p_0\right) - \,^{t+\Delta t}\sigma_m\,^{t+\Delta t}p'_0$$
$$+ \frac{36\,^t G\,^{t+\Delta t}J_{2D}\,d_0}{M^4 D_\lambda} \tag{6.3.30}$$

With $\Delta e^P_{m,j}$ calculated, we have $\Delta\lambda_{,j}$ from (6.2.45), with the coefficients given in (6.3.27), $D_{\lambda,j}$ from (6.3.26), and $^{t+\Delta t}S_{i,j}$ from (6.2.39). Also, the derivatives $^{t+\Delta t}\sigma_{m,j}$ can be calculated from (6.2.51).

Perfect Plasticity. In the case of perfect plasticity (yielding on the critical state line), we have $\Delta e^P_{m,j} = 0$ and $\Delta\lambda_{,j}$ follows from (6.3.24):

$$\Delta\lambda_{,j} = \frac{M\sqrt{3}}{6\,^t G\,^{t+\Delta t}\sigma^E_m}\left(\frac{^t G}{\sqrt{^{t+\Delta t}J^E_{2D}}}\,^{t+\Delta t}S^E_j - \frac{\sqrt{^{t+\Delta t}J^E_{2D}}}{^{t+\Delta t}\sigma^E_m}\,^{t+\Delta t}\sigma^E_{m,j}\right)$$
$$\tag{6.3.31}$$

where (6.2.44) was used. The derivatives $^{t+\Delta t}\sigma^E_{m,j}$ are given in (6.2.51), with $\Delta e^P_{m,j} = 0$. With $\Delta\lambda_j$ determined, we obtain $D_{\lambda,j}$ from (6.3.26) and $^{t+\Delta t}S_{i,j}$ from (6.2.39). Also we have $^{t+\Delta t}\sigma_{m,j} = {}^{t+\Delta t}\sigma^E_{m,j}$.

Finally, we may note that in general the $^{t+\Delta t}\mathbf{C}^{EP}$ matrices are non-symmetric. In practice they may be symmetrized but then quadratic convergence is, in general, lost.

6.3.4 Examples

Example 6.3.1. Conventional Triaxial Test of Cam-Clay Material.
Calculate the material response represented by the Cam-clay model, subjected to the conventional triaxial test shown in Fig. E.6.3-1a. The material data are shown in the same figure, with Young's modulus $E = E_0$.

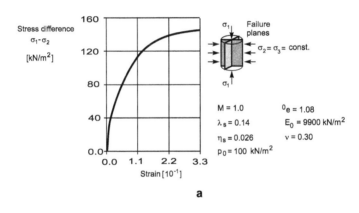

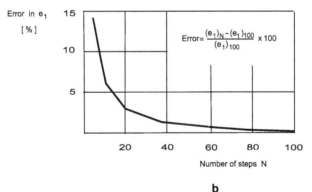

Fig. E.6.3-1. Triaxial compression test of Cam-clay material. **a** Stress difference σ_1-σ_2 versus axial strain; **b** Error in axial strain e_1

We use the computational procedure of Section 6.3.2 for the stress calculation, under the following conditions: two principal stresses are equal

($\sigma_2 = \sigma_3 = p_0 = 100$ kN/m^2) while the axial compressive stress σ_1 is increasing (as published earlier in Kojic et al. 1994). We assume small strain kinematic conditions.

The problem is solved by using the full-Newton iterative method with line searching (Bathe 1996). With use of the arc-length method, the stress level of $\sigma_1 = 214$ shown in Fig. E.6.3-1a is reached in 4 steps with 8 iterations per step. After the next 8 steps, the stress $\sigma_1 = 250$ is reached at which stage the material behaves as perfectly plastic. The results shown in Fig. E.6.3-1a agree with those reported in Desai and Siriwardane (1984). Note that rather large strains correspond to the critical stress, and for this range of strains a large strain formulation is appropriate (see Chapter 7).

In order to investigate the robustness and accuracy of the stress integration algorithm, we have solved this example using equal time (load) steps to reach the axial stress $\sigma_1 = 180$. Figure E.6.3-1b shows the error in the axial strain e_1 with respect to the 100-step solution. Note that the error in e_1 is rather large for a small number of time steps, but decreases rapidly with the number of steps N. The error in e_1 and in the other strain components is due to the fact that the normal to the yield surface changes significantly during the plastic flow.

Example 6.3.2. Triaxial Compression Test of Cam-Clay Material. Calculate the strains in the soil specimen shown in Fig. E.6.3-2a, when the specimen is subjected to axial and radial stresses. The Cam-clay material model is used to model the soil specimen, with the material properties shown in Fig. E.6.3-2a.

The sequence of loads, and the solution response, are as follows:

A0: Axial and radial stresses equal 100 kN/m^2. At this stress level, the bulk modulus K computed from (6.3.5) is equal to the bulk modulus as calculated from the given material properties E_0 and ν. The strains are zero.

A1: Axial stress is increased to 220. The material yields and hardens, with the specific volume decreasing and remaining above the CSL.

A2: Axial stress is decreased to 100, with the material unloading elastically.

A3: Axial and radial stresses are decreased to 50, again the material unloads elastically.

A4/A5: Axial stress is increased to yield and yielding continues with decreasing axial stress. The specific volume stays below the CSL.

The computed axial and radial strains are shown in the figure. Note that the strains become rather large, and therefore a large strain formulation may be more appropriate than the small strain formulation used in this example.

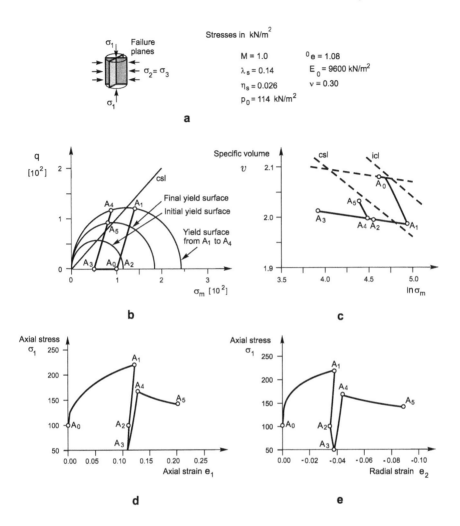

Fig. E.6.3-2. Triaxial compression test of Cam-clay material. **a** Soil specimen and material data; **b** Stress path and yield surfaces in the mean stress σ_m- deviatoric stress q plane; **c** Loading path in the $ln\sigma_m-$ specific volume v plane; **d** The axial stress σ_1- axial strain e_1 dependence; **e** Dependence of σ_1 on radial strain e_2

6.4 A General Soil Plasticity Model

While the material models considered above are quite powerful in representing soil and rock material behavior, there are of course also shortcomings, for example the material model conditions do not depend on the third stress invariant. However, more general models have been developed, see, for example Desai (1980a), (2001); Sture et al. (1989); Jeremic et al. (1999); Wang et al. (2001); Borja et al. (2001).

In this section we briefly develop the computational algorithm for a general soil plasticity model that relies on the theoretical assumptions and experimental results of Desai and Siriwardane (Desai 1980a, 1989; Desai and Siriwardane 1980b, 1984). We present the formulation of the model and the numerical procedure for stress integration as a direct application of the general relations derived in Section 4.2. We consider the associated and nonassociated flow rules and illustrate the application of the algorithm in a numerical example.

6.4.1 Formulation of the Model

A general soil model can be described by a yield criterion dependent on all stress invariants and which also displays work-hardening characteristics. Models of this type were proposed in the above cited references and we present one of the models (Desai 1989). The yield condition for this model is

$$
f_y = p_a^{-2} \, J_{2D} + F_b \, F_s = 0 \tag{6.4.1}
$$

where

$$
F_b = -\alpha \left(\frac{I_1}{p_a} \right)^n + \gamma \left(\frac{I_1}{p_a} \right)^2 \tag{6.4.2}
$$

$$
F_s = (1 - \beta \, s_r)^m \tag{6.4.3}
$$

$$
s_r = \frac{\sqrt{27}}{2} \, J_{3D} \, J_{2D}^{-1.5} \tag{6.4.4}
$$

and I_1, J_{2D} and J_{3D} are, respectively, the first invariant of stress, and the second and third invariants of the stress deviator (see (3.2.11) and (3.2.13)); p_a is the standard atmospheric pressure. The coefficients m, n, γ and β are material constants, while α is the hardening function

$$
\alpha = \alpha \left(\bar{e}^P, \bar{e}_V^P, \bar{e}'^P \right) \tag{6.4.5}
$$

where $\bar{e}^P, \bar{e}_V^P$ and $\bar{e}'^P$ are the effective plastic strain, effective volumetric and deviatoric plastic strains, respectively, defined as

$$
\begin{aligned}
\bar{e}^P &= \int \left(de^P \cdot de^P \right)^{1/2} \\
\bar{e}_V^P &= e_V^P / \sqrt{3} \\
\bar{e}'^P &= \int \left(de'^P \cdot de'^P \right)^{1/2}
\end{aligned} \tag{6.4.6}
$$

Here de^P and de'^P are the increments of plastic strain and deviatoric plastic strain, and e_V^P is the volumetric plastic strain. Note the difference between the definitions of the effective plastic strain in (6.4.6) and in (3.2.46). An analytical form of the hardening function includes the ratios r_V and r_D defined as

$$r_V = \bar{e}_V^P / \bar{e}^P$$
$$r_D = \bar{e}'^P / \bar{e}^P \tag{6.4.7}$$

Figure 6.4.1 shows the yield function in the $(I_1, \sqrt{J_{2D}})$ plane and in the plane whose normal in the principal stress space has the components equal to $\sqrt{3}/3$. The hardening function α of the form (6.4.5) shows that the size of the yield surface depends on the history of the volumetric and deviatoric plastic deformations. A specific form of α is

$$\alpha = a_1(\bar{e}^P)^{-\eta} \tag{6.4.8}$$

where a_1 and η are material constants.

The use of the nonassociated flow rule (6.1.11) may be more appropriate for sand-type soils. The plastic potential Q for this model has the form (6.4.1), with the hardening function α_Q instead of α,

$$\alpha_Q = \alpha + k(\alpha_0 - \alpha)(1 - r_V) \tag{6.4.9}$$

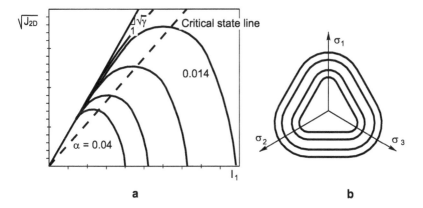

Fig. 6.4.1. A general yield function for geological material, Desai (1989). **a** In $I_1, \sqrt{J_{2D}}$ space; **b** In plane whose normal in the principal stress space $\sigma_1, \sigma_2, \sigma_3$ is given by $\sqrt{3}/3, \sqrt{3}/3, \sqrt{3}/3$

Here α_0 is the hardening function α corresponding to an initial hydrostatic compression (see Desai and Siriwardane 1984), and

$$k = k_1 + k_2 s_r \qquad (6.4.10)$$

where k_1 and k_2 are material constants.

6.4.2 Stress Integration

We perform the stress integration by implementing the computational steps of Table 4.2.2. In order to determine the unit normal $\mathbf{n}$, we express the derivatives $(\mathbf{f}_{y,\sigma})_k$, $k = 1, \ldots, 6$ using the one-index notation for the stresses and strains (see Section 4.4.3 and notation in (4.4.52)). Hence, from (4.2.9) and (4.2.10) we obtain

$$
\begin{aligned}
(f_{y,\sigma})_k &= A_{kj}\, f'_{y,j} + \frac{1}{3} f_{y,m} & j, k = 1, 2, 3 \\
(f_{y,\sigma})_k &= f'_{y,k} & k = 4, 5, 6
\end{aligned}
\qquad (6.4.11)
$$

where the matrix $\mathbf{A}$ is defined in (3.2.70), $f'_{y,j} = \partial f_y/\partial S_j$, and $f_{y,m} = \partial f_y/\partial \sigma_m$. The derivatives of the yield function (6.4.1) can be written as

$$
\begin{aligned}
f'_{y,k} &= p_a^{-2}\, J'_{2D,k} + F_b F'_{s,k} \\
f_{y,m} &= F_s F_{b,m}
\end{aligned}
\qquad (6.4.12)
$$

where

$$
\begin{aligned}
J'_{2D,k} &= S'_k \\
F'_{s,k} &= -\beta\, m\, (1 - s_r)^{m-1} s_{r,k} \\
F_{b,m} &= -\frac{3\alpha}{p_a}\, n \left(\frac{I_1}{p_a}\right)^{n-1} + 6\frac{\gamma}{p_a}\frac{I_1}{p_a}
\end{aligned}
\qquad (6.4.13)
$$

with $F'_{b,k} = 0$. Here we have used the notation: $J'_{2D,k} = \partial J_{2D}/\partial S_k$; $S'_k = S_k$, $k = 1, 2, 3$; $S'_k = 2S_k$, $k = 4, 5, 6$; and $F_{b,m} = \partial F_b/\partial \sigma_m$, $F'_{s,k} = \partial F_s/\partial S_k$, $s'_{r,k} = \partial s_r/\partial S_k$. The derivatives $s'_{r,k}$ follow from (6.4.4),

$$s'_{r,k} = \frac{\sqrt{27}}{2}\left(J_{2D}^{-1.5} J'_{3D,k} - \frac{3}{2} J_{3D}\, J_{2D}^{-2.5} J'_{2D,k}\right) \qquad (6.4.14)$$

The derivatives $J'_{3D,k} = \partial J_{3D}/\partial S_k$ are

$$J'_{3D,1} = S_2 S_3 - S_5^2 \qquad\qquad J'_{3D,2} = S_1 S_3 - S_6^2$$
$$J'_{3D,3} = S_1 S_2 - S_4^2 \qquad\qquad J'_{3D,4} = 2\,(S_5 S_6 - S_3 S_4) \qquad (6.4.15)$$
$$J'_{3D,5} = 2\,(S_4 S_6 - S_1 S_5) \qquad J'_{3D,6} = 2\,(S_4 S_5 - S_2 S_6)$$

With the derivatives (6.4.12) to (6.4.15) calculated, we can determine the derivatives (6.4.11) and the normal $\mathbf{n}$ in (4.2.20) corresponding to the last known trial state in the iterative solution procedure of Table 4.2.2.

In the case of nonassociated plasticity we need the normal $\mathbf{n}_Q = (\partial Q/\partial\boldsymbol{\sigma})/\|\partial Q/\partial\boldsymbol{\sigma}\|$, since the increment of plastic strain $\Delta\mathbf{e}^P$ is

$$\boxed{\Delta\mathbf{e}^P = \|\Delta\mathbf{e}^P\|\,\mathbf{n}_Q} \qquad (6.4.16)$$

This expression is appropriately used in Table 4.2.2. For the material model of Section 6.4.1, and with the correction (6.4.9) of the hardening function α, we have

$$Q = p_a^{-2} J_{2D} + \bar{F}_b F_s \qquad (6.4.17)$$

where

$$\bar{F}_b = F_b - k\,(1 - r_V)(\alpha_0 - \alpha)\left(\frac{I_1}{p_a}\right)^n \qquad (6.4.18)$$

The derivatives of $\bar{F}_b$ differ from the derivatives of F_b:

$$\bar{F}_{b,m} = F_{b,m} - \frac{3n}{p_a}\,k\,(1 - r_V)(\alpha_0 - \alpha)\left(\frac{I_1}{p_a}\right)^{n-1}$$
$$\bar{F}'_{b,j} = -(1 - r_V)(\alpha_0 - \alpha)\left(\frac{I_1}{p_a}\right)^n k_2\,s'_{r,j} \qquad (6.4.19)$$

where we have used the relation (6.4.10). With these and the above expressions for the derivatives $f_{y,k}$ we can determine the derivatives $\partial Q/\partial\boldsymbol{\sigma}$.

6.4.3 Elastic-Plastic Matrix

The general expression for the tangent elastic-plastic matrix in associated plasticity, consistent with the governing parameter method of Section 4.2, is given in (4.2.31) and in Table 4.2.2. The internal variable is represented by the hardening function α, therefore the coefficient a_σ in (4.2.33) for calculation of $\partial(\|\Delta\mathbf{e}^P\|)/\partial^{t+\Delta t}\hat{\mathbf{e}}$ is

$$\boxed{a_\sigma = {}^{t+\Delta t}\mathbf{f}_{y,\sigma}^T\,\Delta\boldsymbol{\sigma}' - \frac{\partial^{t+\Delta t}f_y}{\partial\alpha}\,\alpha'} \qquad (6.4.20)$$

where we calculate the derivatives of the stress increments $\Delta\sigma$ and the function α with respect to $\|\Delta e^P\|$ while finding the solution of the governing equation (see details in Section 4.2.2).

A procedure analogous to the above can be followed for the calculation of $^{t+\Delta t}\mathbf{C}^{EP}$ in the case of nonassociated plasticity.

6.4.4 Example

Example 6.4.1. Conventional Triaxial Test Using the General Soil Model. Calculate the material response for the conventional triaxial compression test (see Section 6.1.2 and Fig. 6.1.7a) of the general soil plasticity model described in Section 6.4.1.

We use the material data given in Desai (1989) and shown in Fig. E.6.4-1. The first regime of plastic deformation corresponds to the volumetric compression, and (6.4.1) reduces to

$$f_y = \alpha(\frac{I_1}{p_a})^n - \gamma(\frac{I_1}{p_a})^2 = 0 \tag{a}$$

with α given by (6.4.8) and the effective plastic strain (6.4.6) is

$$\bar{e}^P = \sqrt{3}\,e_m^P \tag{b}$$

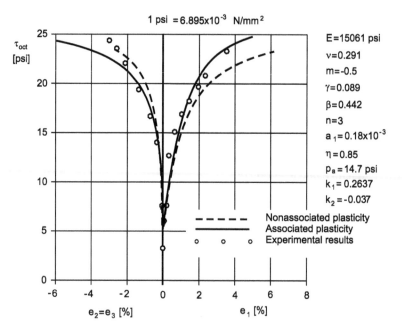

Fig. E.6.4-1. Conventional triaxial test using the general soil model

The hydrostatic pressure for this regime increases to $p = 13$ psi. Substituting this value of pressure into (a), with use of (6.4.8) and (b), we calculate the effective plastic strain $^0\bar{e}^P$ for the conventional triaxial test

$$^0\bar{e}^P = 2.1329 \times 10^{-3} \tag{c}$$

The pressure $p = 13$ is used in the test as a constant lateral stress.

The test is modeled as follows. Since the two principal stresses σ_2 and σ_3 are constant and equal to 13, we have that the constitutive relation (4.2.15) reduces to

$$^{t+\Delta t}\sigma_{xx} = E\,^{t+\Delta t}e''_{xx} + 2\nu\,^{t+\Delta t}\sigma_{yy} - E\,\Delta e^P_{xx} \tag{d}$$

where

$$^{t+\Delta t}e''_{xx} = {}^{t+\Delta t}e_{xx} - {}^{t}e^P_{xx} \tag{e}$$

and $^{t+\Delta t}\sigma_{xx}$ and $^{t+\Delta t}\sigma_{yy}$ are the axial and lateral stresses.

The solution for stresses and strains is obtained using 30 steps with $\Delta e_{xx} = 1.6 \times 10^{-3}$, following the procedure of Table 4.2.2. In solving the governing equation $^{t+\Delta t}f_y = 0$ given by (6.4.1), we have employed the Newton iteration method, with use of the numerical derivatives of the type (4.2.34).

The calculated results are shown in Fig. E.6.4-1 for the associated and nonassociated plasticity, with the experimental results reported in Desai (1989). Here the octahedral shear stress $\tau_{oct} = \sqrt{2}/3\,|\sigma_{xx} - \sigma_{yy}|$.

7. Large Strain Elastic-Plastic Analysis

In this chapter we present a numerical procedure for the analysis of large strain deformations in isotropic plasticity. After short introductory remarks in Section 7.1, we give in Section 7.2 a review of the basic notions of large strain kinematics of deformation. Then in Section 7.3, we give in detail the stress integration procedure in isotropic plasticity based on the multiplicative decomposition of the deformation gradient and the governing parameter method of Section 4.2. The logarithmic strain is employed and the corresponding formulation is called the updated-Lagrangian-Hencky (ULH) formulation. The computational procedure is implemented for the deformations of metals and geomaterials.

7.1 Introduction

In the previous chapters we presented the computational algorithms for stress calculations when considering inelastic deformations assuming small displacements and small strains (within several percent). Hence, the change of geometry was neglected and the inelastic problems are considered to fall into the category of *materially-nonlinear-only* problems (Bathe 1982, 1996).

Frequently, however, in engineering practice the *strains are small* but the *displacements are large.* Such problems are called *geometrically nonlinear.* They are materially linear if the stress-strain relations are linear, and materially nonlinear if these relations are nonlinear, as in inelastic analysis. The corresponding stress and strain measures must reflect these conditions. For example, the Green-Lagrange strain and the second Piola-Kirchhoff stress are effective strain and stress measures used in geometrically nonlinear but small strain analysis (the total Lagrangian formulation is used, Bathe 1982, 1996), see Table 7.1.1.

However, in many types of problems the *strains are also large,* from several percent to hundreds of percent (see Fig.1.3 in Chapter 1). Figure 7.1.1 shows schematically examples representing: (a) small strain and small displacement conditions, (b) small strain and large displacement conditions, (c) large strain conditions.

Table 7.1.1. Stress and strain measures for the three types of problems in inelastic analysis

Problem type	Displacements	Strains	Stresses
Geometrically linear	Small	Infinitesimal	Cauchy
Geometrically nonlinear− small strains	Large	Green-Lagrange	Second Piola-Kirchhoff[1]
Geometrically nonlinear− large strains	Large	Logarithmic	Cauchy

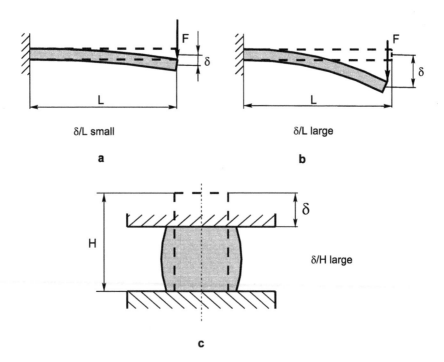

Fig. 7.1.1. Examples of nonlinear problems. **a** Geometrically linear; **b** Geometrically nonlinear, small strains; **c** Geometrically nonlinear, large strains

---

[1] Of course, Cauchy stresses are always the final calculated results

The development of methods for solving large strain inelastic problems is not just an extension of the methodology employed in the small strain case. To solve large strain problems we need additional kinematic quantities to describe the deformations.

In principle, there are two different formulations used to solve large strain problems, namely the rate formulations, and the total formulations.

The *rate formulations* are based on the integration of constitutive relations involving stress and strain rates. This concept has been widely used in large-scale elastic-plastic analyses, see, e.g., Lee and Mallett (1982).

The *total formulations* are based on kinematic quantities defined with respect to an adopted reference configuration and on incremental solution procedures. The total formulations have advantages with respect to the rate formulations (see Bathe et al. 1974, 1975; Kojic and Bathe 1987c; Bathe 1982, 1996).

The solution of large strain engineering problems with plastic deformations has become feasible due to the efficiency of the finite element method and the computational speed and capacity of computers. Various procedures have been proposed for large strain analysis (see Bathe 1996; Simo and Hughes 1998). We present two techniques within the framework of a total formulation, based on the multiplicative decomposition of the deformation gradient and with use of the logarithmic strains.

7.2 Basic Notions in the Kinematics of Large Strain Deformations

In this section we review quantities that describe the kinematics of large strain deformations of a continuous medium. A more detailed description of these quantities is given in Bathe (1996).

Motion. Consider a material body $\mathcal{B}$ consisting of material particles continuously distributed in space, Fig. 7.2.1a. We use a stationary Cartesian coordinate system (x_1, x_2, x_3) to define the motion of the particles. At the initial time $t = 0$, a generic material particle P occupies an infinitely small volume of space, i.e., a geometric point whose position vector is

$$\boxed{^0\mathbf{x} = {}^0\mathbf{x}(0)}$$
(7.2.1a)

or, in component form,

$$^0x_i = {}^0x_i(0) \qquad i = 1, 2, 3$$
(7.2.1b)

Thus we may think of having labeled the material particle with its initial position vector $^0\mathbf{x}$. At time t the position vector is

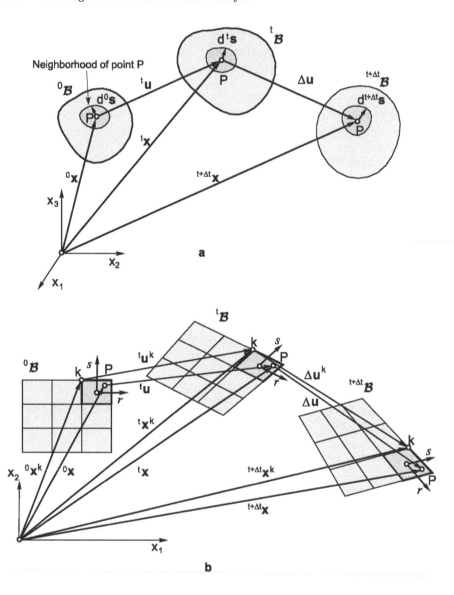

Fig. 7.2.1. Motion. **a** General representation of a body motion; **b** Representation of a body motion by finite elements, 2-D view; r, s axes shown for 4-node element

$$\boxed{{}^{t}\mathbf{x} = {}^{t}\mathbf{x}(t)} \qquad (7.2.2)$$

and the change of the position of the material particle in time defines the *motion of the particle*.

Consider next the set of all material particles of the body B. The set of geometric points defined by the position vectors $^0\mathbf{x}$ of all material particles represents the *initial configuration* ^0B of the body, and the set of geometric points $^t\mathbf{x}$ represents the *configuration* ^tB *at time t*. The volumes in the 3-D space occupied at time 0 and t by the body are ^0V and ^tV, respectively. The vector relation

$$^t\mathbf{x} = {}^t\mathbf{x}\left(^0\mathbf{x}, t\right) \qquad (7.2.3)$$

defines the complete motion of the material body B. Therefore the *motion of the body is given by the change of the body configuration* during time. This description of the body motion is called the *Lagrangian description* of motion.

Note that considering a rigid body, the relative positions of the material particles do not change during time.

In the finite element discretization we select the *nodal points* of the finite elements to represent the motion of the body (see Fig. 7.2.1b). The position vectors of nodal points are defined as

$$^t\mathbf{x}^k = {}^t\mathbf{x}^k(t) \qquad (7.2.4)$$

We do not write $^t\mathbf{x}^k\left(^0\mathbf{x}^k, t\right)$ on the right hand side because the number k (the node number) identifies the material point. Further, using the *isoparametric (natural) coordinates* r, s, t to define a material point P within the finite element (see Section 2.3.2), we have the material point position given by

$$^t\mathbf{x} = \sum_{k=1}^{N} h_k \, {}^t\mathbf{x}^k \qquad (7.2.5)$$

where $h_k(r, s, t)$ [2] are the interpolation functions and N is the number of nodal points of the finite element. Hence, the motion of the whole body B is defined by the motion of the *finite number* of nodal points.

The coordinates used to identify material particles within the body are called the *material coordinates*. Therefore the initial coordinates ^0x_i, or the coordinates r, s, t are the material coordinates. Note that the material coordinates for a particle do not change during the body motion, i.e., for a particle,

$$^0x_i = \text{const.} \qquad (7.2.6)$$

---

[2] We use t for time and for the third coordinate in the natural coordinate system r, s, t. We will point out in the text the meaning of t if there could be any ambiguity.

$$r = \text{const.}, \quad s = \text{const.}, \quad t = \text{const.} \tag{7.2.7}$$

We next define the *displacement* of a material particle P in the time interval 0 to t as

$$^t\mathbf{u}\left(^0\mathbf{x}, t\right) = {}^t\mathbf{x}\left(^0\mathbf{x}, t\right) - {}^0\mathbf{x} \tag{7.2.8}$$

or, using the finite element representation,

$$^t\mathbf{u} = \sum_{k=1}^{N} h_k \left({}^t\mathbf{x}^k - {}^0\mathbf{x}^k \right) = h_k \, {}^t\mathbf{u}^k \tag{7.2.9}$$

where $^t\mathbf{u}^k$ are the nodal point displacements, defined as

$$^t\mathbf{u}^k = {}^t\mathbf{x}^k - {}^0\mathbf{x}^k \tag{7.2.10}$$

It is implied that the interpolation (7.2.5) is used for all times, including for the initial configuration.

Deformation Gradient. The deformation gradient represents the *basic quantity* used to describe the kinematics of deformation of a continuous medium. The deformation gradient at time t, with the reference configuration $^0\mathcal{B}$, is defined as [3]

$$^t_0\mathbf{X} = \frac{\partial \, ^t\mathbf{x}}{\partial \, ^0\mathbf{x}} \tag{7.2.11}$$

with the components

$$^t_0 X_{ij} = \frac{\partial \, ^t x_i}{\partial \, ^0 x_j} \tag{7.2.12}$$

If we define the operator $_0\nabla$ as

$$_0\nabla = \left[\frac{\partial}{\partial \, ^0 x_n}\right] = \frac{\partial}{\partial \, ^0 x_n} \, \mathbf{i}_n \tag{7.2.13}$$

then the deformation gradient can be written as

---

[3] The deformation gradient is by many authors denoted using the letter $\mathbf{F}$. However, we are using $\mathbf{F}$ already to denote the nodal point forces corresponding to the element stresses, see Chapters 1 and 2, Section 4.1 and Bathe (1996), and the notation used here is rather natural when $^0\mathbf{x}$ and $^t\mathbf{x}$ are the material particle coordinates at times 0 and t.

$$\begin{aligned}{}_0^t\mathbf{X} = ({}_0\nabla\,{}^t\mathbf{x}^T)^T = \left[\frac{\partial\,{}^t x_m}{\partial\,{}^0 x_n}\right] = {}_0^t X_{mn}\mathbf{i}_m \otimes \mathbf{i}_n\end{aligned}$$
(7.2.14)

where in addition to the matrix notation we also use tensor notation, see Appendix A2.

If the finite element discretization of the continuum is used, we determine the deformation gradient at a point P from (7.2.5) as

$$\begin{aligned}{}_0^t\mathbf{X} = \left[\frac{\partial\,{}^t x_i}{\partial\,r_m}\frac{\partial\,r_m}{\partial\,{}^0 x_j}\right] = {}^t\mathbf{J}^T\,({}^0\mathbf{J}^{-1})^T\end{aligned}$$
(7.2.15)

where ${}^t\mathbf{J}$ and ${}^0\mathbf{J}$ are the Jacobian matrices at time t and time 0 of the transformation between the Cartesian and natural coordinate systems (see (2.3.36) and Bathe 1982, 1996). We use here r_m, with $m = 1, 2, 3$, to denote the natural coordinates as

$$r_1 = r \qquad r_2 = s \qquad r_3 = t$$
(7.2.16)

and then

$$\begin{aligned}{}^t J_{im} = \frac{\partial\,{}^t x_m}{\partial r_i} = \sum_{k=1}^{N}\frac{\partial h_k}{\partial r_i}\,{}^t x_m^k\end{aligned}$$

(7.2.17)

$$\begin{aligned}{}^0 J_{mj} = \frac{\partial\,{}^0 x_j}{\partial r_m} = \sum_{k=1}^{N}\frac{\partial h_k}{\partial r_m}\,{}^0 x_j^k\end{aligned}$$

We also use

$$\begin{aligned}{}_t^0\mathbf{X} = \frac{\partial\,{}^0\mathbf{x}}{\partial\,{}^t\mathbf{x}} = \left[\frac{\partial\,{}^0 x_i}{\partial\,{}^t x_j}\right]\end{aligned}$$
(7.2.18)

with

$$\begin{aligned}{}_0^t\mathbf{X}\,{}_t^0\mathbf{X} = \left[\frac{\partial\,{}^t x_i}{\partial\,{}^0 x_k}\frac{\partial\,{}^0 x_k}{\partial\,{}^t x_j}\right] = \mathbf{I}\end{aligned}$$
(7.2.19)

Here ${}_t^0\mathbf{X}$ is the *inverse deformation gradient* and clearly

$$\begin{aligned}{}_t^0\mathbf{X} = {}_0^t\mathbf{X}^{-1}\end{aligned}$$
(7.2.20)

The deformation gradient at a material point at time $t + \Delta t$ is

$$\begin{aligned}{}_{0}^{t+\Delta t}\mathbf{X} = \frac{\partial\,{}^{t+\Delta t}\mathbf{x}}{\partial\,{}^0\mathbf{x}} = \left[\frac{\partial\,{}^{t+\Delta t} x_i}{\partial\,{}^0 x_j}\right]\end{aligned}$$
(7.2.21)

and, using configuration ^tB as the reference configuration, we have

$$^{t+\Delta t}_{t}\mathbf{X} = \frac{\partial\,^{t+\Delta t}\mathbf{x}}{\partial\,^t\mathbf{x}} = \left[\frac{\partial\,^{t+\Delta t}x_i}{\partial\,^t x_j}\right] \tag{7.2.22}$$

By employing the chain rule we further have

$$^{t+\Delta t}_{0}X_{ij} = \frac{\partial\,^{t+\Delta t}x_i}{\partial\,^0 x_j} = \frac{\partial\,^{t+\Delta t}x_i}{\partial\,^t x_m}\frac{\partial\,^t x_m}{\partial\,^0 x_j} = \,^{t+\Delta t}_{t}X_{im}\,^t_0X_{mj} \tag{7.2.23}$$

or, in the matrix form

$$\boxed{^{t+\Delta t}_{0}\mathbf{X} = \,^{t+\Delta t}_{t}\mathbf{X}\,^t_0\mathbf{X}} \tag{7.2.24}$$

The quantity $^{t+\Delta t}_{t}\mathbf{X}$ is called the *relative deformation gradient*.

The deformation gradient can be expressed in terms of displacements,

$$\boxed{^t_0\mathbf{X} = \mathbf{I} + \frac{\partial\,^t\mathbf{u}}{\partial\,^0\mathbf{x}}} \tag{7.2.25}$$

where $\partial\,^t\mathbf{u}/\partial\,^0\mathbf{x}$ is the *displacement gradient*. Also, for the inverse deformation gradient

$$\boxed{^0_t\mathbf{X} = \mathbf{I} - \frac{\partial\,^t\mathbf{u}}{\partial\,^t\mathbf{x}}} \tag{7.2.26}$$

Similarly, from (7.2.23) and the relation

$$^{t+\Delta t}\mathbf{x} = \,^t\mathbf{x} + \Delta\mathbf{u} \tag{7.2.27}$$

where $\Delta\mathbf{u}$ is the displacement increment (see Fig. 7.2.1), we have

$$^{t+\Delta t}_{0}\mathbf{X} = \,^t_0\mathbf{X} + \frac{\partial(\Delta\mathbf{u})}{\partial\,^0\mathbf{x}} \tag{7.2.28}$$

and

$$\boxed{^{t+\Delta t}_{t}\mathbf{X} = \mathbf{I} + \frac{\partial(\Delta\mathbf{u})}{\partial\,^t\mathbf{x}}} \tag{7.2.29}$$

If the finite element discretization is employed, the displacement gradient is calculated using the interpolation (7.2.9). Then, for example, $^t_0\mathbf{X}$ defined in (7.2.25) can be expressed as

$$\begin{smallmatrix}t\\0\end{smallmatrix}\mathbf{X} = \mathbf{I} + \left[\frac{\partial\,^t u_i}{\partial r_m}\right] (^0\mathbf{J}^{-1})^T \tag{7.2.30}$$

where

$$\frac{\partial\,^t u_i}{\partial r_m} = \sum_{k=1}^N \frac{\partial h_k}{\partial r_m}\,^t u_i^k \tag{7.2.31}$$

Analogously, we have

$$\begin{smallmatrix}t+\Delta t\\t\end{smallmatrix}\mathbf{X} = \mathbf{I} + \left[\frac{\partial\,(\Delta u_i)}{\partial r_m}\right] (^t\mathbf{J}^{-1})^T \tag{7.2.32}$$

where

$$\frac{\partial\,(\Delta u_i)}{\partial r_m} = \sum_{k=1}^N \frac{\partial h_k}{\partial r_m}\,\Delta u_i^k \tag{7.2.33}$$

Using the deformation gradient we can calculate the volume change of the material from the relation (see Bathe 1982, 1996)

$$\boxed{d\,^t V = \left(\det\,\begin{smallmatrix}t\\0\end{smallmatrix}\mathbf{X}\right) d^0 V} \tag{7.2.34}$$

where $d\,^t V$ and $d^0 V$ are the elementary volumes occupied by the same material particles in the configurations $^t\mathcal{B}$ and $^0\mathcal{B}$, and $(\det\,\begin{smallmatrix}t\\0\end{smallmatrix}\mathbf{X})$ is the determinant of $\begin{smallmatrix}t\\0\end{smallmatrix}\mathbf{X}$ (see (A2.17)).

The deformation gradient has a physical meaning. To describe that meaning, consider the relation

$$d\,^t x_i = \begin{smallmatrix}t\\0\end{smallmatrix}X_{ik}\, d^0 x_k \tag{7.2.35a}$$

which follows from the definition (7.2.12). In matrix form we have that

$$\boxed{d\,^t\mathbf{s} = \begin{smallmatrix}t\\0\end{smallmatrix}\mathbf{X}\, d^0\mathbf{s}} \tag{7.2.35b}$$

Hence, $\begin{smallmatrix}t\\0\end{smallmatrix}\mathbf{X}$ relates the differential material vectors $d\,^t\mathbf{s}$ and $d^0\mathbf{s}$ (consisting of the same material particles) in the configuration $^t\mathcal{B}$ and the initial configuration $^0\mathcal{B}$ at a material point (point P in Fig. 7.2.1). Similarly, if the body moves from the configuration $^t\mathcal{B}$ to the configuration $^{t+\Delta t}\mathcal{B}$, we can write

$$d^{t+\Delta t}\mathbf{s} = \begin{smallmatrix}t+\Delta t\\t\end{smallmatrix}\mathbf{X}\, d\,^t\mathbf{s} \tag{7.2.36}$$

and $\begin{smallmatrix}t+\Delta t\\t\end{smallmatrix}\mathbf{X}$ relates the material vectors in these two configurations.

Note that in case the material undergoes a rigid body translation plus a rotation between two configurations, the corresponding deformation gradient only represents a rotation (see (A2.39)).

Using the deformation gradient we can calculate the stretch of a material element. Namely, using (7.2.35) it follows that

$$
{}_0^t\lambda = \| {}_0^t\mathbf{X}\,{}^0\mathbf{n} \|
\tag{7.2.37}
$$

where $\|\cdot\|$ is the Euclidean vector norm (see (A2.30)), and the stretch ${}_0^t\lambda$ is defined as

$$
{}_0^t\lambda = \frac{\|d\,{}^t\mathbf{s}\|}{\|d\,{}^0\mathbf{s}\|} = \frac{d\,{}^t s}{d\,{}^0 s}
\tag{7.2.38}
$$

Here $d\,{}^0 s$ and $d\,{}^t s$ are the lengths of the material element in the two configurations. The unit vector ${}^0\mathbf{n}$ defines the direction of a material element in the initial configuration ${}^0\mathcal{B}$,

$$
{}^0\mathbf{n} = \frac{d\,{}^0\mathbf{s}}{d\,{}^0 s} = \left[\frac{d\,{}^0 x_k}{d\,{}^0 s} \right]
\tag{7.2.39}
$$

It follows that the stretches of the differential material vectors initially in the directions of the coordinate axes x, y, z are

$$
{}_0^t\lambda_x = \left\| {}_0^t\mathbf{X}^{(1)} \right\| \qquad {}_0^t\lambda_y = \left\| {}_0^t\mathbf{X}^{(2)} \right\| \qquad {}_0^t\lambda_z = \left\| {}_0^t\mathbf{X}^{(3)} \right\|
\tag{7.2.40}
$$

where ${}_0^t\mathbf{X}^{(i)}$, $i = 1, 2, 3$ are the first, second and third column-vectors in the matrix ${}_0^t\mathbf{X}$. The stretch of a material element with the direction of the unit vector ${}^t\mathbf{n}$ in the configuration ${}^t\mathcal{B}$ can be obtained from (7.2.18) and (7.2.35) as

$$
{}_0^t\lambda = \left\| {}_t^0\mathbf{X}\,{}^t\mathbf{n} \right\|^{-1}
\tag{7.2.41}
$$

The angle ${}^t\theta$ between the two material elements having the initial directions ${}^0\mathbf{n}$ and ${}^0\hat{\mathbf{n}}$, ${}^0\mathbf{n}$ and ${}^0\hat{\mathbf{n}}$ being unit vectors, can be calculated using (7.2.35) and (7.2.39), as

$$
{}^t\theta = \cos^{-1}\left[({}_0^t\mathbf{X}\,{}^0\mathbf{n}) \cdot ({}_0^t\mathbf{X}\,{}^0\hat{\mathbf{n}}) / ({}_0^t\lambda\,{}_0^t\hat{\lambda}) \right]
\tag{7.2.42}
$$

The initial angle ${}^0\theta$ between the material elements that have directions ${}^t\mathbf{n}$ and ${}^t\hat{\mathbf{n}}$ can be obtained as

$$
{}^0\theta = \cos^{-1}\left[({}^0_t\mathbf{X}\,{}^t\mathbf{n})\cdot({}^0_t\mathbf{X}\,{}^t\hat{\mathbf{n}})\,{}^t_0\lambda\,{}^t_0\hat{\lambda}\right]
\tag{7.2.43}
$$

Example 7.2.1 illustrates the calculation of the deformation gradient.

Deformation Tensors. Consider first the squared lengths of the material vector $d^t\mathbf{s}$ at a material point P (see Fig.7.2.1a) in the two configurations ${}^t\mathcal{B}$ and ${}^0\mathcal{B}$. It follows from (7.2.35) and (7.2.18) that

$$
\left(d^t s\right)^2 = d^t x_i\, d^t x_i = \frac{\partial^t x_i}{\partial^0 x_m}\frac{\partial^t x_i}{\partial^0 x_n}\, d^0 x_m\, d^0 x_n
\tag{7.2.44a}
$$

and

$$
\left(d^0 s\right)^2 = d^0 x_i\, d^0 x_i = \frac{\partial^0 x_i}{\partial^t x_m}\frac{\partial^0 x_i}{\partial^t x_n}\, d^t x_m\, d^t x_n
\tag{7.2.45a}
$$

or, in matrix (tensor) notation

$$
(d^t s)^2 = d^0\mathbf{s}^T\; {}^t_0\mathbf{C}\, d^0\mathbf{s}
\tag{7.2.44b}
$$

and

$$
(d^0 s)^2 = d^t\mathbf{s}^T\; {}^0_t\mathbf{B}\, d^t\mathbf{s}
\tag{7.2.45b}
$$

where

$$
\boxed{\; {}^t_0\mathbf{C} = {}^t_0\mathbf{X}^T\; {}^t_0\mathbf{X} \;}
\tag{7.2.46a}
$$

and

$$
\boxed{\; {}^0_t\mathbf{B} = {}^0_t\mathbf{X}^T\; {}^0_t\mathbf{X} \;}
\tag{7.2.47a}
$$

are called the *right Cauchy-Green deformation tensor* and the *Finger deformation tensor*, respectively. Their components are

$$
{}^t_0 C_{ij} = \frac{\partial^t x_m}{\partial^0 x_i}\frac{\partial^t x_m}{\partial^0 x_j}
\tag{7.2.46b}
$$

and

$$
{}^0_t B_{ij} = \frac{\partial^0 x_m}{\partial^t x_i}\frac{\partial^0 x_m}{\partial^t x_j}
\tag{7.2.47b}
$$

We will also use the tensor

$$
\boxed{\; {}^t_0\mathbf{B} = {}^t_0\mathbf{X}\; {}^t_0\mathbf{X}^T \;}
\tag{7.2.48}
$$

which is called the *left Cauchy-Green deformation tensor,* where of course

$$\substack{0\\t}\mathbf{B} = \substack{t\\0}\mathbf{B}^{-1} = \left(\substack{t\\0}\mathbf{X}^{-1}\right)^T \, \substack{t\\0}\mathbf{X}^{-1} \tag{7.2.49}$$

The stretch given in (7.2.38) and angle between two material elements defined in (7.2.42) can be obtained using the tensor $\substack{t\\0}\mathbf{C}$ as

$$\substack{t\\0}\lambda = \left(^0\mathbf{n}^T \, \substack{t\\0}\mathbf{C} \, ^0\mathbf{n}\right)^{1/2} \tag{7.2.50}$$

and

$$^t\theta = \cos^{-1}\left[^0\mathbf{n}^T \, \substack{t\\0}\mathbf{C} \, ^0\hat{\mathbf{n}}/ \left(\substack{t\\0}\lambda \, \substack{t\\0}\hat{\lambda}\right)\right] \tag{7.2.51}$$

Also, using the tensor $\substack{0\\t}\mathbf{B}$ we obtain

$$\substack{t\\0}\lambda = \left(^t\mathbf{n}^T \, \substack{0\\t}\mathbf{B} \, ^t\mathbf{n}\right)^{-1/2} \tag{7.2.52}$$

and (see (7.2.43))

$$^0\theta = \cos^{-1}\left[^t\mathbf{n}^T \, \substack{0\\t}\mathbf{B} \, ^t\hat{\mathbf{n}} \, \substack{t\\0}\lambda \, \substack{t\\0}\hat{\lambda}\right] \tag{7.2.53}$$

The deformation tensors are symmetric and positive definite. Symmetry follows from (7.2.46) and (7.2.48). The deformation tensors are positive definite because their eigenvalues are real and positive.[4]

It is useful for further developments to relate the deformation tensors corresponding to the configurations $^0\mathcal{B}$, $^t\mathcal{B}$ and $^{t+\Delta t}\mathcal{B}$. Using (7.2.24) and (7.2.46) we can write

$$^{t+\Delta t}_{0}\mathbf{C} = {}^{t+\Delta t}_{0}\mathbf{X}^T \, {}^{t+\Delta t}_{0}\mathbf{X} = \substack{t\\0}\mathbf{X}^T \, {}^{t+\Delta t}_{t}\mathbf{X}^T \, {}^{t+\Delta t}_{t}\mathbf{X} \, \substack{t\\0}\mathbf{X}$$

and

$$^{t+\Delta t}_{0}\mathbf{C} = \substack{t\\0}\mathbf{X}^T \, {}^{t+\Delta t}_{t}\mathbf{C} \, \substack{t\\0}\mathbf{X} \tag{7.2.54}$$

where

$$^{t+\Delta t}_{t}\mathbf{C} = {}^{t+\Delta t}_{t}\mathbf{X}^T \, {}^{t+\Delta t}_{t}\mathbf{X} \tag{7.2.55}$$

is the *relative right Cauchy-Green deformation tensor* with the reference configuration $^t\mathcal{B}$.

The following important characteristic of the right Cauchy-Green deformation tensor can be obtained from (7.2.54). If the material undergoes only a rigid body motion from the configuration $^t\mathcal{B}$ to $^{t+\Delta t}\mathcal{B}$, then

---

[4] By definition a matrix is positive definite if all eigenvalues are positive.

$$t+\Delta t \atop t} \mathbf{X} = {}^{t+\Delta t}_{t} \mathbf{R} \tag{7.2.56}$$

where ${}^{t+\Delta t}_{t}\mathbf{R}$ is the rotation tensor (see (A2.33)). Substituting into (7.2.54) we obtain

$${}^{t+\Delta t}_{0}\mathbf{C} = {}^{t}_{0}\mathbf{X}^{T} {}^{t+\Delta t}_{t}\mathbf{R}^{T} {}^{t+\Delta t}_{t}\mathbf{R} {}^{t}_{0}\mathbf{X} = {}^{t}_{0}\mathbf{X}^{T} {}^{t}_{0}\mathbf{X} = {}^{t}_{0}\mathbf{C} \tag{7.2.57}$$

since the rotation tensor ${}^{t+\Delta t}_{t}\mathbf{R}$ satisfies the orthogonality condition (A2.41),

$$\boxed{{}^{t+\Delta t}_{t}\mathbf{R}^{T} {}^{t+\Delta t}_{t}\mathbf{R} = \mathbf{I}} \tag{7.2.58}$$

Hence, the right Cauchy-Green deformation tensor is *invariant with respect to a rigid body motion*.

Analogously, from (7.2.48) we obtain

$${}^{t+\Delta t}_{0}\mathbf{B} = {}^{t+\Delta t}_{0}\mathbf{X} {}^{t+\Delta t}_{0}\mathbf{X}^{T} = {}^{t+\Delta t}_{t}\mathbf{X} {}^{t}_{0}\mathbf{X} {}^{t}_{0}\mathbf{X}^{T} {}^{t+\Delta t}_{t}\mathbf{X}^{T}$$

and

$$\boxed{{}^{t+\Delta t}_{0}\mathbf{B} = {}^{t+\Delta t}_{t}\mathbf{X} {}^{t}_{0}\mathbf{B} {}^{t+\Delta t}_{t}\mathbf{X}^{T}} \tag{7.2.59}$$

Substituting for ${}^{t+\Delta t}_{t}\mathbf{X}$ from (7.2.56) into (7.2.59), we find that ${}^{t+\Delta t}_{0}\mathbf{B} \neq {}^{t}_{0}\mathbf{B}$, i.e., the left Cauchy-Green deformation tensor *is not invariant* with respect to a rigid body rotation.

The relations (7.2.54) and (7.2.59) relate the deformation tensors in different configurations, but with respect to the same coordinate system x_i. In Example 7.2.1 we demonstrate that a coordinate transformation changes of course the components of the tensor but does not have any effect on the material deformation (see also Bathe 1996).

Polar Decomposition. Let us summarize some basic relations regarding the *polar decomposition theorem*. A detailed derivation of the theorem is given in Bathe (1982, 1996).

Consider the deformation of the material at a point P, Fig. 7.2.2a. Let the vectors ${}^{t}\mathbf{p}_\alpha$ and ${}^{t}\bar{\mathbf{p}}_\alpha$ (assumed to be of unit length) denote the eigenvectors of ${}^{t}_{0}\mathbf{C}$ and ${}^{t}_{t}\mathbf{B}$ (or ${}^{t}_{0}\mathbf{B}$). The eigenvalues of the tensors ${}^{t}_{0}\mathbf{C}$ and ${}^{t}_{0}\mathbf{B}$ are $({}^{t}_{0}\lambda_\alpha)^2$. The eigenvectors form two orthonormal bases. We call these bases the *right basis* and the *left basis* with the relation (see (A2.40))

$$\boxed{{}^{t}\bar{\mathbf{p}}_\alpha = {}^{t}_{0}\mathbf{R}\,{}^{t}\mathbf{p}_\alpha \qquad \alpha = 1, 2, 3} \tag{7.2.60a}$$

or in component form,

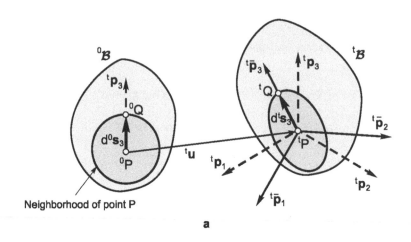

a

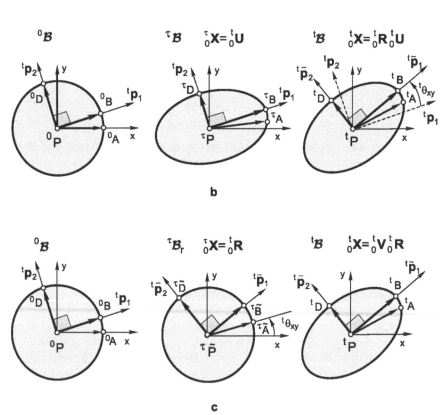

b

c

Fig. 7.2.2. Deformation represented by polar decomposition. **a** Principal directions of $_0^t\mathbf{C}$ and $_0^t\mathbf{B}$; **b** Polar decomposition $_0^t\mathbf{X} = _0^t\mathbf{R}\,_0^t\mathbf{U}$ (two-dimensional problem); **c** Polar decomposition $_0^t\mathbf{X} = _0^t\mathbf{V}\,_0^t\mathbf{R}$ (two-dimensional problem)

$$^t\bar{p}_{(\alpha)m} = {}^t_0 R_{mk}\,{}^t p_{(\alpha)k} \tag{7.2.60b}$$

where the index α in parentheses indicates the vector number and not the component number. The *rotation tensor* ${}^t_0\mathbf{R}$ is given by

$$\boxed{{}^t_0\mathbf{R} = \sum_\alpha {}^t\bar{\mathbf{p}}_\alpha \otimes {}^t\mathbf{p}_\alpha} \tag{7.2.61a}$$

or in component form

$$^t_0 R_{mk} = \sum_\alpha {}^t\bar{p}_{(\alpha)m}\,{}^t p_{(\alpha)k} \tag{7.2.61b}$$

Figure 7.2.2a shows that a material of an initially spherical shape changes into an ellipsoid with the principal axes ${}^t\bar{\mathbf{p}}_1, {}^t\bar{\mathbf{p}}_2, {}^t\bar{\mathbf{p}}_3$ (see Love 1944 for a further discussion).

The polar decomposition theorem gives the relation

$$\boxed{{}^t_0\mathbf{X} = {}^t_0\mathbf{R}\,{}^t_0\mathbf{U}} \tag{7.2.62}$$

where

$$\boxed{{}^t_0\mathbf{U} = \sum_\alpha {}^t_0\lambda_\alpha\,{}^t\mathbf{p}_\alpha \otimes {}^t\mathbf{p}_\alpha} \tag{7.2.63}$$

is the *right stretch tensor*. Here ${}^t_0\lambda_\alpha$ are the principal stretches obtained by the eigenanalysis of ${}^t_0\mathbf{C}$. Hence, we have that *the deformation gradient can be decomposed into a stretch and a rotation*. Physically, this theorem shows that any material element $d^0\mathbf{s}$ deforms into $d^t\mathbf{s}$ by application of the stretch ${}^t_0\mathbf{U}$ and then the rotation ${}^t_0\mathbf{R}$.

Hence we can think of applying a sequence of two deformation gradients, from the initial configuration ${}^0\mathcal{B}$ to a *fictitious*, or *conceptual* configuration ${}^\tau\mathcal{B}$ with the deformation gradient

$$^\tau_0\mathbf{X} = {}^t_0\mathbf{U} \tag{7.2.64}$$

and then from ${}^\tau\mathcal{B}$ to the final configuration ${}^t\mathcal{B}$ by the deformation gradient

$$^t_\tau\mathbf{X} = {}^t_0\mathbf{R} \tag{7.2.65}$$

since

$$^t_0\mathbf{X} = {}^t_\tau\mathbf{X}\,{}^\tau_0\mathbf{X} \tag{7.2.66}$$

Figure 7.2.2b shows this deformation sequence in a two-dimensional deformation (in the x, y plane). Two principal vectors lie in the x, y plane while the third principal vector is normal to the plane of deformation. By applying the stretch ${}_0^t\mathbf{U}$ the circular material surrounding point P deforms into the ellipse in the conceptual configuration ${}^\tau\mathcal{B}$, with the principal directions ${}^t\mathbf{p}_1$ and ${}^t\mathbf{p}_2$. All material vectors containing the point P change lengths and directions (as the vector $\overrightarrow{PA}$ shown in the figure), except the vectors $\overrightarrow{PB}$ and $\overrightarrow{PD}$, lying along the principal axes, which change their lengths only. The shaded square element of the material shown in Fig. 7.2.2b changes its size but the sides remain orthogonal. The final position of the material in the configuration ${}^t\mathcal{B}$ is obtained by the rigid body rotation for the angle ${}^t\theta_{xy}$.

Of course, we can change the sequence of deformations: we first conceptually rotate by the rotation ${}_0^t\mathbf{R}$ and then conceptually stretch by the *left stretch tensor* ${}_0^t\mathbf{V}$,

$$
{}_0^t\mathbf{X} = {}_0^t\mathbf{V}\,{}_0^t\mathbf{R} \tag{7.2.67}
$$

where clearly

$$
{}_0^t\mathbf{V} = {}_0^t\mathbf{R}\,{}_0^t\mathbf{U}\,{}_0^t\mathbf{R}^T \tag{7.2.68}
$$

Figure 7.2.2c shows this sequence of deformations in the case of a two-dimensional deformation. The conceptual configuration ${}^\tau\mathcal{B}_r$ is obtained by the rotation ${}^t\theta_{xy}$ and the final configuration ${}^t\mathcal{B}$ is reached by the stretch tensor ${}_0^t\mathbf{V}$. The stretch tensor ${}_0^t\mathbf{V}$ produces changes of lengths and directions of all material elements, except those lying along the principal directions of the left basis which only change their lengths. Note that the stretch tensor ${}_0^t\mathbf{V}$ has the same principal values ${}_0^t\lambda_\alpha$ as the right stretch tensor ${}_0^t\mathbf{U}$, and is given as

$$
{}_0^t\mathbf{V} = \sum_\alpha {}_0^t\lambda_\alpha\,{}^t\bar{\mathbf{p}}_\alpha \otimes {}^t\bar{\mathbf{p}}_\alpha \tag{7.2.69}
$$

In practical applications we can determine the rotation tensor ${}_0^t\mathbf{R}$ by calculating first the stretches ${}_0^t\lambda_\alpha$ and the right principal base vectors ${}^t\mathbf{p}_\alpha$ by an eigensolution of ${}_0^t\mathbf{C}$ (see Bathe 1996). Then

$$
{}_0^t\mathbf{U}^{-1} = \sum_\alpha {}_0^t\lambda_\alpha^{-1}\,{}^t\mathbf{p}_\alpha \otimes {}^t\mathbf{p}_\alpha \tag{7.2.70}
$$

and we obtain ${}_0^t\mathbf{R}$ as

$$\substack{t\\0}\mathbf{R} = \substack{t\\0}\mathbf{X}\, \substack{t\\0}\mathbf{U}^{-1}$$

(7.2.71)

If the principal values $\substack{t\\0}\lambda_i$ are distinct, we have unique sets of vectors ${}^t\mathbf{p}_1,\, {}^t\mathbf{p}_2,\, {}^t\mathbf{p}_3$, and ${}^t\bar{\mathbf{p}}_1,\, {}^t\bar{\mathbf{p}}_2,\, {}^t\bar{\mathbf{p}}_3$ forming the right and the left bases. In case *two principal values are equal*, say $\substack{t\\0}\lambda_1 = \substack{t\\0}\lambda_2$, the vectors ${}^t\mathbf{p}_3$ and ${}^t\bar{\mathbf{p}}_3$ are uniquely defined, and any sets of orthogonal vectors in the plane normal to ${}^t\mathbf{p}_3$ and ${}^t\bar{\mathbf{p}}_3$, form pairs of base vectors ${}^t\mathbf{p}_1,\, {}^t\mathbf{p}_2$ and ${}^t\bar{\mathbf{p}}_1,\, {}^t\bar{\mathbf{p}}_2$, respectively. Finally, in case *all principal values are equal*, any sets of orthogonal vectors represent admissible right and left bases, and $\substack{t\\0}\mathbf{R}$ can be taken to be the identity matrix (see Bathe 1996).

Generalized Strain Measure. According to Hill (1978), a smooth monotonic function $g(\substack{t\\0}\lambda_i)$ of the principal stretches $\substack{t\\0}\lambda_i$ may define a strain measure. Then the *generalized strains* in the principal basis ${}^t\mathbf{p}_i$ are

$$\substack{t\\0}\mathbf{E}_g = \sum_i {}^t g\left(\substack{t\\0}\lambda_i\right)\, {}^t\mathbf{p}_i \otimes {}^t\mathbf{p}_i$$

(7.2.72a)

or, in matrix form

$$\substack{t\\0}\mathbf{E}_g = {}^t\mathbf{R}_L\, {}^t\mathbf{g}\left(\substack{t\\0}\Lambda\right)\, {}^t\mathbf{R}_L^T$$

(7.2.72b)

where ${}^t\mathbf{g}(\substack{t\\0}\Lambda)$ is the diagonal matrix

$$
{}^t\mathbf{g}\left(\substack{t\\0}\Lambda\right) =
\begin{bmatrix}
{}^t g\left(\substack{t\\0}\lambda_1\right) & & \\
& {}^t g\left(\substack{t\\0}\lambda_2\right) & \\
& & {}^t g\left(\substack{t\\0}\lambda_3\right)
\end{bmatrix}
$$

(7.2.73)

and ${}^t\mathbf{R}_L$ is the matrix defined as

$$
{}^t\mathbf{R}_L = \begin{bmatrix} {}^t\mathbf{p}_1 & {}^t\mathbf{p}_2 & {}^t\mathbf{p}_3 \end{bmatrix}
$$

(7.2.74)

The strain $\substack{t\\0}\mathbf{E}_g$ defined in (7.2.72) corresponds to the *fictitious configuration* ${}^\tau\mathcal{B}$ and the deformation gradient (7.2.64). Since the material vectors with the principal stretches $\substack{t\\0}\lambda_i$ at configuration ${}^t\mathcal{B}$, reached by the stretch and rotation, have directions of the left basis $\bar{\mathbf{p}}_i$, we also can define the generalized strain measure as

$$\substack{t\\0}\hat{\mathbf{E}}_g = \sum_i {}^t g\left(\substack{t\\0}\lambda_i\right)\, {}^t\bar{\mathbf{p}}_i \otimes {}^t\bar{\mathbf{p}}_i$$

(7.2.75a)

or, in matrix form

$$\hat{{}_0^t\mathbf{E}}_g = {}^t\mathbf{R}_E \, {}^t\mathbf{g}\left({}_0^t\Lambda\right) \, {}^t\mathbf{R}_E^T \tag{7.2.75b}$$

where ${}^t\mathbf{R}_E$ is

$$ {}^t\mathbf{R}_E = \begin{bmatrix} {}^t\bar{\mathbf{p}}_1 & {}^t\bar{\mathbf{p}}_2 & {}^t\bar{\mathbf{p}}_3 \end{bmatrix} \tag{7.2.76}$$

It follows from (7.2.60) that

$$ {}^t\mathbf{R}_E = {}_0^t\mathbf{R} \, {}^t\mathbf{R}_L \tag{7.2.77}$$

and

$$ \hat{{}_0^t\mathbf{E}}_g = {}_0^t\mathbf{R} \, {}_0^t\mathbf{E}_g \, {}_0^t\mathbf{R}^T \tag{7.2.78}$$

For a reasonable strain measure we have the conditions that

$$g(1) = 0$$
$$\left(\frac{\partial^t g}{\partial {}_0^t\lambda_i}\right)_{{}_0^t\lambda_i=1} = g'(1) = 1 \tag{7.2.79}$$

since the material is undeformed for the stretch

$$ {}_0^t\lambda_i = 1 \tag{7.2.80}$$

and the generalized strains should reduce to the small strains when considering infinitesimal deformations.

Hill considered a family of functions

$$ {}^t g^{(n)}\left({}_0^t\lambda_i\right) = \frac{1}{2n}\left({}_0^t\lambda_i^{2n} - 1\right) \tag{7.2.81}$$

which satisfy the conditions (7.2.79). For the cases $n = 1$ and $n = -1$ we obtain from (7.2.72) and (7.2.81) that

$$ {}_0^t\mathbf{E}_g^{(1)} = {}_0^t\boldsymbol{\epsilon} = \frac{1}{2}\left({}_0^t\mathbf{C} - \mathbf{I}\right) \tag{7.2.82}$$

$$ {}_0^t\mathbf{E}_g^{(-1)} = {}_t^t\boldsymbol{\epsilon}^A = \frac{1}{2}\left(\mathbf{I} - {}_t^0\mathbf{C}\right) \tag{7.2.83}$$

where ${}_0^t\boldsymbol{\epsilon}$ and ${}_t^t\boldsymbol{\epsilon}^A$ are the *Green-Lagrange* and *Almansi* strains. The Almansi strains are usually used as (Bathe 1982, 1996)

$$
{}^t_0\hat{\mathbf{E}}_g^{(-1)} = {}^t_t\hat{\epsilon}^A = \frac{1}{2}\left(\mathbf{I} - {}^0_t\mathbf{B}\right)
\tag{7.2.84}
$$

If $n \to 0$

$$
{}^tg^{(0)}\left({}^t_0\lambda_i\right) = \ln {}^t_0\lambda_i
\tag{7.2.85}
$$

and we can define the *logarithmic or Hencky strain*, as

$$
{}^t_0\mathbf{E}^{(H)} = \sum_i \ln {}^t_0\lambda_i\, {}^t\mathbf{p}_i \otimes {}^t\mathbf{p}_i
\tag{7.2.86}
$$

$$
{}^t_0\hat{\mathbf{E}}^{(H)} = \sum_i \ln {}^t_0\lambda_i\, {}^t\bar{\mathbf{p}}_i \otimes {}^t\bar{\mathbf{p}}_i
\tag{7.2.87}
$$

in the right and the left bases ("*with the rigid body rotation removed*" and "*not removed*", respectively).

The strain-stretch relations for uniaxial deformation and for the strains defined in (7.2.82), (7.2.83), (7.2.86), and small strains, are given in Example 7.2.2. Small strains are given by

$$
{}^te_{ij} = \frac{1}{2}\left[\partial\,{}^tu_i/\partial\,{}^0x_j + \partial\,{}^tu_j/\partial\,{}^0x_i\right]
\tag{7.2.88}
$$

Assume that the principal directions do not change during the deformations, as, for example, in the cases of triaxial, biaxial or uniaxial extensions or compressions of a specimen. For a principal axis x we have

$$
{}^t_0E^{(H)}_{xx} = \ln {}^t_0\lambda_x = \ln \frac{{}^t\ell_x}{{}^0\ell_x} = \int_{{}^0\ell_x}^{{}^t\ell_x} \frac{d\ell}{\ell}
\tag{7.2.89}
$$

where ${}^0\ell_x$ and ${}^t\ell_x$ are the initial length and the length at time t of a material element in the x-direction. Hence the logarithmic strain, also referred to as the *true* or *natural* strain, represents the integral of $d\ell/\ell$. Note that

$$
d\,{}^t_0E^{(H)}_{xx} = \frac{d\,{}^t\ell_x}{{}^t\ell_x}
\tag{7.2.90}
$$

where $d\,{}^t\ell_x/{}^t\ell_x$ is the infinitesimal strain with respect to the configuration ${}^t\mathcal{B}$. Therefore we can write for the principal directions fixed in space

$$(d_0^t E_{ii}^{(H)})_{\mathbf{p}i=const} = d(\ln {}_0^t\lambda_i) = d^t e_{ii} \quad \text{no sum on i} \qquad (7.2.91)$$

The logarithmic strain measure has been generally used to test materials in elasto-plastic large strain conditions (see, e.g., Crandall et al. 1972; Makinouchi et al. 1993; Lee et al. 1996), and it is therefore the appropriate strain measure to use in the finite element analysis of large strain elastic-plastic response. We shall use this strain measure in the computational algorithms of Section 7.3.2.

Work Conjugate Stress and Strain Measures. For the analysis of large deformation problems, we need to use stress measures that correspond to the above introduced strains. These stress measures are defined using the *stress power* ${}^tP$ *per unit current volume* (at time t, for the configuration ${}^t\mathcal{B}$) given as

$$\boxed{{}^tP = {}^t\tau_{ij}\,{}^tD_{ij}} \qquad (7.2.92)$$

where ${}^t\tau_{ij}$ are the Cauchy stress components, and

$$\boxed{{}^tD_{ij} = \frac{1}{2}\left(\frac{\partial\,{}^tv_i}{\partial\,{}^tx_j} + \frac{\partial\,{}^tv_j}{\partial\,{}^tx_i}\right)} \qquad (7.2.93)$$

are the components of the *rate-of-deformation* (or *velocity strain*) tensor ${}^t\mathbf{D}$, with ${}^tv_i$ the components of the velocity vector ${}^t\mathbf{v}$,

$$\boxed{{}^t\mathbf{v} = \frac{d^t\mathbf{x}}{dt}} \qquad (7.2.94)$$

We say that the Cauchy stresses and velocity strains are *work conjugate* or *energetically conjugate*. Following this definition of work conjugacy, we can determine other stress measures which are work conjugate to given strain measures.

Consider first the generalized strains defined by (7.2.72). The power per unit current volume must be

$$\boxed{{}^tP = {}^t\tau_{ij}^{(g)}\,({}_0^t\dot{\mathbf{E}}_g)_{ij}} \qquad (7.2.95)$$

where the ${}^t\tau_{ij}^{(g)}$ are the components of the generalized stress ${}^t\tau^{(g)}$ which is work conjugate to the generalized strain ${}_0^t\mathbf{E}_g$; and $({}_0^t\dot{\mathbf{E}}_g)_{ij}$ are the components of the time rates of change of the generalized strain ${}_0^t\mathbf{E}_g$. If we let ${}^t\bar{\tau}_{ij}$ be the

Cauchy stress components in the left basis, then it follows from this relation that (for details see Bathe 1996)

$$^t\tau_{ij}^{(g)} = \frac{1}{^tg'\left(^t_0\lambda_i\right)\,^t_0\lambda_j}\,^t\bar{\tau}_{ij} \qquad i = j$$

$$^t\tau_{ij}^{(g)} = \frac{^t_0\lambda_j^2 - ^t_0\lambda_i^2}{2\left[^tg\left(^t_0\lambda_j\right) - ^tg\left(^t_0\lambda_i\right)\right]\,^t_0\lambda_i\,^t_0\lambda_j}\,^t\bar{\tau}_{ij} \qquad (7.2.96)$$

$$i \neq j, \quad ^t_0\lambda_i \neq ^t_0\lambda_j \quad \text{no sum on i,j}$$

These are the general expressions for the stress components $^t\tau_{ij}^{(g)}$, for any function $g(\lambda)$ satisfying the conditions (7.2.79). Note that the stress components $^t\tau_{ij}^{(g)}$ are acting on the planes defined by the right base vectors $^t\mathbf{p}_\alpha$ [5]. The relations between the stress components in the stationary system x_i and the stress components in the left basis follow from the tensorial transformation (see (A.2.24) and (A2.34))

$$\boxed{^t\boldsymbol{\tau} = {}^t\mathbf{R}_E\,{}^t\bar{\boldsymbol{\tau}}\,{}^t\mathbf{R}_E^T} \qquad (7.2.97)$$

where $^t\bar{\boldsymbol{\tau}}$ lists the stress components $^t\bar{\tau}_{ij}$ and the matrix $^t\mathbf{R}_E$ is defined by (7.2.76).

In the case of the logarithmic strain we substitute (7.2.85) into (7.2.96) and obtain

$$^t\tau_{ij}^{(0)} = {}^t\bar{\tau}_{ij} \qquad\qquad i = j \qquad (7.2.98)$$

$$^t\tau_{ij}^{(0)} = \frac{^t_0\lambda_j^2 - ^t_0\lambda_i^2}{2\,^t_0\lambda_i\,^t_0\lambda_j}\ln\left[\frac{^t_0\lambda_i}{^t_0\lambda_j}\right]\,^t\bar{\tau}_{ij} \qquad i \neq j \text{ no sum on i,j}$$

7.2.1 Examples

Example 7.2.1. Large Strain Deformation of a Material Element.
The material element shown in Fig. E.7.2-1 is rotated by the angle α in step 1, and then deformed as shown in the $\bar{x}_1$ direction in step 2. Assume a linear variation of displacements within the element and use $u_B = u_0$.

  a) Determine the deformation gradients $^1_0\mathbf{X}$ and $^0_1\mathbf{X}$ and the line vector $d^1\mathbf{s}$ originally inclined at angle β_0 with respect to the x_1 axis.

  b) Calculate the deformation gradients $^2_0\mathbf{X}$ and $^2_1\mathbf{X}$ and determine the stretches of the line vectors $d\mathbf{a}$ and $d\mathbf{b}$ originally parallel to the coordinate axes $^0x_1$, $^0x_2$. Perform the calculations for a point with the initial coordinates $^0x_1$, $^0x_2$.

---

[5] (7.2.96) is an unusual equation in that the stress components on the left-hand and right-hand sides are measured in different bases, as mentioned above.

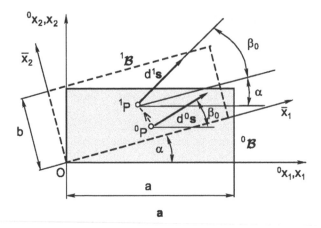

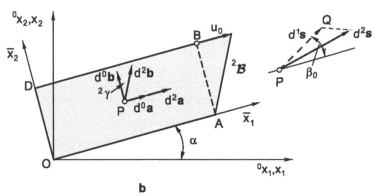

Fig. E.7.2-1. Deformation of a material element in two steps

a) Coordinates of a point P at time $t = t_1$ are (see Fig. E.7.2-1a)

$$^1x_1 = {}^0x_1 \cos\alpha - {}^0x_2 \sin\alpha \qquad {}^1x_2 = {}^0x_1 \sin\alpha + {}^0x_2 \cos\alpha \qquad \text{(a)}$$

Then, according to (7.2.12) the deformation gradient ${}^1_0\mathbf{X}$ is

$$\left[\frac{\partial\,{}^1x_i}{\partial\,{}^0x_j}\right] = {}^1_0\mathbf{X} = {}^1_0\mathbf{R} = \begin{bmatrix} \cos\alpha & -\sin\alpha \\ \sin\alpha & \cos\alpha \end{bmatrix} \qquad \text{(b)}$$

where ${}^1_0\mathbf{R}$ is the rotation matrix. Writing the inverse relations to (a),

$$^0x_1 = {}^1x_1 \cos\alpha + {}^1x_2 \sin\alpha \qquad {}^0x_2 = -{}^1x_1 \sin\alpha + {}^1x_2 \cos\alpha \qquad \text{(c)}$$

we obtain

$$\left[\frac{\partial\,{}^0x_i}{\partial\,{}^1x_j}\right] = {}^0_1\mathbf{X} = {}^1_0\mathbf{R}^T = \begin{bmatrix} \cos\alpha & \sin\alpha \\ -\sin\alpha & \cos\alpha \end{bmatrix} \qquad \text{(d)}$$

The same result for $^0_1\mathbf{X}$ can be obtained by inverting the matrix in (b). Note that

$$^1_0 X_{i3} = \frac{\partial\,^t x_i}{\partial\,^0 x_3} = 0; \quad ^1_0 X_{3i} = 0; \quad i = 1, 2 \qquad ^1_0 X_{33} = 1 \tag{e}$$

where x_3 is the third axis of the Cartesian coordinate system x_1, x_2, x_3.

The components of a material vector $\overrightarrow{PQ}$ of length d^0s at time $t = 0$ and time $t = t_1$ are

$$d\,^0 s_1 = d^0 s \cos \beta_0 \qquad d^0 s_2 = d^0 s \sin \beta_0 \tag{f}$$

and

$$d^1 s_1 = d^0 s \cos(\alpha + \beta_0) \qquad d^1 s_2 = d^0 s \sin(\alpha + \beta_0) \tag{g}$$

The expressions (g) also follow from (7.2.35).

b) We assume the displacement $^2\bar{u}$ in the rotated coordinate system $\bar{x}_1, \bar{x}_2$ to be

$$^2\bar{u}(\bar{x}_1, \bar{x}_2) = \frac{u_0}{ab}\,\bar{x}_1\bar{x}_2 \tag{h}$$

The coordinates of a point in the coordinate system x_1, x_2 are

$$\begin{aligned}
^2 x_1 &= (^0 x_1 + c\,^0 x_1\,^0 x_2) \cos\alpha - {}^0 x_2 \sin\alpha \\
^2 x_2 &= (^0 x_1 + c\,^0 x_1\,^0 x_2) \sin\alpha + {}^0 x_2 \cos\alpha
\end{aligned} \tag{i}$$

where $c = u_0/(ab)$. Now we determine the deformation gradient $^2_0\mathbf{X}$ as follows:

$$\begin{aligned}
^2_0 X_{11} &= (1 + c\,^0 x_2) \cos\alpha & ^2_0 X_{12} &= c\,^0 x_1 \cos\alpha - \sin\alpha \\
^2_0 X_{21} &= (1 + c\,^0 x_2) \sin\alpha & ^2_0 X_{22} &= c\,^0 x_1 \sin\alpha + \cos\alpha
\end{aligned} \tag{j}$$

We next analyze the change of the material vectors $d\mathbf{a}$ and $d\mathbf{b}$ which at $t = 0$ have directions of the coordinate axes x_1 and x_2. By using (7.2.35) we obtain from (j),

$$\begin{aligned}
d^2 a_1 &= {}^2_0 X_{1k}\, d^0 a_k = (1 + c\,^0 x_2) \cos\alpha\, d^0 a \\
d^2 a_2 &= {}^2_0 X_{2k}\, d^0 a_k = (1 + c\,^0 x_2) \sin\alpha\, d^0 a
\end{aligned} \tag{k}$$

where $d^0 a$ is the initial length of the material element. Similarly we have for the line element $d\mathbf{b}$,

$$\begin{aligned}
d^2 b_1 &= {}^2_0 X_{1k}\, d^0 b_k = (c\,^0 x_1 \cos\alpha - \sin\alpha)\, d^0 b \\
d^2 b_2 &= {}^2_0 X_{2k}\, d^0 b_k = (c\,^0 x_1 \sin\alpha + \cos\alpha)\, d^0 b
\end{aligned} \tag{l}$$

From (k) we see that the line elements originally parallel to the x_1 direction extend, and their stretch is

$$\overset{2}{_0}\lambda_{(1)} = \frac{d^2a}{d^0a} = 1 + c^0 x_2 \qquad \text{(m)}$$

where d^2a is the element length at step 2. Note that the line elements along the line OA do not stretch. The line segment $d\mathbf{b}$ changes length and also direction with respect to the material coordinate system $\bar{x}_1, \bar{x}_2$. From (1) we obtain the stretch

$$\overset{2}{_0}\lambda_{(2)} = \frac{d^2b}{d^0b} = \sqrt{1 + c^2 \, ^0 x_1^2} \qquad \text{(n)}$$

and the angle $^2\gamma$ with respect to the material axis $\bar{x}_2$ is determined by

$$\tan\left(^2\gamma\right) = \frac{d^2\,\bar{b}_1}{d^2\,\bar{b}_2} = c^0 x_1 \qquad \text{(o)}$$

where $d^2\,\bar{b}_1$ and $d^2\,\bar{b}_2$ are the components of $d^2\mathbf{b}$ in the $\bar{x}_1, \bar{x}_2$ coordinate system. The line OD remains undeformed, while the stretches and rotations of the material element $d\mathbf{b}$ increases with $^0 x_1$ (see Fig. E.7.2-1b).

For an arbitrary material segment $d\mathbf{s}$ with initial components

$$d^0 s_1 = \cos\beta_0 \, d^0 s \qquad d^0 s_2 = \sin\beta_0 \, d^0 s \qquad \text{(p)}$$

using (7.2.35) and (j) we obtain

$$d^2 s_1 = \left[(1 + c^0 x_2)\cos\alpha\cos\beta_0 + (c^0 x_1 \cos\alpha - \sin\alpha)\sin\beta_0\right] d^0 s$$
$$d^2 s_2 = \left[(1 + c^0 x_2)\sin\alpha\cos\beta_0 + (c^0 x_1 \sin\alpha + \cos\alpha)\sin\beta_0\right] d^0 s \qquad \text{(q)}$$

As shown for the material elements $d\mathbf{a}$ and $d\mathbf{b}$, we can calculate the stretch $\overset{2}{_0}\lambda = d^2 s / d^0 s$ and the direction of the element $d^2\mathbf{s}$. It is found that $\overset{2}{_0}\lambda$ varies between $\overset{2}{_0}\lambda_{(1)}$ and $\overset{2}{_0}\lambda_{(2)}$ and that the rotation angle changes between zero and $^2\gamma$, depending on the initial direction of the line element $d\mathbf{s}$.

We next calculate the relative deformation gradient $\overset{2}{_1}\mathbf{X}$. From (7.2.24) we obtain

$$\overset{2}{_1}\mathbf{X} = \overset{2}{_0}\mathbf{X}\overset{0}{_1}\mathbf{X} \qquad \text{(r)}$$

Substituting (d) and (j) into (r) we obtain

$$\overset{2}{_1}X_{11} = 1 + c\left(^0 x_2 \cos\alpha - {}^0 x_1 \sin\alpha\right)\cos\alpha$$
$$\overset{2}{_1}X_{12} = c\left(^0 x_2 \sin\alpha + {}^0 x_1 \cos\alpha\right)\cos\alpha$$
$$\overset{2}{_1}X_{21} = c\left(^0 x_2 \cos\alpha - {}^0 x_1 \sin\alpha\right)\sin\alpha \qquad \text{(s)}$$
$$\overset{2}{_1}X_{22} = 1 + c\left(^0 x_2 \sin\alpha + {}^0 x_1 \cos\alpha\right)\sin\alpha$$

Finally, we calculate the deformation gradient $\overset{2}{_1}\bar{\mathbf{X}}$ corresponding to the coordinate system $\bar{x}_1, \bar{x}_2$. Using (h) we obtain

$$^2\bar{x}_1 = {}^1\bar{x}_1 + c\,{}^1\bar{x}_1\,{}^1\bar{x}_2 \qquad ^2\bar{x}_2 = {}^1\bar{x}_2 \tag{t}$$

from which follows

$$^2_1\bar{X}_{11} = \frac{\partial\,^2\bar{x}_1}{\partial\,^1\bar{x}_1} = 1 + c\,{}^1\bar{x}_2 \qquad ^2_1\bar{X}_{12} = \frac{\partial\,^2\bar{x}_1}{\partial\,^1\bar{x}_2} = c\,{}^1\bar{x}_1 = c\,{}^0x_1$$

$$^2_1\bar{X}_{21} = \frac{\partial\,^2\bar{x}_2}{\partial\,^1\bar{x}_1} = 0 \qquad ^2_1\bar{X}_{22} = \frac{\partial\,^2\bar{x}_2}{\partial\,^1\bar{x}_2} = 1 \tag{u}$$

The coordinate transformation between the coordinate systems x_1, x_2 and $\bar{x}_1, \bar{x}_2$ can be written in the form

$$x_i = {}^1_0R_{ik}\,\bar{x}_k \tag{v}$$

where the ${}^1_0R_{ik}$ are given by (b). Then, according to the tensorial transformation (A2.24) (see also (A2.34)) we have

$$^2_1\mathbf{X} = {}^1_0\mathbf{R}\,{}^2_1\bar{\mathbf{X}}\,{}^1_0\mathbf{R}^T \tag{w}$$

Substituting $^2_1\bar{\mathbf{X}}$ from (u) into (w) we obtain $^2_1\mathbf{X}$ given by (s).

Example 7.2.2. Strain-Stretch Dependence. Determine the strain-stretch dependence for various strain measures, in uniaxial deformation.

According to the definitions (7.2.88), (7.2.82), (7.2.83) and (7.2.86) for the small strain, Green-Lagrange, Almansi and logarithmic strains, respectively, we obtain the following expressions for these strains

$$^t e_{11} = {}^t_0\lambda - 1 \qquad\qquad ^t_0\epsilon_{11} = \frac{1}{2}({}^t_0\lambda^2 - 1)$$

$$^t_t\epsilon^A_{11} = \frac{1}{2}(1 - {}^t_0\lambda^{-2}) \qquad ^t_0E^{(H)}_{11} = \ln {}^t_0\lambda \tag{a}$$

Figure E.7.2-2 gives a graphical representation of these expressions.

It can be seen from the figure that for stretches around $\lambda = 1$ all strain measures give approximately the same values. Also, in extension the Green-Lagrange strain and logarithmic strain increase to infinity, while the Almansi strain tends to 0.5. In compression, the Green-Lagrange and the Almansi strains change character, while the logarithmic strain tends to minus infinity. The small strain values change linearly with the stretch.

Example 7.2.3. Large Strain Shear. Determine the principal directions, deformation tensors, the spin and the rate-of-deformation tensors for the material subjected to what is referred to as "simple shear", as shown in Fig. E.7.2-3a.

From the geometry in Fig. E.7.2-3a we obtain the deformation gradient as

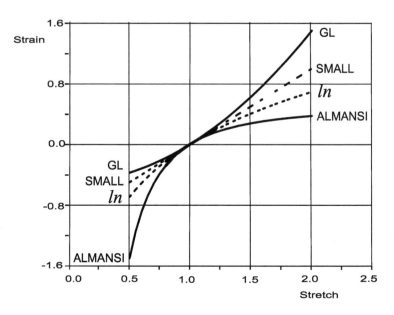

Fig. E.7.2-2. Strain-stretch dependence for various strain measures

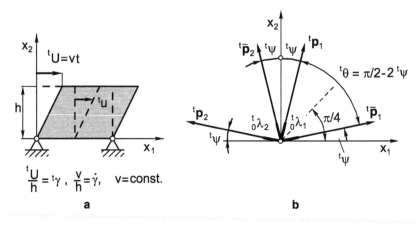

Fig. E.7.2-3. Large strain shear of material. **a** Kinematics of deformation; **b** Principal stretches and principal vectors

$$
{}_{0}^{t}\mathbf{X} = \begin{bmatrix} 1 & {}^{t}\gamma \\ 0 & 1 \end{bmatrix}
\tag{a}
$$

or in tensor notation

$$
{}_{0}^{t}\mathbf{X} = \mathbf{i}_1 \otimes \mathbf{i}_1 + {}^{t}\gamma\, \mathbf{i}_1 \otimes \mathbf{i}_2 + \mathbf{i}_2 \otimes \mathbf{i}_2
\tag{b}
$$

Also $_0^t X_{i3} = _0^t X_{3i} = \delta_{i3}$ and we further consider only components in the plane (x_1, x_2). The right and the left Cauchy-Green deformation tensors are (see (7.2.46) and (7.2.48))

$$_0^t \mathbf{C} = _0^t \mathbf{X}^T \, _0^t \mathbf{X} = \begin{bmatrix} 1 & ^t\gamma \\ ^t\gamma & 1 + ^t\gamma^2 \end{bmatrix} \tag{c}$$

and

$$_0^t \mathbf{B} = _0^t \mathbf{X} \, _0^t \mathbf{X}^T = \begin{bmatrix} 1 + ^t\gamma^2 & ^t\gamma \\ ^t\gamma & 1 \end{bmatrix} \tag{d}$$

The principal stretches, obtained from

$$\det(_0^t \mathbf{C} - _0^t \lambda^2 \mathbf{I}) = 0 \quad \text{or} \quad \det(_0^t \mathbf{B} - _0^t \lambda^2 \mathbf{I}) = 0 \tag{e}$$

are

$$_0^t \lambda_1 = 1/_0^t \lambda_2 = \cot {}^t\psi \tag{f}$$

where

$$^t\psi = \frac{1}{2} \tan^{-1}(2/^t\gamma) \tag{g}$$

is the angle shown in Fig. E.7.2-3b. Solving the eigenvalue problem

$$_0^t \mathbf{C} \, ^t\mathbf{p}_i = _0^t \lambda_i^2 \, ^t\mathbf{p}_i \tag{h}$$

$$_0^t \mathbf{B} \, ^t\bar{\mathbf{p}}_i = _0^t \lambda_i^2 \, ^t\bar{\mathbf{p}}_i \tag{i}$$

we obtain

$$^t\theta_r = \pi/2 - {}^t\psi \qquad ^t\theta_\ell = {}^t\psi \tag{j}$$

where $^t\theta_r$ and $^t\theta_\ell$ are the angles between the $^t\mathbf{p}_1$ and the x_1- axis and between the $^t\bar{\mathbf{p}}_1$ and the x_1-axis, respectively. The first vectors $^t\mathbf{p}_1$ and $^t\bar{\mathbf{p}}_1$ of the right and the left bases start from the initial direction $^0\theta_r = {}^0\theta_\ell = \pi/4$ and then rotate in the opposite directions. For very large stretches, the extended fibers are close to the x_1-direction, while the compressed fibers approach the x_2-direction.

The stretch tensor represented by $_0^t\Lambda$ in the principal directions $\mathbf{p}_i$ and $_0^t\mathbf{U}$ corresponding to the x_1, x_2 coordinate system, are

$$_0^t\Lambda = \begin{bmatrix} _0^t\lambda_1 & 0 \\ 0 & _0^t\lambda_2 \end{bmatrix} \tag{k}$$

$$_0^t\mathbf{U} = \begin{bmatrix} \sin 2\,^t\psi & \cos 2\,^t\psi \\ \cos 2\,^t\psi & ^t\gamma \cos 2\,^t\psi + \sin 2\,^t\psi \end{bmatrix} \tag{l}$$

We have obtained $_0^t\mathbf{U}$ from $_0^t\Lambda$ by the tensorial transformation (A2.24). The rotation tensor $_0^t\mathbf{R}$ can be obtained using (7.2.61), where we substitute

$$^t\mathbf{p}_1 = \sin{}^t\psi\,\mathbf{i}_1 + \cos{}^t\psi\,\mathbf{i}_2 \qquad ^t\mathbf{p}_2 = -\cos{}^t\psi\,\mathbf{i}_1 + \sin{}^t\psi\,\mathbf{i}_2$$
$$^t\bar{\mathbf{p}}_1 = \cos{}^t\psi\,\mathbf{i}_1 + \sin{}^t\psi\,\mathbf{i}_2 \qquad ^t\bar{\mathbf{p}}_2 = -\sin{}^t\psi\,\mathbf{i}_1 + \cos{}^t\psi\,\mathbf{i}_2 \qquad \text{(m)}$$

Then we have

$$_0^t\mathbf{R} = \begin{bmatrix} sin2\,{}^t\psi & \cos 2\,{}^t\psi \\ -\cos 2\,{}^t\psi & \sin 2\,{}^t\psi \end{bmatrix} \qquad \text{(n)}$$

Also, $_0^t\mathbf{R}$ can be obtained from (a) and (l) by employing (7.2.62).
    The rate-of-deformation tensor, defined in (7.2.93), is

$$^t\mathbf{D} = \frac{\dot{\gamma}}{2} \begin{bmatrix} 0 & 1 \\ 1 & 0 \end{bmatrix} \qquad \text{(o)}$$

The spin tensor $^t\mathbf{W}$ defined as

$$^t W_{ij} = \frac{1}{2}\Big(\frac{\partial\,{}^t v_i}{\partial\,{}^t x_j} - \frac{\partial\,{}^t v_j}{\partial\,{}^t x_i}\Big)$$

is in this case

$$^t\mathbf{W} = \frac{\dot{\gamma}}{2} \begin{bmatrix} 0 & 1 \\ -1 & 0 \end{bmatrix} \qquad \text{(p)}$$

    The left deformation tensor $_0^t\mathbf{V}$ follows from (7.2.67), (a) and (n), or (7.2.68), (l) and (n),

$$_0^t\mathbf{V} = \begin{bmatrix} ^t\gamma\cos 2\,{}^t\psi + \sin 2\,{}^t\psi & \cos 2\,{}^t\psi \\ \cos 2\,{}^t\psi & \sin 2\,{}^t\psi \end{bmatrix} \qquad \text{(q)}$$

Also, $_0^t\mathbf{V}$ can be obtained using the tensor transformation

$$_0^t\mathbf{V} = {}^t\mathbf{R}_E\,_0^t\bar{\mathbf{V}}\,{}^t\mathbf{R}_E^T \quad \text{with} \quad {}^t\mathbf{R}_E = [\bar{\mathbf{p}}_1\,\bar{\mathbf{p}}_2] \quad \text{and} \quad _0^t\bar{\mathbf{V}} = {}_0^t\Lambda$$

Note that $_0^t\mathbf{U}$ given by (l) can be obtained from (7.2.63), (f) and (m).
    The rate of change of angles $^t\psi$ and $^t\theta$ follows from the relations (g) and (j),

$$^t\dot{\psi} = -\frac{\dot{\gamma}}{4}\sin^2 2\,{}^t\psi \qquad ^t\dot{\theta} = -2\,{}^t\dot{\psi} = \frac{\dot{\gamma}}{2}\sin^2 2\,{}^t\psi \qquad \text{(r)}$$

Using these relations we obtain from (n),

$$_0^t\dot{\mathbf{R}} = {}^t\dot{\theta} \begin{bmatrix} -\sin{}^t\theta & \cos{}^t\theta \\ -\cos{}^t\theta & -\sin{}^t\theta \end{bmatrix} \qquad \text{(s)}$$

The matrix ${}^{t}\mathbf{R}_{L}$, defined in (7.2.74), is

$$
{}^{t}\mathbf{R}_{L} = [{}^{t}\mathbf{p}_{1}\ {}^{t}\mathbf{p}_{2}] = \begin{bmatrix} \sin {}^{t}\psi & -\cos {}^{t}\psi \\ \cos {}^{t}\psi & \sin {}^{t}\psi \end{bmatrix}
\tag{t}
$$

Differentiating with respect to time we obtain

$$
{}^{t}\dot{\mathbf{R}}_{L} = {}^{t}\dot{\psi} \begin{bmatrix} \cos {}^{t}\psi & \sin {}^{t}\psi \\ -\sin {}^{t}\psi & \cos {}^{t}\psi \end{bmatrix}
\tag{u}
$$

From (t) and (u) we determine the spin tensor Ω_{L} of the right basis rotation (see Bathe 1996)

$$
{}^{t}\Omega_{L} = {}^{t}\mathbf{R}_{L}^{T}\ {}^{t}\dot{\mathbf{R}}_{L} = {}^{t}\dot{\psi} \begin{bmatrix} 0 & 1 \\ -1 & 0 \end{bmatrix}
\tag{v}
$$

In analogy with (t) to (v), from (7.2.76) and (m) follow the expressions:

$$
{}^{t}\mathbf{R}_{E} = \begin{bmatrix} \cos {}^{t}\psi & -\sin {}^{t}\psi \\ \sin {}^{t}\psi & \cos {}^{t}\psi \end{bmatrix} \qquad {}^{t}\dot{\mathbf{R}}_{E} = {}^{t}\dot{\psi} \begin{bmatrix} -\sin {}^{t}\psi & -\cos {}^{t}\psi \\ \cos {}^{t}\psi & -\sin {}^{t}\psi \end{bmatrix}
$$
$$
{}^{t}\Omega_{E} = {}^{t}\mathbf{R}_{E}^{T}\ {}^{t}\dot{\mathbf{R}}_{E} = -{}^{t}\dot{\psi} \begin{bmatrix} 0 & 1 \\ -1 & 0 \end{bmatrix}
\tag{w}
$$

From the expressions for ${}^{t}\Omega_{L}$ and ${}^{t}\Omega_{E}$ we see that the right and the left bases rotate with the same angular velocity, but in the opposite directions.

7.3 Stress Integration in Isotropic Plasticity Using the Logarithmic Strains

In this section we first discuss some basic assumptions used in the stress integration when considering large strain effects in plasticity, and then present a computational algorithm using the Hencky (logarithmic) strain. The stress calculation is performed within the so-called *updated-Lagrangian-Hencky* (ULH) formulation. The derivation of the algorithm is given in a general form and for the general 3-D elastic-plastic deformations, representing an extension of the computational procedures for small strains developed in the previous chapters. Finally, a number of solutions are given, with some comparisons of results to those available in the literature.

7.3.1 Introduction

In the development given below we assume that the elastic deformations are small and therefore we can use the *elastic constitutive law* (see (A1.5))

$$^t\boldsymbol{\tau} = {}^t\mathbf{C}^E \, {}^t_0\mathbf{e}^E \tag{7.3.1}$$

where $^t\boldsymbol{\tau}$ and $^t_0\mathbf{e}^E$ are the Cauchy stress and elastic strain, while $^t\mathbf{C}^E$ is the elastic constitutive matrix (possibly a function of temperature $^t\theta$). In case the elasticity relationship depends on the stress/strain state (see Sections 6.2.2 and 6.3) an incremental form of (7.3.1) is applicable

$$^{t+\Delta t}\boldsymbol{\tau} = f(^t\boldsymbol{\tau}, \, \Delta\mathbf{e}^E, \ldots) \tag{7.3.2}$$

where $^t\boldsymbol{\tau}$ and $^{t+\Delta t}\boldsymbol{\tau}$ are the Cauchy stresses at the start and the end of the time step, and $\Delta\mathbf{e}^E$ is the elastic strain increment in the time step.

Since the stresses are related to the elastic strains, we somehow have to establish these strains from the total deformations. One way to proceed is to use the *additive decomposition* of total strains. Then, analogous to small strain conditions (Chapters 4 and 6) we write

$$^t_0\mathbf{e} = {}^t_0\mathbf{e}^E + {}^t_0\mathbf{e}^P \tag{7.3.3}$$

where $^t_0\mathbf{e}$ is the total strain and $^t_0\mathbf{e}^P$ is the plastic strain. However, this assumption does not have the generality of analysis that can be achieved considering the micromechanics of elastic-plastic material deformation.

A more general approach to obtain the elastic strains is based on the *multiplicative decomposition* of the deformation gradient. The assumption is that the deformation gradient $^t_0\mathbf{X}$ can be written in the form

$$^t_0\mathbf{X} = {}^t_0\mathbf{X}^E \, {}^t_0\mathbf{X}^P \tag{7.3.4}$$

where $^t_0\mathbf{X}^E$ and $^t_0\mathbf{X}^P$ are the *elastic* and *plastic deformation gradients*. This assumption was introduced by Lee and Liu (1967), and Lee (1969), and is based on considerations of micromechanics of plastic flow (Perzyna and Wojno1968). According to (7.3.4), the state of material deformation corresponding to the deformation gradient $^t_0\mathbf{X}$ can conceptually be obtained by first deforming the material plastically for the deformation gradient $^t_0\mathbf{X}^P$, and then elastically for the deformation gradient $^t_0\mathbf{X}^E$. Figure 7.3.1 shows schematically the multiplicative decomposition.

The assumption (7.3.4) leads to the introduction of the so-called *conceptual intermediate local stress-free configuration* as represented in the figure. Namely, since the stresses are proportional to the elastic strains, the state of

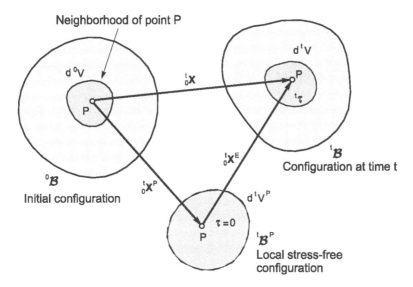

Fig. 7.3.1. Multiplicative decomposition of deformation gradient

deformation corresponding to the gradient ${}_0^t\mathbf{X}^P$ represents the state of *zero stresses*. We use the term *local* stress-free configuration because the stresses are equal to zero at a material point, or in an infinitesimal volume dV surrounding the point. The stress-free configuration ${}^t\mathcal{B}^P$ of the body corresponds to the deformation gradient field ${}_0^t\mathbf{X}^P(x_i)$. This deformation field does not in general represent a field of compatible deformations, and hence the stress-free configuration ${}^t\mathcal{B}^P$ of a body may physically not be possible (and is hence only conceptual). In Fig. 7.3.2a we have a bar stressed in the axial direction by a force which causes the plastic deformation ${}^te_{xx}^P = {}^te^P$ constant throughout the material. The release of the load produces the stress-free state with a field of compatible deformations and constant strain ${}^te_{xx} = {}^te^P$. Therefore the stress-free configuration ${}^t\mathcal{B}^P$ is physically possible. On the other hand, in Fig. 7.3.2b we have a beam plastically deformed by a moment ${}^tM$, assuming that the material is perfectly plastic. The distribution of the total strain in the cross-section corresponds to beam theory (Crandall et al. 1972). We see that a field of incompatible deformations corresponds to the conceptual stress-free state. While the conceptual stress-free configuration is not a compatible displacement configuration, the actual solution of the problem using the decomposition (7.3.4) gives of course the physically correct solution (see Example 7.3.4 where the assumptions on the kinematics of deformation analogous to beam theory are employed and the local stress-free configuration gives an incompatible deformation field). Hence, using the conceptual configuration does not result in a physical contradiction, and indeed as we see below, results into a general solution approach.

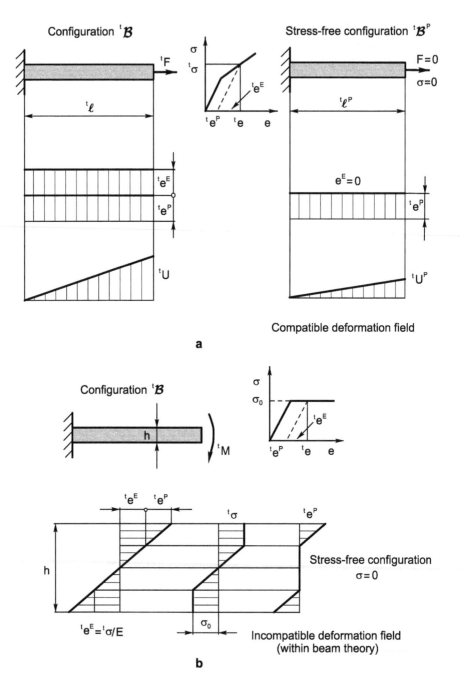

Fig. 7.3.2. Compatible and incompatible strain fields for the conceptual stress-free configurations using multiplicative decomposition of $_0^t\mathbf{X}$. **a** Case of compatible strains; **b** Case of incompatible deformations

7.3.2 Stress Integration in Large Strain Plasticity

We first present the continuum mechanics basis for large strain elastic-plastic deformations assuming isotropic elastic behavior with use of the right basis and employing the Cauchy stresses and the logarithmic strains. Hence we use the "rotated" (with respect to a material) fictitious configuration in which the elastic rigid body motion of the material is removed (see also (7.2.86) and (7.2.87)). Then we give the derivation of the governing relations of the computational procedure for the stress integration in the case of a von Mises material with mixed hardening behavior. Therefore the material data input is the Cauchy stress - logarithmic strain (natural or true strain) relationship. The presentation follows Eterovic and Bathe (1990); Gabriel and Bathe (1995); Bathe (1996); and Montans and Bathe (2003).

The Continuum Mechanics Basis. Let us give some basic assumptions regarding the material deformation and the basic relations used subsequently. We start with the relation

$$\boxed{{}^t J = \det {}^t_0 \mathbf{X} = \det {}^t_0 \mathbf{X}^E \det {}^t_0 \mathbf{X}^P} \tag{7.3.5}$$

from which the volumetric logarithmic strain follows

$$\boxed{{}^t_0 e_V = \ln \left(\det {}^t_0 \mathbf{X} \right) = {}^t_0 e_V^E + {}^t_0 e_V^P} \tag{7.3.6}$$

Hence, for the volumetric strain we have the additive decomposition rule. When the *plastic deformations are incompressible* we have

$$\det {}^t_0 \mathbf{X}^P = 1 \tag{7.3.7}$$

and then it follows from (7.3.5) that

$$^t J = {}^t J^E = \det {}^t_0 \mathbf{X}^E \tag{7.3.8}$$

Therefore, the volumetric deformation is elastic. We can use the polar decomposition (7.2.62) for the elastic deformation gradient ${}^t_0 \mathbf{X}^E$,

$$\boxed{{}^t_0 \mathbf{X}^E = {}^t_0 \mathbf{R}^E {}^t_0 \mathbf{U}^E} \tag{7.3.9}$$

where ${}^t_0 \mathbf{U}^E$ is the elastic stretch tensor and ${}^t_0 \mathbf{R}^E$ is the rotation tensor. The elastic logarithmic strain ${}^t_0 \mathbf{E}^E$, which we assume to correspond to small elastic deformations, in the right basis follows from (7.2.86) and (7.3.9),

$$\,^t_0\mathbf{E}^E = \ln\,^t_0\mathbf{U}^E = \sum_A \ln\,^t_0\lambda^E_A\,\,^t\mathbf{p}^E_A \otimes\,^t\mathbf{p}^E_A \tag{7.3.10}$$

We next define the *modified plastic velocity gradient* $\,^t\bar{\mathbf{L}}^P$ as

$$\,^t\bar{\mathbf{L}}^P = \,^t_0\dot{\mathbf{X}}^P\,^0_t\mathbf{X}^P \tag{7.3.11}$$

and will assume later that the *modified plastic spin vanishes* (for a discussion of this assumption, see Eterovic and Bathe 1991b),

$$\,^t\bar{\mathbf{W}}^P = \mathrm{skw}\,^t\bar{\mathbf{L}}^P = 0 \tag{7.3.12}$$

Using (7.3.11) we have

$$\,^t\bar{\mathbf{D}}^P = \frac{1}{2}\left[\,^t_0\dot{\mathbf{X}}^P\,^0_t\mathbf{X}^P + (\,^t_0\dot{\mathbf{X}}^P\,^0_t\mathbf{X}^P)^T\right] \tag{7.3.13}$$

where $\,^t\bar{\mathbf{D}}^P$ is the *modified plastic velocity strain* tensor.

Since $\,^t_0\mathbf{E}^E$ is associated with "a rotated" conceptual configuration, we define the stress measure $\,^t\bar{\boldsymbol{\tau}}$ corresponding to that configuration as

$$\,^t\bar{\boldsymbol{\tau}} = \,^tJ\,(\,^t_0\mathbf{R}^E)^T\,\,^t\boldsymbol{\tau}\,\,^t_0\mathbf{R}^E \tag{7.3.14}$$

When the elastic strains are small we have

$$\,^tJ = \det\,^t_0\mathbf{X} \approx 1 \tag{7.3.15}$$

Considering a material that in elastic response is isotropic, it can be shown (Bathe 1996) that the following relation holds

$$\,^tJ\,^t\boldsymbol{\tau} \cdot \,^t\mathbf{D} = \,^t\bar{\boldsymbol{\tau}} \cdot \,^t\dot{\bar{\mathbf{E}}}^E + \,^t\bar{\boldsymbol{\tau}} \cdot \,^t\bar{\mathbf{D}}^P \tag{7.3.16}$$

where $\,^t\mathbf{D}$ is the velocity strain tensor (7.2.93). This relation shows that the stress and strain measures are energy conjugate and — furthermore — that the usual small strain plasticity relations can be used with these measures in an incremental step (for more detail see Eterovic and Bathe 1991b).

Note that when the volumetric plastic strains are not equal to zero, as in the case of geological materials, we use the additive decomposition (7.3.6) for the volumetric strains. When the plastic deformations are incompressible, as in the case of metals considered further, the relation (7.3.6) is not used. Also, the derived relations are applicable to plasticity constitutive relations of metals with mixed hardening behavior.

Stress Integration Procedure. As in small strain plasticity (Chapters 4 and 6) we adopt the return mapping method of implicit stress integration. Hence, we first determine the *trial elastic stress state*. The trial elastic deformation gradient $^{t+\Delta t}_0 \mathbf{X}^E_*$ follows from (7.3.4) as

$$
{}^{t+\Delta t}_0 \mathbf{X}^E_* = {}^{t+\Delta t}_0 \mathbf{X} \, {}^0_t \mathbf{X}^P
\tag{7.3.17}
$$

where $^{t+\Delta t}_0 \mathbf{X}$ is known from the displacement field, and $^0_t \mathbf{X}^P$ is known from the history of deformation. We perform the polar decomposition (7.3.9),

$$
{}^{t+\Delta t}_0 \mathbf{X}^E_* = {}^{t+\Delta t}_0 \mathbf{R}^E_* \, {}^{t+\Delta t}_0 \mathbf{U}^E_*
\tag{7.3.18}
$$

and obtain the *trial elastic strain* tensor (7.3.10),

$$
{}^{t+\Delta t}_0 \mathbf{E}^E_* = \ln {}^{t+\Delta t}_0 \mathbf{U}^E_* = \sum_A \ln {}^{t+\Delta t}_0 \lambda^E_{*A} \, {}^{t+\Delta t} \mathbf{p}^E_A \otimes {}^{t+\Delta t} \mathbf{p}^E_A
\tag{7.3.19}
$$

We next use the elastic constitutive law corresponding to the *trial elastic stress* $^{t+\Delta t} \bar{\tau}^E_*$ and elastic logarithmic strain $^{t+\Delta t}_0 \mathbf{E}^E_*$ (Anand 1979, 1985), see (4.4.9) and (4.4.6). The trial elastic mean and deviatoric stresses are (see (A1.12) and (A1.19))

$$
{}^{t+\Delta t} \bar{\sigma}^E_{*m} = c_m \, {}^{t+\Delta t}_0 E^E_{*m}
\tag{7.3.20}
$$

$$
{}^{t+\Delta t} \bar{\mathbf{S}}^E_* = 2G \, {}^{t+\Delta t}_0 \mathbf{E}'^E_*
\tag{7.3.21}
$$

where $^{t+\Delta t}_0 E^E_{*m}$ and $^{t+\Delta t}_0 \mathbf{E}'^E_*$ are the trial elastic mean and deviatoric strains. The deviatoric stress $^{t+\Delta t} \bar{\mathbf{S}}^E_*$ corresponds to the conceptual "rotated" configuration, and the trial elastic stress radius $^{t+\Delta t} \hat{\mathbf{S}}^E_*$ in this configuration can be obtained from (7.3.21) and (4.4.15) as

$$t+\Delta t \hat{\mathbf{S}}_*^E = {}^{t+\Delta t}J^{-1}\, {}^{t+\Delta t}\bar{\hat{\mathbf{S}}}_*^E - {}^t\bar{\alpha} \tag{7.3.22}$$

where ${}^t\bar{\alpha}$ is the back stress in the "rotated" configuration. Next we check for yielding [6] according to (4.4.16), i.e.,

$$t+\Delta t \hat{\sigma}_*^E = \sqrt{\frac{3}{2}} \, \|{}^{t+\Delta t}\hat{\mathbf{S}}_*^E\| \le {}^t\hat{\sigma}_y \tag{7.3.23}$$

The case of no yielding is addressed below, after (7.3.35).

In the case of yielding the plasticity calculations are the same as in Section 4.4.1 for small strains. Namely, the fundamental constitutive relations of plasticity (4.4.18) and (4.4.20) have now the form (Eterovic and Bathe 1991b)

$$\Delta \mathbf{e}^P = \Delta\lambda\, {}^{t+\Delta t}\hat{\mathbf{S}}_\tau \tag{7.3.24}$$

and

$$\Delta\bar{\alpha} = \Delta\lambda\hat{C}\, {}^{t+\Delta t}\hat{\mathbf{S}}_\tau \tag{7.3.25}$$

where the stress radius ${}^{t+\Delta t}\hat{\mathbf{S}}_\tau$ corresponds to the "rotated" configuration. The scalar $\Delta\lambda$ is given by (4.4.24) where

$$\Delta\bar{e}^P = \Delta t\, \sqrt{\frac{2}{3}\, \bar{\mathbf{D}}^P \cdot \bar{\mathbf{D}}^P} \tag{7.3.26}$$

and

$$t+\Delta t \hat{\sigma} = {}^{t+\Delta t}\hat{\sigma}_y = \sqrt{\frac{3}{2}\, {}^{t+\Delta t}\hat{\mathbf{S}}_\tau \cdot {}^{t+\Delta t}\hat{\mathbf{S}}_\tau} \tag{7.3.27}$$

Then the governing equation (4.4.29) is

$$f(\Delta\bar{e}^P) = \frac{{}^{t+\Delta t}\hat{\sigma}_*^E}{{}^{t+\Delta t}\hat{\sigma}_y + \frac{3}{2}\left(2G + \hat{C}\right)\Delta\bar{e}^P} - 1 = 0 \tag{7.3.28}$$

The stress radius ${}^{t+\Delta t}\hat{\mathbf{S}}_\tau$ follows from (4.4.23)

---

[6] We here measure yielding using Cauchy stresses (in the rotated configuration). If the experimental data refers to Kirchhoff stresses, these stresses should be used, but the difference in solution results is small as long as the elastic strains are small (assumed in this Section).

$$t+\Delta t \hat{\mathbf{S}}_\tau = \frac{t+\Delta t \hat{\mathbf{S}}_*^E}{1 + \left(2G + \hat{C}\right) \Delta\lambda} \tag{7.3.29}$$

and the Cauchy stress deviator $t+\Delta t \mathbf{S}_\tau$ can be obtained from (4.4.22),

$$t+\Delta t \mathbf{S}_\tau = {}^t\bar{\alpha} + (1 + \Delta\lambda\hat{C}) {}^{t+\Delta t}\hat{\mathbf{S}}_\tau \tag{7.3.30}$$

In order to determine the stress deviator $t+\Delta t \mathbf{S}$ corresponding to the Cauchy stress $t+\Delta t \tau$, the transformation to the configuration $t+\Delta t \mathcal{B}$ (see (7.3.14)) must be performed. Therefore we have

$$\boxed{t+\Delta t \mathbf{S} = {}^{t+\Delta t}_0 \mathbf{R}_*^E \, {}^{t+\Delta t}\mathbf{S}_\tau \, ({}^{t+\Delta t}_0 \mathbf{R}_*^E)^T} \tag{7.3.31}$$

This transformation takes into account the fact that the true (Cauchy) stresses are acting on the material planes that rotate during the material deformations. We add the mean stress $t+\Delta t \sigma_m$ calculated from (7.3.20) as

$$\boxed{t+\Delta t \sigma_m = {}^{t+\Delta t} J^{-1} \, {}^{t+\Delta t}\bar{\sigma}_{*m}^E} \tag{7.3.32}$$

to the deviator $t+\Delta t \mathbf{S}$, to obtain the Cauchy stress

$$\boxed{t+\Delta t \tau = {}^{t+\Delta t}\mathbf{S} + {}^{t+\Delta t}\sigma_m} \tag{7.3.33}$$

Of course $t+\Delta t \sigma_m$ is a diagonal tensor corresponding to the normal stresses only.

As shown by Gabriel and Bathe (1995), the accuracy of the solution may be improved by use of the α-method ($0 \le \alpha \le 1$).

Finally, the plastic deformation gradient $t+\Delta t \atop 0 \mathbf{X}^P$ must be calculated for use in the next time step. We integrate (7.3.11) (with use of (7.3.12) and (7.3.24)) and obtain

$$\boxed{{}^{t+\Delta t}_0 \mathbf{X}^P = \exp(\Delta \mathbf{e}^P) \, {}^t_0 \mathbf{X}^P} \tag{7.3.34}$$

where

$$\Delta \mathbf{e}^P = \int_t^{t+\Delta t} \bar{\mathbf{D}}^P d\tau = \Delta\lambda \, {}^{t+\Delta t}\hat{\mathbf{S}}_\tau \tag{7.3.35}$$

with initially ${}^{0}_{0}\mathbf{X}^{P} = \mathbf{I}$.

In the case of no yielding in the time step, we calculate the mean stress ${}^{t+\Delta t}\sigma_{m}$ from (7.3.32), and the deviatoric stress ${}^{t+\Delta t}\mathbf{S}$ from (7.3.31) where we use the trial elastic deviatoric stress ${}^{t+\Delta t}\bar{\mathbf{S}}^{E}_{*}$ (see (7.3.21)) instead of ${}^{t+\Delta t}\mathbf{S}_{\tau}$. Then the stress ${}^{t+\Delta t}\tau$ follows from (7.3.33). Of course, the back stress and the plastic deformation gradient remain unchanged.

Table 7.3.1 summarizes the computational steps for the stress calculation. The formulation is of course directly applicable to plane strain and axisymmetric conditions, and the consistent tangent matrix is obtained by differentiation, see Chapter 4, and Pantuso and Bathe (1997). Of course, this formulation can also be extended to include thermal and creep effects (see Pantuso and Bathe 1997).

Table 7.3.1. Stress integration for large strain plastic deformations of a von Mises material

---

1. **Trial elastic state**

$${}^{t+\Delta t}_{0}\mathbf{X}^{E}_{*} = {}^{t+\Delta t}_{0}\mathbf{X} \, {}^{0}_{t}\mathbf{X}^{P}$$

$${}^{t+\Delta t}_{0}\mathbf{E}^{E}_{*} = \ln {}^{t+\Delta t}_{0}\mathbf{U}^{E}_{*} = \sum_{A} \ln {}^{t+\Delta t}_{0}\lambda^{E}_{*A} \, {}^{t+\Delta t}\mathbf{p}^{E}_{A} \otimes {}^{t+\Delta t}\mathbf{p}^{E}_{A}$$

$${}^{t+\Delta t}\bar{\mathbf{S}}^{E}_{*} = 2G \, {}^{t+\Delta t}_{0}\mathbf{E}'^{E}_{*} \qquad {}^{t+\Delta t}\bar{\sigma}^{E}_{*m} = c_{m} \, {}^{t+\Delta t}_{0}E^{E}_{*m}$$

2. **Elastic deformation**
   - Calculate mean stress ${}^{t+\Delta t}\sigma_{m}$
   $${}^{t+\Delta t}\sigma_{m} = {}^{t+\Delta t}J^{-1} \, {}^{t+\Delta t}\bar{\sigma}^{E}_{*m}$$
   - Calculate stress deviator ${}^{t+\Delta t}\mathbf{S}$
   $${}^{t+\Delta t}\mathbf{S} = {}^{t+\Delta t}_{0}\mathbf{R}^{E}_{*} \, {}^{t+\Delta t}\bar{\mathbf{S}}^{E}_{*} ({}^{t+\Delta t}_{0}\mathbf{R}^{E}_{*})^{T}$$
   - Calculate Cauchy stress ${}^{t+\Delta t}\tau$
   $${}^{t+\Delta t}\tau = {}^{t+\Delta t}\mathbf{S} + {}^{t+\Delta t}\sigma_{m}$$
   Go to step 5

3. **Plastic deformation − stress integration** in the "rotated" configuration
   $${}^{t+\Delta t}\hat{\mathbf{S}}^{E}_{*} = {}^{t+\Delta t}J^{-1} \, {}^{t+\Delta t}\bar{\mathbf{S}}^{E}_{*} - {}^{t}\bar{\alpha}$$
   - Solve the governing equation
   $$f(\Delta \bar{e}^{P}) = 0$$
   - Calculate the Cauchy stress deviator ${}^{t+\Delta t}\mathbf{S}_{\tau}$
   - Calculate stress deviator ${}^{t+\Delta t}\mathbf{S}$
   $${}^{t+\Delta t}\mathbf{S} = {}^{t+\Delta t}_{0}\mathbf{R}^{E}_{*} \, {}^{t+\Delta t}\mathbf{S}_{\tau} ({}^{t+\Delta t}_{0}\mathbf{R}^{E}_{*})^{T}$$

4. **Cauchy stress and updated plastic deformation gradient**
   $${}^{t+\Delta t}\tau = {}^{t+\Delta t}\mathbf{S} + {}^{t+\Delta t}J^{-1} \, {}^{t+\Delta t}\bar{\sigma}^{E}_{*m}$$
   $${}^{t+\Delta t}_{0}\mathbf{X}^{P} = \exp(\Delta \mathbf{e}^{P}) \, {}^{t}_{0}\mathbf{X}^{P}$$

5. **End**

Instead of using the right elastic Cauchy-Green deformation tensor (see (7.2.46))

$$
{}_0^t\mathbf{C}^E = ({}_0^t\mathbf{X}^E)^T\,{}_0^t\mathbf{X}^E
$$
(7.3.36)

we can also use – in a second closely related approach – the left elastic Cauchy-Green deformation tensor (see (7.2.48))

$$
{}_0^t\mathbf{B}^E = {}_0^t\mathbf{X}^E({}_0^t\mathbf{X}^E)^T
$$
(7.3.37)

In this case the elastic strains are measured in the configuration in which the elastic rotations have not been removed (the "unrotated" configuration). Then, in the stress integration procedure the history of plastic deformations is followed by appropriate updating of the tensor ${}_0^t\mathbf{B}^E$ instead of the updating of the plastic deformation gradient according to (7.3.34). This approach has been proposed by Simo (1988), (1993a), and Simo and Meschke (1993b); see also Peric and Owen (1992); Simo and Hughes (1998); Kojic et al. (1995d); Kojic et al. (2002b); Kojic (2002c); and Montans and Bathe (200x) for further developments and a variety of applications.

7.3.3 Examples

Example 7.3.1. Large Strain Isochoric Deformation of Material.
Calculate the stresses in case of two large strain isochoric deformations. Consider
  a) "Simple shear" of the material, according to the straining shown in Fig. E.7.3-1a;
  b) Mixed motion, defined by the stretch tensor (Gabriel and Bathe 1995), see (7.2.63) and (7.2.71),

$$
{}_0^t\mathbf{U} = {}^t\mathbf{R}\,{}_0^t\Lambda\,{}^t\mathbf{R}^T
$$
(a)

where the rotation matrix is defined as

$$
{}^t\mathbf{R} = \begin{bmatrix} \cos\varphi(t) & -\sin\varphi(t) & 0 \\ \sin\varphi(t) & \cos\varphi(t) & 0 \\ 0 & 0 & 1 \end{bmatrix}
$$
(b)

and the stretch matrix is

$$
{}_0^t\Lambda = \begin{bmatrix} \lambda(t) & 0 & 0 \\ 0 & 1/\lambda(t) & 0 \\ 0 & 0 & 1 \end{bmatrix}
$$
(c)

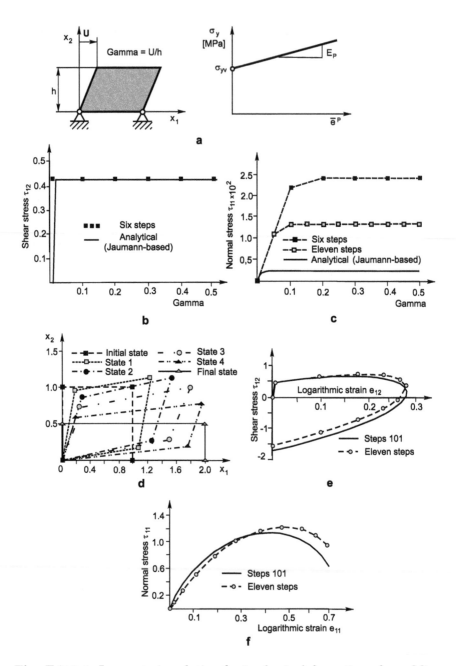

Fig. E.7.3-1. Large strain solution for isochoric deformation of von Mises material. **a** Kinematics of deformation in the case of "simple shear", and yield curve; **b** Shear stress for case **a**; **c** Normal stress τ_{11} for case **a**; **d** Mixed motion, deformation patterns; **e** Shear stress in the case of the mixed motion; **f** Normal stress τ_{11} in the case of the mixed motion

The functions $\varphi(t)$ and $\lambda(t)$ change linearly with time, from $\pi/4$ to 0, and from 1.0 to 2.0, respectively. The material data for the von Mises material is (in MPa):

Shear modulus $G = 76.92$ Bulk modulus $K = 166.67$

Yield stress $\sigma_{yv} = 0.75$ Hardening modulus $E_P = 0.00$ for case **a**

$E_P = 2.00$ for case **b**

The stress calculation is performed using the two approaches described in Section 7.3.2, i. e., with use of the "rotated" and "unrotated" configurations.

a) We assume a perfectly plastic material in this case, i.e., $E_P = 0$. For the stress integration using the left basis (Section 7.3.3) we need the relative deformation gradient, which now is

$$^{t+\Delta t}_{t}\mathbf{X} = \begin{bmatrix} 1 & \Delta\gamma \\ 0 & 1 \end{bmatrix}$$ (d)

The results for the stresses are shown in Figs. E.7.3-1b,c. The same results are obtained with the algorithms using the "rotated" and "unrotated" configurations. The analytical solution based on the Jaumann rate formulation (Weber et al. 1990b) is also given in the figure. Note that the numerical results for the shear (dominant) stress are very close to the analytical solution, even with a small number of steps. Some deviations from the analytical solution are noticeable for the normal stress, but the deviations diminish with an increase in the number of steps.

b) In this case we assume isotropic hardening (see Section 3.2) of the material, with the constant plastic modulus given above. Several deformed shapes of the material element are shown in Fig. E.7.3-1**d,** and the results for the shear and normal stresses are shown in Figs. E.7.3-1e,f.

The numerical solution accuracy of the simple isochoric large strain deformations is presented in Gabriel and Bathe (1995).

Example 7.3.2. Plastic Deformation of Thick-Walled Cylinder. Calculate the stresses and displacements for a thick-walled cylinder when subjected to radial displacements at the inner surface. The material data are given in Fig. E.7.3-2.

The solution is obtained using 20 initially equal 9-node axisymmetric finite elements subjected to prescribed radial displacements at the inner surface of the cylinder.

Fifteen equal steps are employed, with the full Newton-Raphson iterative method including line searches. The number of iterations per step is 3. We give here the unbalanced energies $E_{(i)}$ for the first step:

Iteration	1	2	3
Energy $E_{(i)}$ (Nmm)	$1.39E07$	$5.72E02$	$3.16E$-6

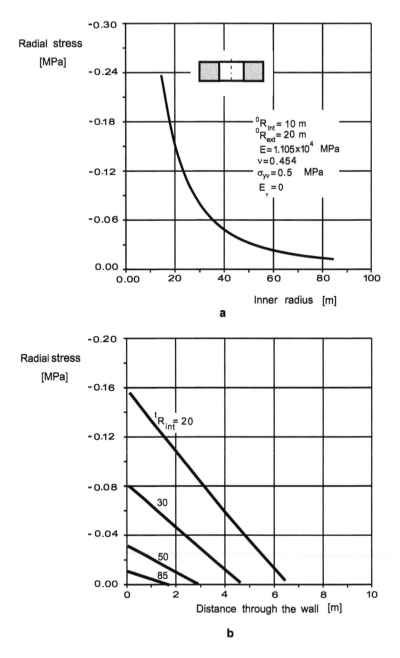

Fig. E.7.3-2. Plastic deformation of the thick-walled cylinder. **a** Compressive radial stress at the inner surface versus inner radius; **b** Radial stress distribution through the wall (measured from the inner surface) for several values of internal radius

The initial unbalanced energy is $E_{(0)} = 2.74 \times 10^9$.

Figure E.7.3-2**a** shows the radial compressive stress at the inner surface as a function of the inner radius. The radial stress distribution through the wall for several values of inner radius is shown in Fig. E.7.3-2**b**. Here the radial coordinate is measured from the cylinder internal surface. Note that the radial stress is equal to zero at the external surface. The results agree with those reported in Simo and Hughes (1998).

Example 7.3.3. Compression and Tension of a Plane Strain Element. A von Mises isotropic metal with the material data given in Fig. E.7.3-3 is subjected to compression and tension under plane strain conditions and free lateral surfaces. Friction at the contact surfaces is neglected. Calculate dependence of the axial stress on the displacement of the top surface.

The solution is obtained using a uniform mesh of plane strain finite elements and prescribed displacements at the top surface. The compressive (and tensile) stress τ_{yy} varies with displacements as shown in the figure (see also Eterovic and Bathe 1990).

Note that the stress in compression is larger than in tension because for the same absolute value of axial displacement, the absolute value of the logarithmic strain e_{yy} is larger in compression (see Fig. E.7.2-2).

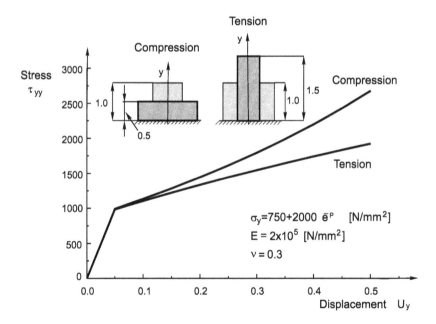

Fig. E.7.3-3. Compression and tension of a plane strain element

Example 7.3.4. Plastic bulging of circular plate. A circular plate connected to a rigid wall at the outer radius, is subjected to a normal pressure increasing with time. The geometric and material data are taken from Ibrahimbegovic (1994); Kojic (2002c):

$$R = 24 \text{ mm}, \quad \delta = 0.314 \text{ mm}$$
$$E = 6.867 \times 10^4 \text{ MPa}, \quad \nu = 0.314, \quad \sigma_y = 177 \, (\bar{e}^P)^{0.29} \text{ MPa}$$

Calculate the deflection of the plate under large strain conditions.

Two finite element models are employed: (a) an axisymmetric solid element model, and (b) a shell model (shown in Fig E.7.3-4a). The change of the shell thickness δ in the time step is taken into account using

$$^{t+\Delta t}\delta = {}^{t}\delta \, \exp(\Delta e_{zz})$$

where Δe_{zz} is the increment of the normal strain through the shell thickness.

The solution is obtained using 105 constant increments of pressure, and the final deformed configuration is shown in Fig E.7.3-4a. The dependences of the pressure and plate thickness at the central point on the displacement of the central point are shown in Figs. E.7.3-4b,c. The results obtained using the two finite element models are practically the same and agree with those reported in Ibrahimbegovic (1994).

Example 7.3.5. Tension of a Circular Bar (Gurson Material Model). A standard smooth tensile circular specimen with the dimensions given in Fig. E.7.3-5a is used to characterize the material behavior and to identify the critical damage parameters for ductile tearing at room temperature, Brocks (1995). The material is assumed to be given by the Gurson material model (see Example 4.5.11). The material constants are as follows:
Young's modulus $E = 250$ Gpa
Poisson's ratio $\nu = 0.3$
The uniaxial yield curve (see (3.2.7) and (3.2.51))

$$\sigma_y = 468.5 + 445.4 \, (\bar{e}^P)^{0.361}$$

Material constants of the Gurson model: $q_1 = 1.5$, $q_2 = 1.0$, $q_3 = 1.5$.
The initial porosity $f_0 = 0.002$; the failure porosity $f_f = 0.315$; the critical porosity $f_c = 0.05$.

One half of the bar is modeled due to symmetry with the boundary conditions shown in the figure. Two-dimensional 8-node axisymmetric elements (168 elements) are used and the loading is applied through the prescribed displacements (Kojic et al. 2002b).

The final end displacement of 3.625 mm (half of the total specimen elongation) is reached in 29 equal load steps. The force-elongation and force-change of diameter relations are shown in Figs. E.7.3-5b,c. The experimental results reported by Brocks (1995) are also shown in the figures.

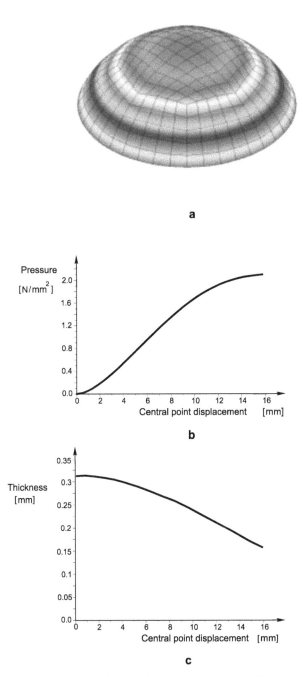

Fig. E.7.3-4. Plastic bulging of circular plate. **a** Final shape and finite element mesh; **b** Pressure - central point displacement dependence; **c** Dependence of plate thickness on displacement at central point

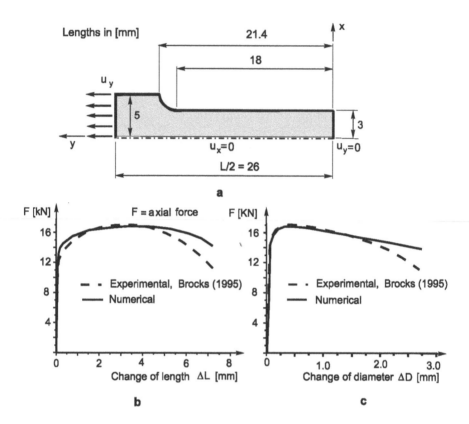

Fig. E.7.3-5. Tension of a circular bar (Gurson material model). **a** Geometry of the specimen and boundary conditions; **b** Axial force - elongation dependence; **c** Axial force - change of diameter dependence

Appendix A1

A Summary of Elastic and Thermoelastic Constitutive Relations

The elastic and thermoelastic constitutive relations are applicable when the material deformations are elastic or inelastic. The expressions are written in the form used in this book. More details are given in, e.g., Timoshenko and Goodier (1951); Wang (1953); Sokolnikoff (1956); Mendelson (1968); Fung (1969); Crandall et al. (1972); Jones (1975); Desai and Siriwardane (1984); Bathe (1982, 1996).

Isotropic Elastic Material. We use the indices 1,2,3 and x, y, z for the Cartesian components. Here we will use constitutive matrices, with two indices (not constitutive tensors of the fourth-order which correspond to the second-order stress and strain tensors), hence we write the stress and strain tensors $\boldsymbol{\sigma}$ and $\mathbf{e}$ using the one-index matrix notation[1] :

$$\boldsymbol{\sigma}^T = [\sigma_1, \ \sigma_2, \ \ldots, \ \sigma_6] \tag{A1.1}$$

where

$$\sigma_1 = \sigma_{11}, \ \sigma_2 = \sigma_{22}, \ \sigma_3 = \sigma_{33}$$
$$\sigma_4 = \sigma_{12}, \ \sigma_5 = \sigma_{23}, \ \sigma_6 = \sigma_{31} \tag{A1.2a}$$

or, with use of the indices x, y, z,

$$\sigma_1 = \sigma_{xx}, \ \sigma_2 = \sigma_{yy}, \ \sigma_3 = \sigma_{zz}$$
$$\sigma_4 = \sigma_{xy}, \ \sigma_5 = \sigma_{yz}, \ \sigma_6 = \sigma_{zx} \tag{A1.2b}$$

and

$$\hat{\mathbf{e}}^T = [e_1, \ e_2, \ \ldots, \ e_6] \tag{A1.3}$$

where

$$e_1 = e_{11}, \ e_2 = e_{22}, \ e_3 = e_{33}$$
$$e_4 = \gamma_{12}, \ e_5 = \gamma_{23}, \ e_6 = \gamma_{31} \tag{A1.4a}$$

or, with use of the indices x, y, z,

---

[1] This notation is also referred to as the Voigt notation

$$e_1 = e_{xx}, \ e_2 = e_{yy}, \ e_3 = e_{zz}$$
$$e_4 = \gamma_{xy}, \ e_5 = \gamma_{yz}, \ e_6 = \gamma_{zx} \tag{A1.4b}$$

Note that instead of the tensorial shear components e_{12}, e_{23} and e_{31}, the engineering strains $\gamma_{12} = 2e_{12}$, $\gamma_{23} = 2e_{23}$ and $\gamma_{31} = 2e_{31}$ (or $\gamma_{xy}, \gamma_{yz}, \gamma_{zx}$) are used in the vector $\hat{\mathbf{e}}$. The engineering shear strain components are used because in engineering practice the elastic constitutive matrix is formed assuming use of the engineering shear strains.

We call $\hat{\mathbf{e}}$ the strain vector. The "overhat" sign is also used for other strain vectors to indicate that the engineering shear strain components are used, such as: the elastic strain $\hat{\mathbf{e}}^E$, the plastic strain $\hat{\mathbf{e}}^P$, the creep strain $\hat{\mathbf{e}}^C$ and the viscoplastic strain $\hat{\mathbf{e}}^{VP}$.

The elastic constitutive relations for an isotropic material in general three-dimensional deformations are given as

$$\sigma = \mathbf{C}^E \hat{\mathbf{e}}^E \tag{A1.5}$$

and

$$\hat{\mathbf{e}}^E = (\mathbf{C}^E)^{-1} \sigma \tag{A1.6}$$

where $\mathbf{C}^E$ is the *elastic constitutive matrix*, and $(\mathbf{C}^E)^{-1}$ is the *elastic compliance matrix*. The coefficients of these matrices can be expressed in terms of two independent material constants, e.g., Young's modulus E and Poisson's ratio ν :

$$\mathbf{C}^E = \frac{E(1-\nu)}{(1+\nu)(1-2\nu)}$$

$$\times \begin{bmatrix} 1 & \dfrac{\nu}{1-\nu} & \dfrac{\nu}{1-\nu} & 0 & 0 & 0 \\[2mm] \dfrac{\nu}{1-\nu} & 1 & \dfrac{\nu}{1-\nu} & 0 & 0 & 0 \\[2mm] \dfrac{\nu}{1-\nu} & \dfrac{\nu}{1-\nu} & 1 & 0 & 0 & 0 \\[2mm] 0 & 0 & 0 & \dfrac{1-2\nu}{2(1-\nu)} & 0 & 0 \\[2mm] 0 & 0 & 0 & 0 & \dfrac{1-2\nu}{2(1-\nu)} & 0 \\[2mm] 0 & 0 & 0 & 0 & 0 & \dfrac{1-2\nu}{2(1-\nu)} \end{bmatrix} \tag{A1.7}$$

and

$$(\mathbf{C}^E)^{-1} = \begin{bmatrix} 1/E & -\nu/E & -\nu/E & 0 & 0 & 0 \\ -\nu/E & 1/E & -\nu/E & 0 & 0 & 0 \\ -\nu/E & -\nu/E & 1/E & 0 & 0 & 0 \\ 0 & 0 & 0 & 1/G & 0 & 0 \\ 0 & 0 & 0 & 0 & 1/G & 0 \\ 0 & 0 & 0 & 0 & 0 & 1/G \end{bmatrix} \qquad (\text{A1.8})$$

where G is the shear modulus

$$G = 0.5E/(1+\nu) \qquad (\text{A1.9})$$

If we introduce the mean stress σ_m and the mean elastic strain e_m^E as

$$\sigma_m = (\sigma_1 + \sigma_2 + \sigma_3)/3 \qquad (\text{A1.10})$$

$$e_m^E = (e_1^E + e_2^E + e_3^E)/3 \qquad (\text{A1.11})$$

then

$$\sigma_m = c_m e_m^E \qquad \text{no sum on } m \qquad (\text{A1.12})$$

where

$$c_m = \frac{E}{1 - 2\nu} \qquad (\text{A1.13})$$

Also,

$$\sigma_m = K e_V^E \qquad (\text{A1.14})$$

where

$$K = \frac{1}{3} c_m \qquad (\text{A1.15})$$

is called the *bulk modulus* of the material, and e_V^E is the elastic volumetric strain

$$e_V^E = e_{xx}^E + e_{yy}^E + e_{zz}^E = 3e_m^E \qquad (\text{A1.16})$$

We can relate the deviatoric stresses S_{ij},

$$S_{ij} = \sigma_{ij} - \sigma_m \delta_{ij} \qquad (\text{A1.17})$$

to the elastic deviatoric strains $e_{ij}^{\prime E}$,

$$\begin{aligned} e_{ij}^{\prime E} &= e_{ij}^E - e_m^E \qquad & i = j \\ e_{ij}^{\prime E} &= \frac{1}{2}\gamma_{ij}^E \qquad & i \neq j \end{aligned} \qquad (\text{A1.18})$$

as

$$S_{ij} = 2G\, e'^E_{ij} \tag{A1.19}$$

The above elastic constitutive matrix $\mathbf{C}^E$ can be directly used for the plane strain and axisymmetric conditions shown in Fig. A1.1. For plane strain deformations we need the relations between the stress components σ_{xx}, σ_{yy} and σ_{xy} and the corresponding strain components. The matrix $\mathbf{C}^E$ reduces to

$$\mathbf{C}^E = \frac{E(1-\nu)}{(1+\nu)(1-2\nu)}
\begin{array}{ccc}
xx & yy & xy \\
\begin{bmatrix}
1 & \dfrac{\nu}{1-\nu} & 0 \\[2mm]
\dfrac{\nu}{1-\nu} & 1 & 0 \\[2mm]
0 & 0 & \dfrac{1-2\nu}{2(1-\nu)}
\end{bmatrix}
\end{array} \tag{A1.20}$$

Note that the normal stress σ_{zz} can be obtained from the relations (assuming elastic deformation)

$$e^E_{zz} = 0 = \left[-\nu\left(\sigma_{xx} + \sigma_{yy}\right) + \sigma_{zz}\right]/E$$

and consequently

$$\sigma_{zz} = \nu(\sigma_{xx} + \sigma_{yy}) \tag{A1.21}$$

In the case of axisymmetric deformations (the radial plane is x, y) we have

$$\mathbf{C}^E = \frac{E(1-\nu)}{(1+\nu)(1-2\nu)}
\begin{array}{cccc}
xx & yy & xy & zz \\
\begin{bmatrix}
1 & \dfrac{\nu}{1-\nu} & 0 & \dfrac{\nu}{1-\nu} \\[2mm]
\dfrac{\nu}{1-\nu} & 1 & 0 & \dfrac{\nu}{1-\nu} \\[2mm]
0 & 0 & \dfrac{1-2\nu}{2(1-\nu)} & 0 \\[2mm]
\dfrac{\nu}{1-\nu} & \dfrac{\nu}{1-\nu} & 0 & 1
\end{bmatrix}
\end{array} \tag{A1.22}$$

We have interchanged here the third and fourth columns as it is usually done in applications (e.g., Bathe 1996).

In the case of plane stress conditions (see Fig. A1.1b) we impose the condition $\sigma_{zz} = 0$ in (A1.5) and obtain

$$e^E_{zz} = -\frac{\nu}{1-\nu}\left(e^E_{xx} + e^E_{yy}\right) \tag{A1.23}$$

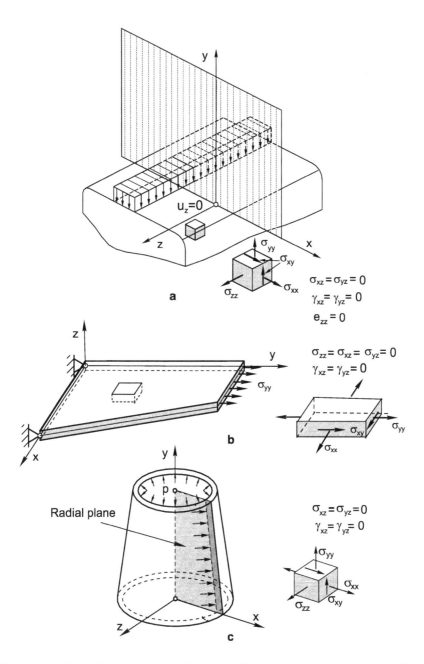

Fig. A1.1. Two-dimensional problems. **a** Plane strain deformation; **b** Plane stress conditions; **c** Axially symmetric deformation

Substituting this expression for e_{zz}^E we obtain the constitutive matrix

$$
\begin{array}{ccc}
xx & yy & xy
\end{array}
$$

$$
\mathbf{C}^E = \frac{E}{1-\nu^2}
\begin{bmatrix}
1 & \nu & 0 \\
\nu & 1 & 0 \\
0 & 0 & \dfrac{1-\nu}{2}
\end{bmatrix}
\tag{A1.24}
$$

This corresponds to *static condensation* of the e_{zz} "degree of freedom" (see Bathe 1996).

The static condensation for plane stress analysis is directly applicable to the shell conditions. Assuming a local orthogonal Cartesian coordinate system r, s, t, with the axis t in the direction of the shell normal (Fig. A1.2a), and imposing the condition $\sigma_{tt} = 0$, we obtain

$$
\begin{array}{cccccc}
rr & ss & tt & rs & st & tr
\end{array}
$$

$$
\mathbf{C}^E = \frac{E}{1-\nu^2}
\begin{bmatrix}
1 & \nu & 0 & 0 & 0 & 0 \\
\nu & 1 & 0 & 0 & 0 & 0 \\
0 & 0 & 0 & 0 & 0 & 0 \\
0 & 0 & 0 & \dfrac{1-\nu}{2} & 0 & 0 \\
0 & 0 & 0 & 0 & \dfrac{1-\nu}{2} & 0 \\
0 & 0 & 0 & 0 & 0 & \dfrac{1-\nu}{2}
\end{bmatrix}
\tag{A1.25}
$$

The zero-row and zero-column is a convenience in the applications of this matrix.

In the case of a curved membrane, the transversal shear terms are equal to zero,

$$
\begin{aligned}
\gamma_{rt} &= 0 & \gamma_{st} &= 0 \\
\sigma_{rt} &= 0 & \sigma_{st} &= 0
\end{aligned}
\tag{A1.26}
$$

Hence the constitutive matrix (A1.25) reduces to the matrix (A1.24) but corresponding to the tangential r, s plane.

The beam conditions can be considered to be a special case of the shell conditions. In the case of the beam shown in Fig. A1.2b we have

$$
\sigma_{ss} = \sigma_{tt} = \sigma_{st} = 0
\tag{A1.27}
$$

We use the additional condition $\sigma_{ss} = 0$ and the matrix (A1.25) reduces to

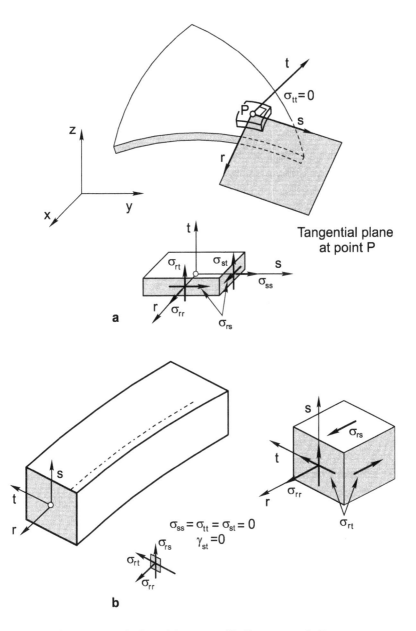

Fig. A1.2. Stresses in shell and beam. **a** Shell stresses; **b** Beam stresses

$$
\begin{array}{ccc}
rr & rs & rt
\end{array}
$$

$$
\mathbf{C}^E = E
\begin{bmatrix}
1 & 0 & 0 \\
0 & \dfrac{1}{2(1+\nu)} & 0 \\
0 & 0 & \dfrac{1}{2(1+\nu)}
\end{bmatrix}
\tag{A1.28}
$$

where (again) we assumed an orthogonal coordinate system.

Isotropic Thermoelastic Material. A temperature change affects the elastic material constants, and causes thermal strains. Hence

$$
E = E(\theta)
$$
$$
\nu = \nu(\theta)
\tag{A1.29}
$$

and for a temperature ${}^t\theta$ at a time t we have that

$$
{}^tE = E|_{\theta = {}^t\theta}
$$
$$
{}^t\nu = \nu|_{\theta = {}^t\theta}
\tag{A1.30}
$$

are the known values.

In an isotropic material the *thermal strains* ${}^t e_{ij}^{TH}$ are

$$
{}^t e_{ij}^{TH} = {}^t\alpha \left({}^t\theta - \theta_{ref}\right) \delta_{ij}
\tag{A1.31}
$$

where ${}^t\alpha$ is the *coefficient of thermal expansion* of the material, and θ_{ref} is the *reference temperature* for which the thermal strains are equal to zero. Note that only normal thermal strains are created by the temperature change, that is, shear thermal strains do not exist.

The *principle of superposition* of thermal strains with other strains in the material is applicable. Therefore, the total strain ${}^t\mathbf{e}$ at a material point of a thermoelastic material is

$$
{}^t\mathbf{e} = {}^t\mathbf{e}^E + {}^t\mathbf{e}^{TH}
\tag{A1.32}
$$

Consider next the *thermoelastic constitutive relations*. Using (A1.6) and (A1.32) we obtain

$$
{}^t\hat{\mathbf{e}} = \left({}^t\mathbf{C}^E\right)^{-1}\,{}^t\boldsymbol{\sigma} + {}^t\mathbf{e}^{TH}
\tag{A1.33}
$$

and

$$
{}^t\boldsymbol{\sigma} = {}^t\mathbf{C}^E \left({}^t\hat{\mathbf{e}} - {}^t\mathbf{e}^{TH}\right)
\tag{A1.34}
$$

where ${}^t\mathbf{C}^E$ corresponds to temperature ${}^t\theta$. Note that, since the shear components of the thermal strain are equal to zero, it follows that $\hat{\mathbf{e}}^{TH} = \mathbf{e}^{TH}$. All

previously written elastic constitutive relations are applicable. For example,
(A1.12) and (A1.19) are now

$$ {}^t\sigma_m = {}^tc_m \left({}^te_m - {}^te^{TH} \right) \tag{A1.35} $$

and

$$ {}^tS_{ij} = 2\,{}^tG\,{}^te'_{ij} \tag{A1.36} $$

where the elastic constants are

$$ {}^tc_m = \frac{{}^tE}{1 - 2\,{}^t\nu} \qquad {}^tG = \frac{{}^tE}{2(1 + {}^t\nu)} \tag{A1.37} $$

Constitutive Relations for Elastic Orthotropic Materials. For an
orthotropic material with the three principal, mutually orthogonal axes a, b, c,
also called the *material axes*, the elastic compliance matrix has the following
form, using the convention in Jones (1975),

$$
\begin{array}{ccccccc}
aa & bb & cc & ab & bc & ca
\end{array}
$$

$$
\left(C^E \right)^{-1} =
\begin{bmatrix}
1/E_a & -\nu_{ba}/E_b & -\nu_{ca}/E_c & 0 & 0 & 0 \\
-\nu_{ab}/E_a & 1/E_b & -\nu_{cb}/E_c & 0 & 0 & 0 \\
-\nu_{ac}/E_a & -\nu_{bc}/E_b & 1/E_c & 0 & 0 & 0 \\
0 & 0 & 0 & 1/G_{ab} & 0 & 0 \\
0 & 0 & 0 & 0 & 1/G_{bc} & 0 \\
0 & 0 & 0 & 0 & 0 & 1/G_{ca}
\end{bmatrix}
\tag{A1.38}
$$

Here E_a, E_b and E_c are the Young's moduli for the material axes a, b and c,
and the ν_{ij} are the Poisson ratios for these axes. The Poisson ratio ν_{ba}, for
example, represents the ratio between the elastic strains e_{aa}^E and e_{bb}^E when
the material is subjected to the uniaxial stress in the b-direction, i.e.,

$$ \nu_{ba} = - \frac{e_{aa}^E}{e_{bb}^E} \bigg|_{\sigma_{bb}=\sigma}, \quad \text{other } \sigma_{ij}=0 $$

Or, in general

$$ \nu_{ji} = - \frac{e_{ii}^E}{e_{jj}^E} \bigg|_{\sigma_{jj}=\sigma}, \quad \text{other } \sigma_{ij}=0 \tag{A1.39} $$

where the indices i, j stand for a, b, c. The coefficients G_{ab}, G_{bc} and G_{ca} represent the shear moduli for the $a - b$, $b - c$ and $a - c$ planes, respectively. The
compliance matrix is symmetric, and therefore the following relations must
be satisfied

$$\frac{\nu_{ab}}{E_a} = \frac{\nu_{ba}}{E_b} \qquad \frac{\nu_{bc}}{E_b} = \frac{\nu_{cb}}{E_c} \qquad \frac{\nu_{ac}}{E_a} = \frac{\nu_{ca}}{E_c} \qquad (A1.40)$$

Hence, the compliance matrix (A1.38) contains nine independent material constants.

By inverting the compliance matrix (A1.38) we obtain the nonzero terms C_{ij}^E,

$$C_{11}^E = \frac{1 - \nu_{bc}\nu_{cb}}{E_b E_c D} \qquad\qquad C_{33}^E = \frac{1 - \nu_{ab}\nu_{ba}}{E_a E_b D}$$

$$C_{12}^E = \frac{\nu_{ba} + \nu_{ca}\nu_{bc}}{E_b E_c D} \qquad\qquad C_{44}^E = G_{ab}$$

$$C_{13}^E = \frac{\nu_{ca} + \nu_{ba}\nu_{cb}}{E_b E_c D} \qquad\qquad C_{55}^E = G_{bc} \qquad (A1.41)$$

$$C_{22}^E = \frac{1 - \nu_{ac}\nu_{ca}}{E_a E_c D} \qquad\qquad C_{66}^E = G_{ca}$$

$$C_{23}^E = \frac{\nu_{cb} + \nu_{ab}\nu_{ca}}{E_a E_c D}$$

where

$$D = \frac{1 - \nu_{ab}\nu_{ba} - \nu_{bc}\nu_{cb} - \nu_{ac}\nu_{ca} - 2\nu_{ba}\nu_{ac}\nu_{cb}}{E_a E_b E_c} \qquad (A1.42)$$

is the determinant of the 3×3 left-hand side top matrix in (A1.38).

The constitutive matrix and the compliance matrix must be positive definite. From this condition some restrictions on the values of the material constants can be derived, see, e.g., Jones (1975).

In the case of orthotropic shell deformations we assume that the c-material axis is in the direction of the normal t, as shown in Fig. A1.3. The stress through the shell thickness must be zero, and using (A1.38) we obtain

$$e_{aa}^E = \frac{1}{E_a}\sigma_{aa} - \frac{\nu_{ba}}{E_b}\sigma_{bb} \qquad e_{bb}^E = -\frac{\nu_{ab}}{E_a}\sigma_{aa} + \frac{1}{E_b}\sigma_{bb} \qquad (A1.43)$$

Solving for σ_{aa} and σ_{bb} and with use of (A1.40) we obtain the constitutive matrix for a shell,

$$
\mathbf{C}^E =
\begin{array}{c}
\begin{array}{cccccc}
aa & bb & cc & ab & bc & ca
\end{array} \\
\left[
\begin{array}{cccccc}
\dfrac{E_a}{1-\nu_{ab}\nu_{ba}} & \dfrac{\nu_{ab}E_b}{1-\nu_{ab}\nu_{ba}} & 0 & 0 & 0 & 0 \\[2ex]
\dfrac{\nu_{ab}E_b}{1-\nu_{ab}\nu_{ba}} & \dfrac{E_b}{1-\nu_{ab}\nu_{ba}} & 0 & 0 & 0 & 0 \\[1ex]
0 & 0 & 0 & 0 & 0 & 0 \\
0 & 0 & 0 & G_{ab} & 0 & 0 \\
0 & 0 & 0 & 0 & G_{bc} & 0 \\
0 & 0 & 0 & 0 & 0 & G_{ca}
\end{array}
\right]
\end{array}
\qquad (A1.44)
$$

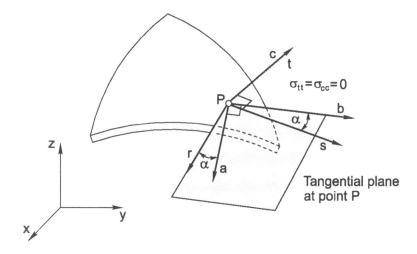

Fig. A1.3. Local orthogonal shell axes r, s, t and material axes a, b, c of orthotropic shell

In practical applications the strains are calculated in the local shell coordinate system r, s, t, and the material axes a, b, c differ by the rotation α around the t-axis, as shown in Fig. A1.3. Therefore, either the transformation of the strains to the material axes, or the transformation of the constitutive matrix to the shell local system, is necessary. If the stresses and constitutive matrix are required in the coordinate system x, y, z, the additional transformations from the r, s, t to the x, y, z coordinate system are required.

In the case of a *membrane*, the constitutive matrix has the form (A1.44), except that the columns and rows corresponding to the transversal shear actions ac and bc contain the zero terms. Hence, we have a *plane stress* deformation in the plane a, b and the constitutive matrix is

$$
\mathbf{C}^E = \begin{array}{c}
\begin{array}{ccc}
aa & \quad bb & \quad ab
\end{array} \\
\left[
\begin{array}{ccc}
\dfrac{E_a}{1 - \nu_{ab}\nu_{ba}} & \dfrac{\nu_{ab}E_b}{1 - \nu_{ab}\nu_{ba}} & 0 \\[2mm]
\dfrac{\nu_{ab}E_b}{1 - \nu_{ab}\nu_{ba}} & \dfrac{E_b}{1 - \nu_{ab}\nu_{ba}} & 0 \\[2mm]
0 & 0 & G_{ab}
\end{array}
\right]
\end{array}
\qquad (A1.45)
$$

The constitutive matrix for the *beam* conditions can be derived from the general constitutive matrix $\mathbf{C}^E$ with the coefficients (A1.41). Then the transformation to the beam coordinate system shown in Fig. A1.2 must be performed, followed by the static condensation to satisfy the conditions (A1.27).

Thermoelastic Constitutive Relations for Orthotropic Materials. The constitutive relations (A1.33) and (A1.34) for the material axes are

$$^t\hat{\mathbf{e}} = \left(^t\mathbf{C}^E\right)^{-1} {}^t\boldsymbol{\sigma} + {}^t\mathbf{e}^{TH}\Big|_{\text{material axes}} \qquad (A1.46)$$

and

$$^t\boldsymbol{\sigma} = {}^t\mathbf{C}^E \left(^t\hat{\mathbf{e}} - {}^t\mathbf{e}^{TH}\right)\Big|_{\text{material axes}} \qquad (A1.47)$$

with $\left(^t\mathbf{C}^E\right)^{-1}$ and $^tC_{ij}^E$ defined by (A1.38) and (A1.41). The material constants are known functions of temperature and the thermal strains are

$$^te_{aa}^{TH} = {}^t\alpha_a \left(^t\theta - \theta_{ref}\right) \quad ^te_{bb}^{TH} = {}^t\alpha_b \left(^t\theta - \theta_{ref}\right) \quad ^te_{cc}^{TH} = {}^t\alpha_c \left(^t\theta - \theta_{ref}\right)$$

$$^te_{ij}^{TH} = 0 \qquad i \neq j \qquad (A1.48)$$

where $^t\alpha_a$, $^t\alpha_b$ and $^t\alpha_c$ are the coefficients of linear expansion in the directions of the material axes.

Note that (A1.46) and (A1.47) are the constitutive relations from which the constitutive relations for other coordinate systems, or for the special cases of shell, plane stress, plane strain, axisymmetry and beam deformations can be derived.

Transformation of the Constitutive Matrix. It is important for practical applications to know how to transform a given constitutive matrix for a change of coordinate system. The constitutive matrix transformation follows from the fact that the stress and strain quantities are tensors, hence their components change according to the tensorial rules (see (A2.24) in Appendix A2).

Let x_i and $\bar{x}_i$ represent two Cartesian coordinate systems. Then the tensor transformations of stresses and strains (see, e.g., Wang 1953; Bathe 1996) can be written as

$$\bar{\sigma}_i = T_{ik}^\sigma \sigma_k \qquad (A1.49)$$

and

$$\bar{e}_i = T_{ik}^\epsilon e_k \qquad (A1.50)$$

where $\mathbf{T}^\sigma$ and $\mathbf{T}^\epsilon$ are the transformation matrices. Here the repeated indices k imply summation, $k = 1, 2, ..., 6$. Also,

$$\sigma_i = \bar{T}_{ik}^\sigma \bar{\sigma}_k = T_{ki}^\epsilon \bar{\sigma}_k \qquad (A1.51)$$

and

$$e_i = \bar{T}_{ik}^\epsilon \bar{e}_k = T_{ki}^\sigma \bar{e}_k \qquad (A1.52)$$

where $\bar{\mathbf{T}}^\sigma$ and $\bar{\mathbf{T}}^\epsilon$ correspond to these transformations. We next define the transformation matrix $\mathbf{T}$ as

$$
\begin{array}{cccc}
 & x_1 & x_2 & x_3 \\
\mathbf{T} = \begin{array}{c} \bar{x}_1 \\ \bar{x}_2 \\ \bar{x}_3 \end{array} & \left[\begin{array}{ccc} \ell_1 & m_1 & n_1 \\ \ell_2 & m_2 & n_2 \\ \ell_3 & m_3 & n_3 \end{array}\right]
\end{array}
\tag{A1.53}
$$

with components

$$
T_{jk} = \cos(\bar{\mathbf{i}}_j, \mathbf{i}_k)
\tag{A1.54}
$$

where $\mathbf{i}_k$ and $\bar{\mathbf{i}}_j$ are the unit vectors of the coordinate axes x_k and $\bar{x}_j$. Then the matrices $\mathbf{T}^\sigma$ and $\mathbf{T}^\epsilon$ are

$$\mathbf{T}^\sigma =$$

$$
\left[\begin{array}{cccccc}
\ell_1^2 & m_1^2 & n_1^2 & 2\ell_1 m_1 & 2m_1 n_1 & 2n_1 \ell_1 \\
\ell_2^2 & m_2^2 & n_2^2 & 2\ell_2 m_2 & 2m_2 n_2 & 2n_2 \ell_2 \\
\ell_3^2 & m_3^2 & n_3^2 & 2\ell_3 m_3 & 2m_3 n_3 & 2n_3 \ell_3 \\
\ell_1 \ell_2 & m_1 m_2 & n_1 n_2 & \ell_1 m_2 + m_1 \ell_2 & m_1 n_2 + n_1 m_2 & n_1 \ell_2 + \ell_1 n_2 \\
\ell_2 \ell_3 & m_2 m_3 & n_2 n_3 & \ell_2 m_3 + m_2 \ell_3 & m_2 n_3 + n_2 m_3 & n_2 \ell_3 + \ell_2 n_3 \\
\ell_3 \ell_1 & m_3 m_1 & n_3 n_1 & \ell_3 m_1 + m_3 \ell_1 & m_3 n_1 + n_3 m_1 & n_3 \ell_1 + \ell_3 n_1
\end{array}\right]
$$

$$
\tag{A1.55}
$$

$$\mathbf{T}^\epsilon =$$

$$
\left[\begin{array}{cccccc}
\ell_1^2 & m_1^2 & n_1^2 & \ell_1 m_1 & m_1 n_1 & n_1 \ell_1 \\
\ell_2^2 & m_2^2 & n_2^2 & \ell_2 m_2 & m_2 n_2 & n_2 \ell_2 \\
\ell_3^2 & m_3^2 & n_3^2 & \ell_3 m_3 & m_3 n_3 & n_3 \ell_3 \\
2\ell_1 \ell_2 & 2m_1 m_2 & 2n_1 n_2 & \ell_1 m_2 + m_1 \ell_2 & m_1 n_2 + n_1 m_2 & n_1 \ell_2 + \ell_1 n_2 \\
2\ell_2 \ell_3 & 2m_2 m_3 & 2n_2 n_3 & \ell_2 m_3 + m_2 \ell_3 & m_2 n_3 + n_2 m_3 & n_2 \ell_3 + \ell_2 n_3 \\
2\ell_3 \ell_1 & 2m_3 m_1 & 2n_3 n_1 & \ell_3 m_1 + m_3 \ell_1 & m_3 n_1 + n_3 m_1 & n_3 \ell_1 + \ell_3 n_1
\end{array}\right]
$$

$$
\tag{A1.56}
$$

Of course, (A1.51) and (A1.52) tell that

$$
\bar{\mathbf{T}}^\sigma = (\mathbf{T}^\epsilon)^T
\tag{A1.57}
$$

$$
\bar{\mathbf{T}}^\epsilon = (\mathbf{T}^\sigma)^T
\tag{A1.58}
$$

and we have that

$$
(\mathbf{T}^\sigma)^T \mathbf{T}^\epsilon = \mathbf{I}_6
\tag{A1.59}
$$

where $\mathbf{I}_6$ is the 6×6 identity matrix.

The transformation (A1.49) can also be written in the form

$$[\bar{\sigma}_{ij}] = [T_{mn}] [\sigma_{kl}] [T_{rs}]^T \tag{A1.60}$$

according to tensor transformation (A2.24). Also, using the tensor components e_{ij} of strain, the same transformation (A1.60) is applicable

$$[\bar{e}_{ij}] = [T_{mn}] [e_{kl}] [T_{rs}]^T \tag{A1.61}$$

Note that the transformation matrix $\mathbf{T}^\sigma$ and the transformation (A1.49) can be applied to any symmetric second-order tensor, while the transformation (A1.50) contains the definition of doubled shear terms (A1.4).

From the transformations (A1.49) to (A1.52) and the constitutive relations (A1.5), follows

$$\bar{C}^E_{js} = T^\sigma_{jk} C^E_{kr} T^\sigma_{sr} \tag{A1.62a}$$

or, in matrix form,

$$\bar{\mathbf{C}}^E = \mathbf{T}^\sigma \, \mathbf{C}^E \, (\mathbf{T}^\sigma)^T \tag{A1.62b}$$

Also,

$$\mathbf{C}^E = (\mathbf{T}^\epsilon)^T \bar{\mathbf{C}}^E \, \mathbf{T}^\epsilon \tag{A1.63}$$

where $\mathbf{C}^E$ and $\bar{\mathbf{C}}^E$ correspond to two coordinate systems. In the case of an isotropic elastic material, see the constitutive matrix (A1.7), we have

$$\bar{\mathbf{C}}^E = \mathbf{C}^E \tag{A1.64}$$

for any coordinate system. This is also true for the matrices (A1.20) and (A1.22) and any coordinate system obtained by rotation about the z-axis. Also, for the constitutive matrices (A1.24), (A1.25) and (A1.28) obtained by the static condensations, the relation (A1.62) holds only for the rotations about the z-axis, t-axis and r-axis, respectively.

The relations (A1.62) and (A1.63) are also applicable to a general constitutive matrix. For example, in the case of the constitutive relations

$$d\boldsymbol{\sigma} = \mathbf{C} \, d\hat{\mathbf{e}} \tag{A1.65}$$

the constitutive matrix $\mathbf{C}$ transforms according to (A1.62) and (A1.63),

$$\bar{\mathbf{C}} = \mathbf{T}^\sigma \, \mathbf{C} \, (\mathbf{T}^\sigma)^T \tag{A1.66}$$

and

$$\mathbf{C} = (\mathbf{T}^\epsilon)^T \, \bar{\mathbf{C}} \, \mathbf{T}^\epsilon \tag{A1.67}$$

Appendix A2

Notation - Matrices and Tensors

In this Appendix we list some basic relations for matrices and tensors used in the book only to introduce the notation employed. For the theory and use of matrices and tensors, see, e.g., Green and Zerna (1960); Malvern (1969); Bathe (1996); Chapelle and Bathe (2003).

Vectors/Matrices. Let $\mathbf{b}$ be a vector of dimension n

$$\mathbf{b} = \begin{bmatrix} b_1 \\ b_2 \\ \cdot \\ \cdot \\ \cdot \\ b_n \end{bmatrix} \tag{A2.1}$$

where $b_1, b_2, ..., b_n$ are the components of the vector. The transpose of $\mathbf{b}$, denoted by $\mathbf{b}^T$, is the $1 \times n$ matrix,

$$\mathbf{b}^T = [b_1 b_2 ..... b_n] \tag{A2.2}$$

The vectors in (A.2.1) and in (A.2.2) are called a column vector and a row vector, respectively. A vector with three components has the geometric meaning shown in Fig. A2.1.

A two-dimensional square matrix $\mathbf{B}$ of dimension n is

$$\mathbf{B} = \begin{bmatrix} B_{11} & B_{12} & \cdot & \cdot & \cdot & B_{1n} \\ B_{21} & B_{22} & \cdot & \cdot & \cdot & B_{2n} \\ \cdot & \cdot & & & & \cdot \\ \cdot & \cdot & & & & \cdot \\ \cdot & \cdot & & & & \cdot \\ B_{n1} & B_{n2} & \cdot & \cdot & \cdot & B_{nn} \end{bmatrix} \tag{A2.3}$$

where B_{ij} are the components of the matrix. The square matrix is symmetric when

$$B_{ij} = B_{ji} \qquad i, j = 1, 2, ..., n \tag{A2.4}$$

The basic matrix operations used are as follows.

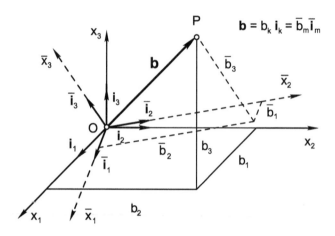

Fig. A2.1. Geometric representation of a vector in 3-D space

The *addition (subtraction)* of vectors **a** and **b** results in vector **c**,

$$\mathbf{c} = \mathbf{a} + \mathbf{b} \tag{A2.5a}$$

or, in component form,

$$c_i = a_i + b_i \qquad i = 1, 2, ..., n \tag{A2.5b}$$

The addition of matrices **A** and **B** results in the matrix **C**,

$$\mathbf{C} = \mathbf{A} + \mathbf{B} \tag{A2.6a}$$

or, in component form,

$$C_{ij} = A_{ij} + B_{ij} \qquad i, j = 1, 2, ..., n \tag{A2.6b}$$

The *multiplication of vectors* **a** and **b** gives the scalar c,

$$c = \mathbf{a}^T \mathbf{b} = \mathbf{b}^T \mathbf{a} = a_k b_k = \sum_{k=1}^{n} a_k b_k \tag{A2.7}$$

which is called the *scalar product* of vectors **a** and **b**. We use, as indicated, the *summation convention*: repeated indices in an expression imply summation. We also use

$$\mathbf{C} = \mathbf{a} \mathbf{b}^T \tag{A2.8a}$$

with components

$$C_{ij} = a_i b_j \tag{A2.8b}$$

The *multiplication of a matrix* **A** by the *vector* **b** gives the vector **c**,

$$\mathbf{c} = \mathbf{Ab} \tag{A2.9a}$$

with components

$$c_i = A_{ik} b_k = \sum_{k=1}^{n} A_{ik} b_k \tag{A2.9b}$$

The *multiplication of matrices* **A** and **B** gives the matrix **C** as

$$\mathbf{C} = \mathbf{AB} \tag{A2.10a}$$

with components

$$C_{ij} = A_{ik} B_{kj} \tag{A2.10b}$$

Note that

$$\mathbf{C}^T = \mathbf{B}^T \mathbf{A}^T \tag{A2.11a}$$

or in component form

$$(\mathbf{C}^T)_{ij} = B_{ki} A_{jk} \tag{A2.11b}$$

The *scalar product of matrices* **A** and **B** results in the scalar c as

$$c = \mathbf{A} \cdot \mathbf{B} = \mathbf{A}_{ij} B_{ij} \tag{A2.12}$$

The *inverse matrix* of a matrix **B**, denoted as $\mathbf{B}^{-1}$, is the matrix that satisfies the relation

$$\mathbf{B}^{-1}\mathbf{B} = \mathbf{BB}^{-1} = \mathbf{I} \tag{A2.13}$$

or in component form

$$B_{ik}^{-1} B_{kj} = I_{ij} \tag{A2.14}$$

where

$$I_{ij} = \delta_{ij} \tag{A2.15}$$

are the components of the *identity matrix* **I**. Here the δ_{ij} is the Kronecker delta symbol which is defined as

$$\begin{aligned} \delta_{ij} &= 1 \qquad i = j \\ \delta_{ij} &= 0 \qquad i \neq j \end{aligned} \tag{A2.16}$$

The *determinant* of a 3×3 matrix **B**, denoted as det **B**, is defined as

$$\det \mathbf{B} = e_{ijk} B_{1i} B_{2j} B_{3k} \qquad (A2.17)$$

where e_{ijk} is the permutation symbol, with

$$
\begin{aligned}
e_{ijk} &= 0 && \text{for} && i = j, \text{ or } j = k, \text{ or } k = i \\
e_{ijk} &= 1 && \text{for} && \text{even permutation of } 1,2,3 \qquad (A2.18) \\
e_{ijk} &= -1 && \text{for} && \text{odd permutation of } 1,2,3
\end{aligned}
$$

Tensors. Tensors are represented by *components* corresponding to the *base vectors* of a coordinate system. We use Cartesian systems only and tensors are represented by components corresponding to the *unit vectors* of the Cartesian system (see Fig. A2.1).

A *first order tensor* (or vector) **b** is defined as

$$\mathbf{b} = b_k \mathbf{i}_k \qquad k = 1, 2, 3 \qquad (A2.19)$$

where the b_k are the components, and the $\mathbf{i}_k$ are the unit vectors of the Cartesian system. A geometric representation of the the relation (A2.19) and of the vector components is shown in Fig. A2.1.

A *second order tensor* **B** (in this book called tensor) is defined as, using the tensor product symbol $\otimes$,

$$\mathbf{B} = B_{mn} \mathbf{i}_m \otimes \mathbf{i}_n \qquad i, m = 1, 2, 3 \qquad (A2.20)$$

Here, a repeated index implies summation. This representation of tensors is called the *tensor (or direct) notation*. Comparing (A2.1) with (A2.19), and (A2.3) with (A2.20), it can be seen that tensor components can be represented in matrix form.

In a coordinate system with unit vectors $\bar{\mathbf{i}}$ (see Fig. A2.1), the vector **b** and tensor **B** are

$$\mathbf{b} = \bar{b}_k \bar{\mathbf{i}}_k \qquad (A2.21)$$

and

$$\mathbf{B} = \bar{B}_{mn} \bar{\mathbf{i}}_m \otimes \bar{\mathbf{i}}_n \qquad (A2.22)$$

The relations between tensorial components in the two coordinate systems are given as

$$\bar{b}_m = T_{mk} b_k \qquad (A2.23a)$$

and

$$\bar{B}_{ij} = T_{ik} B_{km} T_{jm} \qquad (A2.24a)$$

where T_{km} gives the cosine of the angle between the unit vectors $\bar{\mathbf{i}}_k$ and $\mathbf{i}_m$, see (A1.54). The relations (A2.23a) and (A2.24a) *represent the tensorial transformations* of first order and second order tensors. Using matrix notation we can write the relations (A2.23a) and (A2.24a) as

$$\{\bar{b}_i\} = [T_{mn}] \{b_k\} \tag{A2.23b}$$

and

$$[\bar{B}_{ij}] = [T_{mn}] [B_{kl}] [T_{rs}]^T \tag{A2.24b}$$

In analogy with the above mathematical relations for matrices, we give some basic relations for tensors (Malvern 1969, Chapelle and Bathe 2003). The scalar (inner) product of two vectors **a** and **b** is

$$c = \mathbf{a} \cdot \mathbf{b} = a_k b_k \tag{A2.25}$$

giving the scalar c (see (A2.7)). The dot product (multiplication) of tensor **A** and vector **b** gives vector **c** according to the relation (see (A2.9))

$$\mathbf{c} = c_m \mathbf{i}_m = \mathbf{Ab} = A_{mp} \mathbf{i}_m \otimes \mathbf{i}_p \cdot b_k \mathbf{i}_k = A_{mk} b_k \mathbf{i}_m \tag{A2.26}$$

where we have used the orthogonality condition

$$\mathbf{i}_p \cdot \mathbf{i}_k = \delta_{pk} \tag{A2.27}$$

The dot product (multiplication) of tensors **A** and **B** gives tensor **C** as (see (A2.10))

$$\mathbf{C} = \mathbf{C}_{mn} \mathbf{i}_m \otimes \mathbf{i}_n = \mathbf{AB} = A_{mk} B_{kn} \mathbf{i}_m \otimes \mathbf{i}_n \tag{A2.28}$$

where again we have used the orthogonality condition (A2.27). We also use the notation that the scalar (inner) product between two tensors **A** and **B** is (see (A2.12))

$$c = \mathbf{A} \cdot \mathbf{B} = A_{ij} B_{ij} \tag{A2.29}$$

The *Euclidean norm* of a vector **b** is

$$\|\mathbf{b}\| = (b_k b_k)^{1/2} \tag{A2.30}$$

and of a tensor **B** is

$$\|\mathbf{B}\|_2 = (B_{ij} B_{ij})^{1/2} \tag{A2.31}$$

The rotation tensor **R** corresponding to two sets of orthonormal vectors $\mathbf{i}_m$ and $\bar{\mathbf{i}}_m$ is defined from the relation

$$\bar{\mathbf{i}}_m = \mathbf{R} \mathbf{i}_m \tag{A2.32a}$$

or in component form, with respect to the coordinate system x_k with the unit vectors $\mathbf{i}_k$,

$$(\bar{\mathbf{i}}_m)_k = R_{kp}\,\delta_{pm} = R_{km} \tag{A2.32b}$$

From this relation follows that the components of the rotation tensor in the coordinate system x_k are

$$R_{jk} = \cos(\mathbf{i}_j, \bar{\mathbf{i}}_k) \tag{A2.33}$$

Comparing this expression with (A1.54) we obtain

$$R_{ij} = T_{ji} \tag{A2.34a}$$

hence

$$\mathbf{R} = \mathbf{T}^T \tag{A2.34b}$$

where $\mathbf{T}$ is the transformation matrix. A geometric interpretation of the relations (A2.32) and (A2.34) is shown in Fig. A2.2a. Also, if we multiply a vector $\mathbf{b}$ by the rotation tensor, we obtain

$$\hat{\mathbf{b}} = \mathbf{b}_{rot} = \mathbf{R}\mathbf{b} \tag{A2.35}$$

where $\hat{\mathbf{b}}$ is the rotated vector.

To illustrate the difference between the transformation of vector components and a vector rotation, we consider the example shown in Fig. A2.2b. The transformation matrix and the rotation tensor are

$$\mathbf{T} = \mathbf{R}^T = \begin{bmatrix} \cos\alpha & \sin\alpha \\ -\sin\alpha & \cos\alpha \end{bmatrix} \tag{A2.36}$$

The multiplication $\mathbf{Tb}$ according to (A2.23) gives

$$\bar{b}_1 = b\cos(\phi - \alpha), \quad \bar{b}_2 = b\sin(\phi - \alpha) \tag{A2.37}$$

while the multiplication $\mathbf{Rb}$ gives

$$\hat{b}_1 = b\cos(\phi + \alpha), \quad \hat{b}_2 = b\sin(\phi + \alpha) \tag{A2.38}$$

where b is the Euclidean norm (modulus) of the vector $\mathbf{b}$, see (A2.30). Obviously, in (A2.37) the *same vector* $\mathbf{b}$ is simply expressed with its components in the rotated coordinate system $\bar{x}_1, \bar{x}_2$; while the multiplication $\mathbf{Rb}$ leads to *another vector* $\hat{\mathbf{b}}$, which is *rotated* with respect to the vector $\mathbf{b}$, and the components are given in the coordinate system x_1, x_2.

In general, if two sets of orthonormal vectors are defined as $\mathbf{p}_\alpha$ and $\bar{\mathbf{p}}_\alpha$, then the rotation tensor is given as

$$\mathbf{R} = \sum_{\alpha=1}^{3} \bar{\mathbf{p}}_\alpha \otimes \mathbf{p}_\alpha \tag{A2.39a}$$

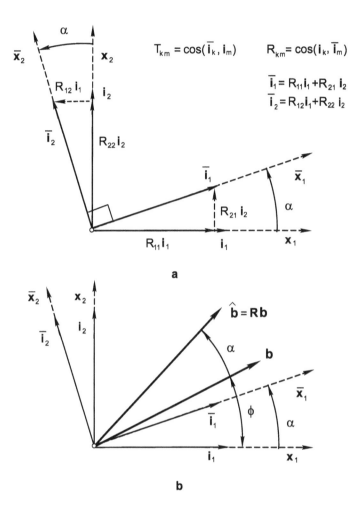

Fig. A2.2. Geometric representation of change of coordinate basis and rotation (2-D). **a** With the definition of the rotation tensor **R**; **b** Rotation of a vector

or in component form

$$R_{ij} = \sum_{\alpha=1}^{3} (\bar{\mathbf{p}}_\alpha)_i \, (\mathbf{p}_\alpha)_j \qquad \text{(A2.39b)}$$

Also, we have that

$$\bar{\mathbf{p}}_\alpha = \mathbf{R}\mathbf{p}_\alpha \qquad \text{(A2.40a)}$$

or in component form

$$(\bar{\mathbf{p}}_\alpha)_i = R_{ik}(\mathbf{p}_\alpha)_k \qquad \text{(A2.40b)}$$

The rotation tensor is *orthogonal* because it satisfies the orthogonality condition

$$\mathbf{R}\mathbf{R}^T = \mathbf{I} \qquad (A2.41)$$

Obviously, the transformation matrix $\mathbf{T}$ is also an orthogonal matrix.

Appendix A3

List of Main Symbols

In this Appendix we give a list of the most frequently used symbols in the book. For convenience of reading, these symbols are also defined in the text when they appear first time.

a) Alphabetic Symbols

${}_0^t\mathbf{B}$	left Cauchy-Green deformation tensor
${}_t^t\mathbf{B_L}$	linear strain-displacement matrix
${}_t^t\mathbf{B_{NL}}$	nonlinear strain-displacement matrix
$\mathcal{B}$	material body, configuration
$\mathbf{C}$	constitutive matrix
$\mathbf{C}^E$	elastic
$\mathbf{C}^{EP}$	elastic-plastic
$\mathbf{C}^{VP}$	viscoplastic
$\mathbf{C}^{EPC}$	elastic-plastic-creep
$\mathbf{C}^P$	plastic modulus tensor in hardening
${}_0^t\mathbf{C}$	right Cauchy-Green deformation tensor
$\mathbf{D}$	strain rate tensor
$\mathbf{e}$	strain tensor (tensorial shear strain components)
$\mathbf{e}^C$	creep
$\mathbf{e}^E$	elastic
$\mathbf{E}_g$	generalized
$\mathbf{E}^{(H)}$	logarithmic
$\mathbf{e}^{IN}$	inelastic
$\mathbf{e}^P$	plastic
$\mathbf{e}^{TH}$	thermal
$\mathbf{e}^{VP}$	viscoplastic
$\mathbf{e}''$	trial elastic at end of time step
$\hat{\mathbf{e}}$	strain vector (engineering shear strain components)
$\hat{\mathbf{e}}^C$	creep
$\hat{\mathbf{e}}^E$	elastic
$\hat{\mathbf{e}}^P$	plastic
$\hat{\mathbf{e}}^{VP}$	viscoplastic
$\hat{\mathbf{e}}''$	trial elastic at end of time step
e	void ratio
e_m	mean strain

$\bar{e}^P$	effective plastic strain (von Mises yield condition)
e_V	volumetric strain
E	Young's modulus of elasticity
E_P	plastic modulus
E_T	tangent modulus
f	function, porosity for Gurson material model
$f(p)$	function of the governing parameter p
f_y	yield function, in general and for metals
f_C	for cap yield conditions
f_{DP}	for Drucker-Prager yield conditions
$\mathbf{F}$	force vector
G	elastic shear modulus
i	iteration counter in equilibrium iterations
$\mathbf{I}$	identity matrix
$\mathbf{J}$	Jacobian matrix
J_{2D}	second invariant of stress tensor
k	proportionality coefficient for viscoplastic models, material constant for cap plasticity models, trial number in solving the governing equation
$\mathbf{K}$	stiffness matrix
M	mixed hardening parameter (von Mises yield condition), material parameter (Cam-clay model)
$\mathbf{n}$	normal to yield surface
N	number of nodal points of a finite element
$\mathbf{N}$	shape coefficient matrix (Hill's material model)
p	governing parameter
p_0	hardening function (axis of ellipse) in Cam-clay model
$\mathbf{p}_i$	principal vectors of the right basis
$\bar{\mathbf{p}}_i$	of the left bases
q	deviatoric stress in triaxial test of soil
Q	plastic potential in nonassociated plasticity
r	natural coordinate of a finite element
r_0	Lankford coefficient (orthotropic sheet metals), angle $= 0^0$
r_{45}	$= 45^0$
r_{90}	$= 90^0$
R	ratio of semiaxes in Cam-clay model, radius (size) of yield surface (von Mises yield condition)
$\mathbf{R}$	rotation tensor
s	natural coordinate of a finite element
$\mathbf{S}$	deviatoric stress
$\hat{\mathbf{S}}$	radius of yield surface (von Mises yield condidtion)
$^{t+\Delta t}_t\mathbf{S}$	Second Piola-Kirchhoff stress
t	time (real time) or parameter to denote load or strain level, natural coordinate of a finite element
$\mathbf{T}$	transformation matrix

$\mathbf{T}^\epsilon$	for strains
$\mathbf{T}^\sigma$	for stresses
$\mathbf{u}$	displacement vector, at a material point
$\mathbf{U}$	at a nodal finite element point
${}_0^t\mathbf{U}$	right stretch tensor
v	specific volume of soil
V	volume of material
${}_0^t\mathbf{V}$	left stretch tensor
W	internal work per unit volume
W^E	elastic
W^P	plastic
$\mathbf{x}$	position vector
X	position of cap for cap soil plasticity model
$\mathbf{X}$	deformation gradient
$\mathbf{X}^E$	elastic
$\mathbf{X}_*^E$	trial elastic
$\mathbf{X}^P$	plastic

b) Greek Symbols

α	material constant for cap soil plasticity models, hardening function for a general soil plasticity model, thermal expansion coefficient
$\boldsymbol{\alpha}$	back stress (von Mises yield condition)
$\boldsymbol{\beta}$	internal variables
γ	proportionality coefficient in stress - creep strain relations
δ_{ij}	Kronecker delta symbol ($= 1$, for $i = j$; $= 0$, for $i \neq j$)
$\Delta(*)$	increment in time step of a quantity $(*)$
η	viscosity coefficient
θ	temperature
λ	proportionality coefficient in stress - plastic strain relations
${}_0^t\lambda$	stretch from initial to the current state of deformation
ν	Poisson's ratio
σ_y	yield stress
σ_{yv}	initial
$\bar{\sigma}$	effective stress (in von Mises yield condition)
$\hat{\sigma}$	reduced
$\bar{\sigma}_a$	equivalent stress (in Hill's yield condition)
σ_m	mean stress
$\boldsymbol{\sigma}$	engineering stress
$\boldsymbol{\sigma}^E$	elastic stress solution in time step
$\boldsymbol{\sigma}^*$	conjugate stress in viscoplasticity, effective stress in soil
$\boldsymbol{\tau}$	true (Cauchy) stress
$\boldsymbol{\tau}^{(g)}$	generalized stress measure

References

Al-Tabbaa, A. (1987), **Permeability and Stress-Strain Response of Speswhite Kaolin**, Ph.D. thesis, Cambridge Univ., U.K.

Aifantis, E. C. (1987), The physics of plastic deformation, Int. J. Plasticity, Vol. 3, pp. 211-247.

Anand, L. (1979), On H. Hencky's approximate strain-energy functions for moderate deformations, J. Appl. Mechanics, Vol. 46, pp. 78-82.

Anand, L. (1985), Constitutive equations for hot-working of metals, Int. J. Plasticity, Vol. 1, pp. 213-231.

Asaro, R. J. (1983), Micromechanics of crystals and polycrystals, Adv. Appl. Mech., Vol. 23.

Atkinson, J. (1993), **The Mechanics of Soils and Foundations**, McGraw-Hill Book Co., New York.

Barlat, F., and Lian, J. (1989), Plastic behavior and stretchability of sheet metals, Part I: A yield function for orthotropic sheets under plane stress conditions, Int. J. Plasticity, Vol. 5, pp. 51-66.

Barlat, F., Lege, D. J., and Brem, J. C. (1991), A six-component yield function for anisotropic materials, Int. J. Plasticity, Vol. 7, pp. 693-712.

Bathe, K. J., and Wilson, E. L. (1973), NONSAP - A general finite element program for nonlinear dynamic analysis of complex structures, Paper No. M3-1, Proc. Second Conf. Struct. Mech. in Reactor Technology.

Bathe, K. J., Ozdemir, H., and Wilson, E. L. (1974), Static and dynamic geometric and material nonlinear analysis, Technical Report UCSESM 74-4, University of California, Berkeley.

Bathe, K. J., Ramm, E., and Wilson, E. L. (1975), Finite element formulations for large deformation dynamic analysis, Int. J. Num. Meth. Engng., Vol. 9, pp. 353-386.

Bathe, K. J., and Ozdemir, H. (1976), Elastic-plastic large deformation static and dynamic analysis, Comp. Structures, Vol. 6, pp. 81-92.

Bathe, K. J., and Cimento, A. P. (1980), Some practical procedures for the solution of nonlinear finite element equations, Comp. Meth. Appl. Mech. Engng., Vol. 22, pp. 59-85.

Bathe, K. J. (1982), **Finite Element Procedures in Engineering Analysis**, Prentice-Hall, Englewood Cliffs, N.J.

Bathe, K. J., Almeida, C. A., and Ho, L. W. (1983), A simple and effective pipe elbow element - some nonlinear capabilities, Comp. Structures, Vol. 17, pp. 659-667.

Bathe, K. J., Chaudhary, A. B., Dvorkin, E. N., and Kojic, M. (1984), On the solution of nonlinear finite element equations, in: Damjanic, F. et al., eds., **Proc. Int. Conference on Computer-Aided Analysis and Design of Concrete Structures I,** Pineridge Press, Swansea, U. K., pp. 289-299.

Bathe, K. J. (1996), **Finite Element Procedures**, Prentice-Hall, Englewood Cliffs, N.J.

Bathe, K. J. (1998), Crush simulation of cars with finite element analysis, ASME, Mechanical Engineering.

Bathe, K. J., ed. (1999), **Nonlinear Finite Element Analysis and ADINA**, Comp. Structures, Vol. 72.

Bathe, K. J., Rugonyi, S., and De, S. (2000a), On the current state of finite element methods - solids and structures with full coupling to fluid flows, ICIAM 99, Proc. plenary lectures (Ball JM and Hunt JCR, eds.), Oxford Univ. Press.

Bathe, K. J., Iosilevich, A., and Chapelle, D. (2000b), An evaluation of the MITC shell elements, Comp. Structures, Vol. 75, pp. 1-30.

Bathe, K. J., ed. (2001a), **Computational Fluid and Solid Mechanics**, Proc. First M. I. T. Conference on Computational Fluid and Solid Mechanics, Elsevier Science.

Bathe, K. J. (2001b), The inf-sup condition and its evaluation for mixed finite element methods, Comp. Structures, Vol. 79, pp. 243-252, 971.

Bathe, K. J., ed. (2003), **Computational Fluid and Solid Mechanics 2003**, Proc. Second M. I. T. Conference on Computational Fluid and Solid Mechanics, Elsevier Science.

Bathe, K. J. and Montans, F. J. (2004), On modeling mixed hardening in computational plasticity, Comp. Structures, Vol. 82, pp. 535-539.

Bishop, A. W., and Henkel, D. J. (1957), **The Measurements of Soil Properties in Triaxial Test**, W. Arnold, London.

Borja, R. I., and Lee, S. R. (1990), Cam-clay plasticity, Part I: Implicit integration of elasto-plastic constitutive relations, Comp. Meth. Appl. Mech. Engng., Vol. 78, pp. 49-72.

Borja, R. I. (1991), Cam-clay plasticity, Part II: Implicit integration of constitutive equation based on nonlinear state predictor, Comp. Meth. Appl. Mech. Engng., Vol. 88, pp. 225-240.

Borja, R. I., Lin, C. H., and Montans, F. J. (2001), Cam-clay plasticity, Part IV: Implicit integration of anisotropic, bounding surface model with nonlinear hyperelasticity and ellipsoidal loading function, Comp. Meth. Appl. Mech. Engng., Vol. 190 (26-27), pp. 3293-3323.

Britto, A. M., and Gunn, M. J. (1987), **Critical State Soil Mechanics Via Finite Elements**, Ellis Horwood Ltd., Chichester, England.

Brocks, W. (1995), **Numerical Round Robin on Micromechanical Models**, IWM-Bericht T8/95.

Brown, G. M. (1970), A self-consistent polycrystalline model for creep under combined stress states, J. Mech. Phys. Solids, Vol. 18, pp. 367-381.

Brunet, M., and Sabourin, F. (1996), Simulation of necking using damage mechanics in 3-D sheet metal forming analysis, in: Lee, J. K., Kinzel, G. L., and Wagoner, R H., eds., **NUMISHEET'96, Proc. The 3rd Int. Conference**, Dearborn, Michigan, Sept. 29 - Oct. 3, pp. 212-219.

Bucalem, M. L., and Bathe, K. J. (200x), **Hierarchical Modeling in Finite Element Analysis**, Springer-Verlag, in preparation.

Burland, J. B. (1965), The yielding and dilatation of clay; correspondence, Geotechnique, Vol. 15, No. 2, pp. 211-219.

Bushnell, D. (1977), A Strategy for the solution of problems involving large deflections, plasticity and creep, Int. J. Num. Meth. Engng., Vol. 11, pp. 683-708.

Casagrande, A. (1936), Characteristics of cohesionless soils affecting the stability of slopes and earth fills, J. Boston Soc. Civil Engng., pp. 257-276.

Chapelle, D., and Bathe, K. J. (1998), Fundamental considerations for the finite element analysis of shell structures, Comp. Structures, Vol. 66, No. 1, 19-36, pp. 711-712.

Chapelle, D., and Bathe, K. J. (2000), The mathematical shell model underlying general shell elements, Int. J. Num. Meth. Engng., Vol. 48, No. 2, pp. 289-313.

Chapelle, D., and Bathe, K. J. (2003), **The Finite Element Analysis of Shells - Fundamentals**, Springer Verlag.

Chen, W. F., and Saleeb, A. F. (1982), **Constitutive Equations for Engineering Materials, Vol. 1: Elasticity and Modelling**, J. Wiley & Sons, New York.

Chen, W. F., and Baladi, G. Y. (1985), **Soil Plasticity - Theory and Implementation,** Elsevier, Amsterdam.

Chen, W. F., and Han, D. J. (1988), **Plasticity for Structural Engineers,** Springer-Verlag, New York.

Chen, W. F., and Mizuno, E. (1990), **Nonlinear Analysis in Soil Mechanics,** Elsevier, Amsterdam.

Christian, J. T., and Desai, C. S. (1977), Constitutive laws for geologic media, in: Christian, J. T., and Desai, C. S., eds., **Numerical Methods in Geotechnical Engineering**, McGraw-Hill Book Co., New York.

Chung, K., and Shah, K. (1992), Finite element simulation of sheet metal forming for planar anisotropic metals, Int. J. Plasticity, Vol. 8, pp. 453-476.

Collatz, L. (1966), **The Numerical Treatment of Differential Equations**, 3rd ed., Springer-Verlag, New York.

Crandall, S., Dahl, N., and Lardner, T. (1972), **An Introduction to the Mechanics of Solids,** Second Edition, McGraw-Hill Book Co., New York.

Crisfield, M. A. (1991), **Non-Linear Finite Element Analysis of Solids and Structures**, J. Wiley & Sons, Chichester, England.

Cristescu, N. D., and Goida, G. (1994), **Visco-Plastic Behaviour of Geomaterials**, Springer-Verlag, Wien - New York.

Dafalias, Y. F., and Popov, E. P. (1975), A model for nonlinearly hardening materials for complex loading, Acta Mech., Vol. 21, No. 3, pp. 173-192.

Dafalias, Y. F., and Popov, E. P. (1977), Cyclic loading for materials with a vanishing elastic region, Nucl. Eng. Des., Vol. 41, No. 2, pp. 293-302.

Dafalias, Y. F. (1979), A bounding surface plasticity model, Proc. 7th Can. Congr. Appl. Mech., Sherbrooke, Canada.

Das, B. M. (1997), **Advanced Soil Mechanics**, Taylor & Francis, Washington D. C.

Davis, E. A. (1943), Increase of stress with permanent strain and stress-strain relations in the plastic state for copper under combined stresses, Trans. ASME, Vol. 65, pp. A187-A196.

Davis, E. H. (1968), Theories of plasticity and failure of soil masses, in: Lee, I. K., ed., **Soil Mechanics: Selected Topics**, Butterworts, London, pp. 341-389.

De Jong, G. De J. (1959), **Statics and Kinematics in the Failable Zone of a Granular Material**, Waltman, Delft.

Desai, C. S. (1980a), A general basis for yield, failure and potential functions in plasticity, Int. J. Num. Anal. Meth. Geomech., Vol. 4, pp. 361-375.

Desai, C. S., and Siriwardane, H. J. (1980b), A concept of correction functions to account for non-associative characteristics of geologic media, Int. J. Num. Anal. Meth. Geomech., Vol. 4, pp. 377-387.

Desai, C. S., and Siriwardane, H. J. (1984), **Constitutive Laws for Engineering Materials,** Prentice-Hall, Englewood Cliffs, NJ.

Desai, C. S. (1989), Notes for advanced school - Numerical methods in geomechanics including constitutive modelling, Udine, Italy.

Desai, C. S. (2001), **Mechanics of Materials and Interfaces, The Disturbed State Concept**, CRC Press, Boca Raton, FL, USA.

DiMaggio, F. L., and Sandler, I. S. (1971), Material model for granular soils, J. Engng. Mech. Div., ASCE 97 (EM3), pp. 935-950, Proc. Paper 8212.

Djapic-Oosterkamp, L., Ivankovic, A., and Venizelos, G. (2000), High strain rate properties of selected aluminum alloys, Materials Sc. and Engng., Vol. A278, pp. 225-235.

Drucker, D. C. (1951), A more fundamental approach to plastic stress-strain relations, Proc. 1st National Congress of Applied Mechanics, ASME, Chicago, pp. 487-491.

Drucker, D. C., and Prager, W. (1952), Soil mechanics and plastic analysis or limit design, Q. Appl. Math., Vol. 10, No. 2, pp. 157-164.

Drucker, D. C., Gibson, R. E., and Henkel, D. J. (1957), Soil mechanics and work hardening theories of plasticity, Trans. ASCE, Vol. 122, pp. 338-364.

Eterovic, A. L., and Bathe, K. J. (1990), A hyperelastic-based large strain elasto-plastic constitutive formulation with combined isotropic- kinematic hardening using the logarithmic stress and strain measures, Int. J. Num. Meth. Engng., Vol. 30, pp. 1099-1114.

Eterovic, A. L., and Bathe, K. J. (1991a), An interface interpolation scheme for quadratic convergence in the finite element analysis of contact problems, in: Wriggers, P. and Wagner, W., eds., **Nonlinear Computational Mechanics**, Springer-Verlag, Berlin.

Eterovic, A. L., and Bathe, K. J. (1991b), On large strain elasto-plastic analysis with frictional contact conditions, Rend. Sem. Mat., Univers. Politecn. Torino.

Finnie, I., and Heller, W. R. (1959), **Creep of Engineering Materials**, McGraw-Hill Book Co., New York.

Fung, Y. C. (1969), **Foundations of Solid Mechanics,** Prentice-Hall, Englewood Cliffs, N. J.

Gabriel, G., and Bathe, K. J. (1995), Some computational issues in large strain elasto-plastic analysis, Comp. Structures, Vol. 56, No. 2/3, pp. 249-267.

Graham, J., Noonan, M. L., and Lew, K. V. (1983), Yield states and stress-strain relationships in a natural plastic clay, Canadian Geotech. J., Vol. 20, No. 3, pp. 502-516.

Green, A. E., and Zerna, W. (1960), **Theoretical Elasticity,** 2nd ed., Clarendon Press, Oxford, England.

Gurson, L. (1977), Continuum theory of ductile rupture by void nucleation and growth: Part I - yield criteria and flow rules for porous ductile media, J. Engng. Mater. Technol., Trans. ASME, Vol. 99, pp. 2-15.

Hill, R. (1948), A theory of the yielding and plastic flow of anisotropic metals, Proc. Royal Soc. London, Series A, Vol. 193, Issue 1033, pp. 281-297.

Hill, R. (1950), **The Mathematical Theory of Plasticity,** Oxford University Press, London.

Hill, R. (1968), On constitutive inequalities for simple materials - II, J. Mech. Phys. Solids, Vol. 16, pp. 315-322.

Hill, R. (1978), Aspects of invariance in solid mechanics. In: Yih, C.-S, ed., **Advances in Applied Mechanics**, Vol. 18, pp. 1-75, Academic Press, New York.

Hill, R. (1990), Constitutive modeling of orthotropic plasticity in sheet metals, J. Mech. Phys. Solids, Vol. 38, No. 3, pp. 405-417.

Hill, R. (1993), A user-friendly theory of orthotropic plasticity in sheet metals, Int. J. Mech. Sci., Vol. 35, No. 1, pp. 19-25.

Hiller, J. F., and Bathe, K. J. (2003), Measuring convergence of mixed finite element discretizations: An application to shell structures, Comp. Stuctures, Vol. 81, pp. 639-654.

Hinton, E., and Owen, D. R. J. (1980), **Finite Elements in Plasticity: Theory and Practice**, Pineridge Press, Swansea, U.K.

Hodge, P. G., Jr. (1957), Discussion of paper: W. Prager, A new method of analyzing stresses and strains in work-hardening solids, J. Appl. Mech., Trans. ASME, Vol. 23, pp. 482-484.

Huber M. T. (1904), Specific strain work as a measure of material effort, Czasopismo Techniczne, Lwow, vol. 22, pp. 34 - 40, 49-50, 61-62, 80-81.

Hvorslev, M. J. (1969), Physical properties of remolded cohesive soils, transl. 69-5, U. S. Army Corps of Engineers Waterways Exper. Stn., Vicksburg, Miss.

Ibrahimbegovic, A. (1994), Finite elastoplastic deformations of space-curved membranes, Comp. Meth. Appl. Mech. Engng., Vol. 119, pp. 371-394.

Ilyushin, A. A. (1946), Some problems in the theory of plastic deformation, Prikladnaia Mathematika and Mechanika, Vol. 7, pp. 245-272; ((1943) English translation, RMB-12, Grad. Div. Appl. Math., Brown Univ.).

Ingham, T. J. (2001), Issues in the seismic analysis of bridges, in: Bathe, K. J., ed., **Computational Fluid and Solid Mechanics**, Elsevier Science.

Iwan, W. D. (1967), On a class of models for the yielding behavior of continuous and composite systems, J. Appl. Mech., Trans ASME, Vol. 34, No. E3, pp. 612-617.

Jeremic, B., Runesson, K., and Sture, S. (1999), A model for elastic-plastic pressure sensitive materials subjected to large deformations, Int. J. Solids and Structures, Vol. 36, pp. 4901-4918.

Johnson, W., and Mellor, P. (1983), **Engineering Plasticity**, Ellis Horwood Ltd., Chichester, England.

Jones, R. M. (1975), **Mechanics of Composite Materials,** McGraw-Hill Book Co., New York.

Kawka, M., and Bathe, K. J. (2001), Implicit integration for solution of metal forming processes, in: Bathe, K. J., ed., **Computational Fluid and Solid Mechanics**, Elsevier Science.

Keedwell, M. J., ed. (1988), **Rheology and Soil Mechanics**, Proc. Intern. Conf., Coventry, U.K., Elsevier Applied Science, London.

Kohata, Y., Tatsuoka, Y., Wang, L., Jiang, G. L., Hoque, E., and Kodaka, T. (1997), Modelling the non-linear deformation properties of stiff geomaterials, Geotechnique, Vol. 47, No. 3, pp. 563-580.

Koiter, W. T. (1953), Stress-strain relations, uniqueness and variational theorems for elastic, plastic materials with singular yield surface, Q. Appl. Math., Vol. 11, pp. 350-354.

Kojic, M., and Cheatham, J. B., Jr. (1974a), Theory of plasticity of porous media with fluid flow, Soc. Petrol. Engng. J., Vol. 14, No. 3, pp. 263-270.

Kojic, M., and Cheatham, J. B., Jr. (1974b), Analysis of the influence of fluid flow on the plasticity of porous rock under axially-symmetric punch, Soc. Petrol. Engng. J., Vol. 14, No. 3, pp. 271-278.

Kojic M., and Bathe, K. J. (1987a), The effective-stress-function algorithm for thermo-elasto-plasticity and creep, Int. J. Num. Meth. Engng., Vol. 24, pp. 1509-1532.

Kojic, M., and Bathe, K. J. (1987b), Thermo-elastic-plastic and creep analysis of shell structures, Comp. Structures, Vol. 26, No. 1/2, pp. 135-143.

Kojic, M., and Bathe, K. J. (1987c), Studies of finite element procedures - stress solution of a closed elastic strain path with stretching and shearing using the updated Lagrangian Jaumann formulation, Comp. Structures, Vol. 26, No. 1/2, pp. 175-179.

Kojic, M. (1992), An implicit procedure for stress integration of a general anisotropic von Mises material, in: Owen, D. R. J., and Onate, E., eds., **Computational Plasticity**, Pineridge Press, Swansea, U.K.

Kojic, M. (1993), Implicit stress integration for elastic-plastic deformation of von Mises material with mixed hardening, Theoretical and Appl. Mech. (Yugoslavian), Vol. 19, pp. 121-140.

Kojic, M., Slavkovic, R., Grujovic, N., and Vukicevic, M. (1994), Implicit stress integration algorithm for modified Cam-clay material, Theoretical and Appl. Mech. (Yugoslavian), Vol. 20, pp. 95-118.

Kojic, M., Grujovic, N., Slavkovic, R., and Kojic, A. (1995a), Elastic-plastic orthotropic pipe deformation under external load and internal pressure, AIAA Journal, Vol. 33, No. 12, pp. 2354-2358.

Kojic, M., Zivkovic, M., and Kojic, A. (1995b), Elastic-plastic analysis of orthotropic multilayered beam, Comp. Structures, Vol. 57, pp. 205-211.

Kojic, M., Begovic, D., and Grujovic, N. (1995c), A computational procedure for implicit stress integration of anisotropic thermo-plastic and/or anisotropic creep constitutive relations of metals, in: Owen, D. R. J., and Onate, E., eds., **Computational Plasticity**, pp. 249-259, Pineridge Press, Swansea, U.K.

Kojic, M., Slavkovic, R., Grujovic, N., and Zivkovic, M. (1995d), A solution procedure for large strain plasticity of the modified Cam-clay material, in: Theocaris, P. S., and Gdoutos, E. E., eds., Proc. 4th Greek National Congress on Mechanics, pp. 511-518.

Kojic, M., Slavkovic, R., Grujovic, N., and Zivkovic, M. (1995e), Implicit stress integration procedure for the generalized cap model in soil plasticity, in: Owen, D. R. J., Onate, E., eds., **Computational Plasticity**, Pineridge Press, Swansea, U. K., pp. 1809-1820.

Kojic, M. (1996a), The governing parameter method for implicit integration of viscoplastic constitutive relations for isotropic and orthotropic metals, Computational Mechanics, Vol. 19, pp. 49-57.

Kojic, M., Grujovic, N., Slavkovic, R., and Zivkovic, M. (1996b), A general orthotropic von Mises plasticity material model with mixed hardening: model definition and implicit stress integration procedure, J. Appl. Mech., Trans. ASME, Vol. 63, pp. 376-382.

Kojic, M. (2002a), Stress integration procedures for inelastic material models within the finite element method, Appl. Mech. Reviews, Vol. 55, No. 4, pp. 389-414.

Kojic, M., Vlastelica, I., and Zivkovic, M. (2002b), Implicit stress integration for large strain elastic-plastic deformation of metal and Gurson material model, Int. J. Num. Meth. Engng., Vol. 53, pp. 2701-2720.

Kojic, M. (2002c), An extension of 3-D procedure to large strain analysis of shells, Comp. Meth. Appl. Mech. Engng., Vol. 191, pp. 2447-2462.

Kojic, M., Vlastelica, I., and Zivkovic, M. (2004), Implicit stress integration algorithm for Gurson model in case of large strain shell deformation, Facta Universitatis, Nis, Serbia, Vol. 3, No. 16, pp. 85-100.

Kotsovos M., and Pavlovic, M. (1995), **Structural Concrete, Finite-Element Analysis for Limit-State Design**, Thomas Telford, London.

Kraus, H. (1980), **Creep Analysis,** J. Wiley & Sons, New York.

Krieg, R. D. (1975), A practical two-surface plasticity theory, J. Appl. Mech., Vol. 42, pp. 641-646.

Krieg R. D., and Key, S. W. (1976), Implementation of a time independent plasticity theory into structural computer programs, in: **Constitutive Equations in Viscoplasticity: Computational and Engineering Aspects**, The Winter Annual Meeting of ASME, New York.

Krieg, R. D., and Krieg, D. B. (1977), Accuracies of numerical solution methods for the elastic-perfectly plastic model, ASME J. Press. Vess. Tech., Vol. 99, pp. 510-515.

Lee, E. H., and Liu, D. T. (1967), Finite strain elastic-plastic theory particularly for plane wave analysis, J. Appl. Phys., Vol. 38, pp. 19-28.

Lee, E. H. (1969), Elastic-plastic deformations at finite strains, J. Appl. Mech., Trans. ASME, Vol. 36, pp. 1-6.

Lee, E. H., and Mallett, R. L., eds. (1982), **Plasticity of Metals at Finite Strain: Theory, Computation and Experiment**, Rensselaer Polytechnic Institute, Troy.

Lee, J. K., Kinzel, G. L., and Wagoner, R. H., eds. (1996), **NUMISHEET'96, Proc. The 3rd Int. Conference**, Dearborn, Michigan, Sept.29 - Oct.3, pp. 476-477.

Lege, D. J., Barlat, F., and Brem, J. C. (1989), Characterization and modeling of the mechanical behavior and formability of a 2008-T4 sheet sample, Int. J. Mech. Sci., Vol. 31, No. 7, pp. 549-563.

Lévy, M. (1870), Memoire sur les équations générales des mouvements intérieurs des corps solides ductiles au delà des limites où l'élasticité pourrait les ramener à leur premier état, Comptes Rendus, Vol. 70, pp. 1323-1325.

Lewis, R. W., and Schrefler, B. A. (1987), **The Finite Element Method in the Deformation and Consolidation of Porous Media**, J. Wiley & Sons, Chichester, England.

Li, K. P., and Cescotto, S. (1996), Numerical simulations of 3-D sheet metal forming by mixed brick element, in: Lee, J. K., Kinzel, G. L., and Wagoner, R H., eds., **NUMISHEET'96, Proc. The 3rd Int. Conference**, Deaborn, Michigan, Sept. 39-Oct. 3, pp. 80-86.

Love, A. E. H. (1944), **A Treatise on the Mathematical Theory of Elasticity,** Dover Publ. Co., New York.

Lubliner, J. (1984), A maximum-dissipation principle in generalized plasticity, Acta Mechanica, Vol. 52, pp. 225-237.

Lubliner, J. (1990), **Plasticity Theory,** Macmillan Publ. Co., New York.

Lush, A. M., Weber, G., and Anand, L. (1989), An implicit time-integration procedure for a set of internal variable constitutive equations for isotropic elasto-viscoplasticity, Int. J. Plasticity, Vol. 5, pp. 521-549.

Makinouchi, A., Nakamachi, E., Onate, E., and Wagoner, R. H., eds. (1993), **NUMISHEET'93, Proc. The 2nd Int. Conference**, Isehara, Japan, 31 Aug. - 2 Sept., p. 403.

Malvern, L. E. (1969), **Introduction to the Mechanics of a Continuous Medium,** Prentice-Hall, Englewood Cliffs, N. J.

Mandel, J. (1964), Contribution théoretique à l'etude de l'écrouissage et des lois de l'écoulement plastique, Proc. 11th Int. Congr. Appl. Mech., pp. 502-509.

Marcal, P. V. (1965), A stiffness method for elastic-plastic problems, Int. J. Mech. Sci., pp. 229-238.

McClintock, F., and Argon, A., eds. (1966), **Mechanical Behaviour of Materials,** Addison-Wesley Publ. Co., Reading, MA.

Mendelson, A. (1968), **Plasticity: Theory and Application**, The Macmillan Co., New York.

Montans, F. J, and Bathe K. J. (2003), On the stress integration in large strain elasto-plasticity, in: Bathe, K. J., ed., **Computational Fluid and Solid Mechanics**, Proc. Second M. I. T. Conference on Computational Fluid and Solid Mechanics, Elsevier Science.

Montans, F. J, and Bathe K. J. (200x), Computational issues in large strain elasto-plastic analysis: An algorithm for mixed hardening including plastic spin, submitted for publication.

Mroz, Z. (1967), On the description of anisotropic work hardening, J. Mech. Phys. Solids, Vol. 15, pp. 163-175.

Mroz, Z., Norris, V. A., and Zienkiewicz, O. C. (1978), An anisotropic hardening model for soils and its application to cyclic loading, Int. J. Num. Anal. Meth. Geomech., Vol. 2, pp. 203-221.

Mroz, Z., Norris V. A., and Zienkiewicz, O. C. (1979), Application of an anisotropic hardening model in the analysis of elasto-plastic deformation of soils, Geotechnique, Vol. 29, No. 1, pp. 1-34.

Nakamachi, E. (1993), Sheet forming process characterization by static-explicit anisotropic elastic-plastic finite element simulation, in: Makinouchi, A., Nakamachi, E., Onate, E., and Wagoner, R. H., eds., **NUMISHEET'93, Proc. The 2nd Int. Conference**, Isehara, Japan, 31 Aug. - 2 Sept., pp. 109-124.

Naylor, D. J., and Pande, G. N. (1981), **Finite Elements in Geotechnical Engineering**, Pineridge Press, Swansea, U. K.

Odquist, F. K. G. (1974), **Mathematical Theory of Creep and Creep Rupture**, Clarendon Press, Oxford.

Ortiz, M., Pinsky P. M., and Taylor, R. L. (1983), Operator split methods for the numerical solution of the elasto-plastic dynamic problem, Comp. Meth. Appl. Mech. Engng., Vol. 39, pp. 137-157.

Ortiz, M., and Popov, E. P. (1985), Accuracy and stability of integration algorithms for elastoplastic constitutive relations, Int. J. Num. Meth. Engng., Vol. 21, 1561-1576.

Pantuso, D., and Bathe, K. J. (1997), Finite element analysis of thermo-elasto-plastic solids in contact, in: Owen, D. R. J., Onate, E., and Hinton, E., eds., **Computational Plasticity**, pp. 72-87, CIMNE, Barcelona.

Penny, R. K., and Marriott, D. L. (1971), **Design for Creep**, McGraw-Hill Book Co., Ltd., London.

Peric, D., and Owen, D. R. J. (1992), A model for finite strain elasto-plasticity based on logarithmic strains: computational issues, Comp. Meth. Appl. Mech. Engng., pp. 35-61.

Perzyna, P. (1966), Fundamental problems in viscoplasticity, Advances in Appl. Mechanics, Vol. 9, pp. 243-377.

Perzyna, P., and Wojno, W. (1968), Thermodynamics of a rate sensitive plastic material, Archwm. Mech. Stosow, Vol. 20, pp. 499-510.

Pietruszczak, S., and Pande, G. N., eds. (1989), **Numerical Models in Geomechanics,** Numog III, Proc. Int. Symp., Niagara Falls, Canada, Elsevier Applied Science, London.

Prager, W. (1955), The Theory of Plasticity: A survey of recent achievements (James Clayton Lecture), Inst. of Mech. Engng., Vol. 162, pp. 41-57.

Prager, W. (1956), A new method of analyzing stresses and strains in work-hardening plastic solids, J. Appl. Mech., Trans. ASME, Vol. 23, pp. 493-496.

Prager, W. (1959), **Introduction to Plasticity,** Addison-Wesley, Reading, MA.

Prager, W., and Hodge, P. (1968), **Theory of Perfectly Plastic Solids,** Dover Publications, Inc., New York.

Prandtl, L. (1924), Spannungsverteilung in plastischen Körpern, Proc. 1st Int. Congress on Appl. Mechanics, Delft, pp. 43-54.

Pugh, C. E., Corum, J. M., Lin, K. C., and Greenstreet, W. L. (1972), Report TM-3602, Oak Ridge National Laboratory, Oak Ridge, TN.

Pugh, C. E., et al. (1974), Report ORNL-5014, Oak Ridge National Laboratory, Oak Ridge, TN.

Pugh, C. E. (1975), Constitutive equations for creep analysis of liquid moderated fast breeder reactor (LMFBR) components, in: Zamrik, S. Y., and Jetter, R. I., eds., **Advances in Design for Elevated Temperature Environment**, pp. 1-16, ASME, New York.

Pugh, C. E., and Robinson, D. N. (1978), Some trends in constitutive equation model development for high-temperature behaviour of fast-reactor structural alloys, Nucl. Eng. Des., Vol. 48, pp. 269-276.

Rammerstorfer, F. G., Fischer, D. F., Mitter, W., Bathe K. J., and Snyder, M. D. (1981), On thermo-elastic-plastic analysis of heat treatment processes including creep and phase changes, Comp. Structures, Vol. 13, pp. 771-779.

Rees, D. W. A. (1983), The potential theory of creep, in: Wilshire, B., and Owen, D. R. J., eds., **Engineering Approaches to High Temperature Design,** pp. 57-84, Pineridge Press, Swansea, U.K..

Rendulic, L. (1936), Pore index and pore water pressure, Bauingenieur, Vol. 17, No. 559.

Reuss, E. (1930), Berücksichtigung der elastischen Formänderungen in der Plastizitätstheorie, Z. Angew. Math. Mech., Vol. 10, pp. 266-274.

Rice, J. R. (1971), Inelastic constitutive relations for solids: An internal-variable theory and its application to metal plasticity, J. Mech. Phys. Solids, Vol. 19, pp. 433-455.

Rice, J. R., and Tracey, D. M. (1973), Computational fracture mechanics, in: Fendes, S. J., ed., **Numerical and Computer Methods in Structural Mechanics**, Academic Press, N. Y.

Rice, J. R. (1975), Continuum mechanics and thermodynamics of plasticity in relation to microscale deformation mechanisms, in: Argon, A. S., ed., **Constitutive Equations in Plasticity**, pp. 23-79, The MIT Press, Cambridge, MA.

Roscoe, K. H., Shofield, A. N., and Wroth, C. P. (1958), On the yielding of soils, Geotechnique, Vol. 8, No. 1, pp. 22-52.

Roscoe, K. H., Shofield, A. N., and Thurairajah, A. (1963a), Yielding of soils in states wetter than critical, Geotechnique, Vol. 13, No. 2, pp. 211-240.

Roscoe, K. H., and Schofield, A. N. (1963b), Mechanical behaviour of an idealized "wet" clay, Proc. European Conf. on Soil Mechanics and Foundation Engineering, Wiesbaden, Vol. 1, pp. 47-54.

Roscoe, K. H., and Burland, J. B. (1968), On the generalized stress-strain behaviour of wet clay, in: Heyman, J., and Leckie, F. A., eds., **Engineering Plasticity**, pp. 535-609, Cambridge Univ. Press, London.

Saint-Venant B. de (1870), Mémoire sur l'établissement des équations différentielles des mouvements intérieurs opérés dans les corps solides ductiles au delà des limites où l'élasticité pourrait les ramener à leur premier état, Comptes Rendus, Vol. 70, pp. 473-480.

Sandler, I. S., DiMaggio F. L., and Baladi, G. Y. (1976), Generalized cap model for geological materials, J. Geotech. Engng. Div., ASCE 102 (GT7), pp. 683-699.

Sautter, W., Kochendörfer, A., and Dehlinger, U. (1953), Über die Gesetzmäßigkeiten der plastischen Verformung von Metallen unter einem mehrachsigen Spannungszustand, Z. Metallk., Vol. 44, pp. 553-565.

Shreyer, H. L., Kulak R. F., and Kramer, J. M. (1979), Accurate numerical solutions for elastic-plastic models, ASME J. Press. Vess. Tech., Vol. 101, pp. 226-234.

Simo, J. C., and Taylor, R. L. (1985a), Consistent tangent operators for rate independent elasto-plasticity, Comp. Meth. Appl. Mech. Engng., Vol. 48, pp. 101-118.

Simo, J. C., and Ortiz, M. (1985b), A unified approach to finite deformation elastoplasticity based on use of hyperelastic constitutive equations, Comp. Meth. Appl. Mech. Engng., Vol. 49, pp. 221-245.

Simo, J. C., and Taylor, R. L. (1986), A return mapping algorithm for plane stress elastoplasticity, Int. J. Num. Meth. Engng., Vol. 22, pp. 649-670.

Simo, J. C. (1988), A framework for finite strain elastoplasticity based on maximum plastic dissipation and the multiplicative decomposition, Part II. Computational aspects, Comp. Meth. Appl. Mech. Engng., Vol. 68, pp. 1-31.

Simo, J. C. (1993a), Recent developments in the numerical analysis of plasticity, in Stein, E. (ed.), **Progress in Computational Analysis of Inelastic Structures**, Springer-Verlag, Wien, pp. 114-173.

Simo, J. C., and Meschke, G. (1993b), A new class of algorithms for classical plasticity extended to finite strains. Application to geomaterials, Computational Mechanics, Vol. 11, pp. 253-278.

Simo, J. C., and Hughes, T. J. R. (1998), **Computational Inelasticity**, Springer-Verlag, New York.

Smith, J. O., and Sidebottom, O. M. (1965), **Inelastic Behaviour of Load - Carrying Members**, J. Wiley & Sons, Inc., New York.

Snyder, M. D., and Bathe, K. J. (1980), Finite element analysis of thermo-elastic-plastic and creep response, Report 82448-10, Acoustic and Vibration Lab., Mech. Eng. Dept., M.I.T.

Snyder, M., and Bathe, K. J. (1981), A solution procedure for thermo-elastic-plastic and creep problems, J. Nucl. Eng. Des., Vol. 64, pp. 49-80.

Sokolnikoff, S. (1956), **Mathematical Theory of Elasticity,** McGraw-Hill, New York.

Sture, S., Runesson, K., and Macari-Pasqualino, E. J. (1989), Analysis and calibration of a three-invariant plasticity model for granular materials, Ing. Arch., Vol. 59, pp. 253-266.

Tatsuoka, F. (1972), **Shear Tests in Triaxial Apparatus - a Fundamental Study of the Deformation of Sand** (in Japanese), Ph.D. thesis, (cited in Wood (1990)), Tokyo University.

Taylor, G. I., and Quinney, H. (1931), The plastic distortion of metals, Phil. Trans. Roy. Soc., London, Vol. A230, pp. 323-362.

Terzaghi, K. (1936), The shearing resistance of saturated soil and the angle between the planes of shear, **Proceedings 1st Int. SMFE Conference**, Harvard, Cambridge, MA, Vol. 1, pp. 54-56.

Terzaghi, K., Peck, R. B., and Mesri, G. (1996), **Soil Mechanics in Engineering Practice**, third edition, J. Wiley & Sons, New York.

Timoshenko, S., and Goodier, J. N. (1951), **Theory of Elasticity**, McGraw-Hill Book Co., New York.

Timoshenko, S. (1953), **History of Strength of Materials**, McGraw-Hill Book Co., New York.

Tresca, H. (1864), Sur l'ecoulement des corps solids soumis à de fortes pression, Comptes Rendus, Vol. 59, p. 754.

Tvergaard, V. (1981), Influence of voids on shear band instabilities under plane strain conditions, Int. J. Fracture, Vol. 17, pp. 389-407.

Tvergaard, V. (1987), Effect of yield surface curvature and void nucleation on plastic flow localization, J. Mech. Phys. Solids, Vol. 35, pp. 43-60.

Ulm, F. J., and Coussy, O. (2003), **Mechanics and Durability of Solids, Vol. 1, Solid Mechanics**, Prentice-Hall, Englewood Cliffs, N.J.

Vesic, A. S., and Clough, G. W. (1968), Behaviour of granular materials under high stresses, Proc. ASCE, J. Soil Mech. and Foundations Div., Vol. 94 (SM3), pp. 661-688.

von Mises, R. (1913), Mechanik der festen Körper im plastisch deformablen Zustandt, Göttinger Nach., Math.-Phys. Kl., pp. 582-592.

Walczak, J., Sieniawski, J., and Bathe, K. J. (1983), On the analysis of creep stability and rupture, Comp. Structures, Vol. 17, pp. 783-792.

Wang, C. T. (1953), **Applied Elasticity**, McGraw-Hill Book Co., New York.

Wang, J., Bathe, K. J., and Walczak, J. (2001), A stress integration algorithm for J3-dependent elasto-plasticity models, in: Bathe, K. J. , ed., **Computational Fluid and Solid Mechanics**, Elsevier Science.

Washizu, K. (1975), **Variational Methods in Elasticity and Plasticity**, second edition, Pergamon Press, Oxford.

Weber, G., G., and Anand, L. (1990a), Finite deformation constitutive equations and a time integration procedure for isotropic, hyperelastic-viscoplastic solids, Comp. Meth. Appl. Mech. Engng., Vol. 79, pp. 173-202.

Weber, G. G., Lush, A. M., Zavaliangos, A., and Anand, L. (1990b), An objective time-integration procedure for isotropic rate-independent and rate dependent elastic-plastic constitutive equations, Int. J. Plasticity, Vol. 6, pp. 701-744.

Wilkins, M. L. (1964), Calculation of elastic-plastic flow, in: Alder, B., Fernback, S., and Rotenberg, M., eds., **Methods of Computational Physics,** Vol. 3, Academic Press, N. Y.

Wood, D. M. (1990), **Soil Behaviour and Critical State Soil Mechanics,** Cambridge University Press, Cambridge.

Worswick, M. J., and Pick, R. J. (1991), Void growth in plastically deformed free-cutting brass, J. Appl. Mech., Vol. 58, pp. 631-638.

Życzkowski, M. (1981), **Combined Loadings in the Theory of Plasticity**, Polish Scientific.

Index

additive decomposition, 350, 353
algorithm characteristics
– accuracy, 108, 147, 182
– efficiency, 108
– robustness, 106
– stability, 118
associated flow rule, 68
associated hardening law, 72
associated plasticity, 319
associated viscoplasticity, 220
axial
– force, 78, 162, 236, 237
– strain, 74
– stress, 82, 155, 233, 273, 298
axisymmetric conditions, 370
axisymmetric problem, 138

back stress, 64, 129, 250
Barlat's models for sheet metals
– six-component model, 96
– tri-component model, 97
base vectors
– left, 333
– right, 333
Bauschinger effect, 43, 67
beam
– bending, 156
– elastic constitutive matrix, 372
– elastic-plastic deformation, 154
– orthotropic, elastic-plastic deformation, 194
– stiffness, 241
– thermoelastic-plastic and creep
  deformation, 232
– viscoplastic deformation, 257
bilinear stress-strain relation, 42,
  50, 74, 153
bisection method, 13
bulk modulus, 130, 286, 302, 369

Cam-clay

– model formulation, 300
– stress integration, 305
– triaxial compression test, 312
– yield condition, 304
cantilever, 240
cap model, 280
– generalized, 284
– with plane cap, 281
cap yielding, 289
Cauchy stress, 28, 39, 340, 350
closest point projection, 115
coefficient of thermal expansion,
  216, 374
compatible deformations, 351
complementarity condition, 73
compliance matrix, 195, 198, 368,
  375
compressibility, 265
configuration of a body (system)
– at end of time step, 28, 112, 329
– at start of time step, 28, 112, 329
– fictitious (conceptual), 335, 337,
  354
– initial, 325
– local stress-free, 350
– reference, 28, 323
– rotated, 353, 361
– undeformed, 40
– unrotated, 359, 361
conjugate stress, 248
consolidation, 268
consolidation time, 268
constitutive relations
– creep, 213
– elastic, 130, 189, 368
– elastic orthotropic, 375
– for the back stress, 132
– orthotropic thermoelastic, 377
– plastic, 278
– tangent uniaxial, 151

– thermoelastic, 374
constitutive tensor (matrix), 29
– beam (elastic), 372
– consistent tangent, 109
– consistent tangent elastic-plastic, 141, 193, 290, 317
– consistent tangent elastic-plastic-creep, 231, 244
– consistent tangent viscoplastic, 254, 257
– elastic, 113, 350, 368
– elastic (axial symmetry), 370
– elastic (plane strain), 370
– elastic (plane stress), 372
– elastic (shell), 372
– elastic deviatoric, 139, 253
– elastic-plastic, 118, 122
– orthotropic elastic shell, 376
– orthotropic membrane, 377
– tangent, 30
conventional triaxial compression, 273
conventional triaxial test, 311
convergence criterion, 25
– displacement, 25
– energy, 25
– force, 25
convex yield surface, 73, 86, 97, 118
Coulomb friction slider, 217
creep curve, 203
creep deformations, 201
creep law
– eight-parameter, 206
– exponential, 206
– power, 206
creep model
– uniaxial, 204
– multiaxial, 213
creep recovery, 205
creep relaxation, 205
creep stage
– primary, 203
– secondary, 203
– tertiary, 203

critical
– deviatoric stress, 276, 304
– shear stress, 270
– specific volume, 276
– state, 270, 275, 304, 308
– state line, 276
– void ratio, 276
cutting plane procedure, 127

Darcy's law, 267
dashpot, 217
deformation gradient, 326
– elastic, 350
– inverse, 327
– plastic, 350, 357
– relative, 328
deformation tensor
– Cauchy-Green left, 331
– Cauchy-Green right, 331
– Finger, 331
deformation theory of plasticity, 50
determinant, 383
deviatoric plane, 56, 165, 250
deviatoric strain, 131, 251, 369
deviatoric stress, 46, 130, 251, 369
dilatancy, 270
displacement vector, 23, 112, 326
distortion strain energy, 46, 55
drained conditions, 267
Drucker-Prager model, 281

effective creep strain, 214
effective creep strain rate, 213
effective deviatoric plastic strain, 315
effective plastic strain, 58, 130, 225, 315
effective stress, 56, 214, 225
effective stress (for geological materials), 265
effective stress principle, 266
effective viscoplastic strain, 250
effective volumetric plastic strain, 315
effective-stress-function, 134

effective-stress-function algorithm, 108
eigenvalue, 333
eigenvector, 333
elastic limit, 40
elastic predictor, 114
elastic unloading, 61, 73
elliptical cap, 285
equation-of-state method, 206
equilibrium equations, 32, 178, 242
equilibrium iterations, 24, 29, 107, 169, 179
equivalence of plastic work, 54, 185
equivalent plastic strain, 88, 120
equivalent stress, 87
estimated creep strain increment, 242, 244
Euclidean norm, 57, 225, 330, 385
Euler integration method, 111, 132, 224, 247
explicit
– solution algorithm, 3
– stress integration method, 124

failure surface, 287
finite element discretization, 325
flow rule, 68, 132, 185
fluidity parameter, 219
force vector
– external, 3, 23, 28, 106
– internal, 3, 23
– nodal, 32, 107

general soil plasticity model
– conventional triaxial test, 318
– model formulation, 314
– stress integration, 316
geological material, 263
geometrically nonlinear problem, 21, 27, 321
governing parameter, 109, 134, 145, 153, 186, 225, 252, 288, 289, 308
– equation, 110, 134, 141, 155, 191, 236, 252, 288, 308, 356
– function, 109, 116, 137

– method application in plasticity, 115, 316
– method application in thermo-plasticity and creep, 225
– method application in viscoplasticity, 249
– method formulation, 108
Gurson material model, 184, 364

hardening
– cap, 284
– function, 315
– in viscoplasticity, 219
– isotropic, 43, 67, 135, 176
– kinematic, 43, 67, 136, 176
– law, 97, 114, 285, 301
– mixed, 44, 64, 66, 67, 135, 176, 250
– regime, 308
hardening in creep
– strain, 208
– time, 208
Hill's basic model for sheet metals, 91
Hill's orthotropic plasticity model, 85, 188
Hill's plasticity model for sheet metals, 94
hoop stress, 82, 155

identity matrix, 383
impermeability, 267
implicit, 3
– solution algorithm, 3
– stress integration algorithm, 109, 125, 184, 247
incompatible strain field, 352
incompressible, 223, 353
incremental analysis, 106
inelastic deformations, 106, 201
– time-dependent, 201
– time-independent, 201
initial (reference) state, 269
integration parameter, 111, 132
integration point, 106
integration weight, 35, 106

internal variables, 72, 106, 109, 222, 248
interpolation function, 31
invariant
– of deviatoric stress, 46, 314
– of stress, 46, 314
inverse matrix, 383
isochoric, 49, 231, 263, 359
isoparametric coordinates, 325
isoparametric finite element, 30, 33
isotropic compression, 271
isotropic compression line, 271
isotropy condition, 45

Jacobian matrix, 34, 327

kinematic hardening modulus, 68
Kirchhoff stress, 39
Koiter flow rule, 74
Kronecker delta symbol, 46, 255, 383
Kuhn-Tucker conditions, 73

Lagrangian description of motion, 325
Lankford coefficients, 89
large strain
– plastic deformation, 358
large strains, 321
Levy-von Mises equations, 53
linearization, 4, 21, 29, 127
Lipschitz continuity condition, 19
load step, 23
loading
– cyclic, 44, 209
– history, 50, 51
– hydrostatic, 300
– hydrostatic compression, 271
– inverse, 172
– multiaxial, 44, 48
– non-radial, 149, 181
– proportional, 51, 170
– proportional cyclic, 235, 238
– radial, 149, 181
– reverse radial, 181

– uniaxial, 39, 151

macromechanical theories, 37
material body, 323
material coordinates, 325
material model, 1
material particle, 323
material vector, 329
materially-nonlinear-only (MNO) problem, 23, 321
matrix, 381
mean stress, 46, 223
mean-value theorem, 15
membrane, 139, 141, 372
micromechanical theories, 37
mixed hardening parameter, 67
mixed motion, 359
modified creep strain, 211
modified effective creep strain, 215
modified plastic spin, 354
modified plastic velocity gradient, 354
modified plastic velocity strain, 354
monotonic function, 135, 337
motion, 324
multiplicative decomposition, 350

neutral loading, 73
Newton-Raphson iterative solution method, 4, 16, 19, 24, 242
– full, 25
– modified, 25, 169
nodal point displacements, 23, 30, 107, 326
nonassociated flow rule, 68, 278, 315
nonassociated plasticity, 319
nonlinear
– finite element analysis, 106
– inelastic analysis, 322
– material behavior, 106
– strain increments, 28
– structural analysis, 1
– structural problems, 19
– inelastic analysis, 1
nonsingular matrix, 19

normal to yield surface, 68, 118, 149, 172, 248, 255, 278, 317
normality principle, 68
numerical integration, 35, 106

O.R.N.L. hardening rule, 209
octahedral shear stress, 319
oedometer test, 269
one-index notation, 141, 254, 316, 367
origin (in creep), 210
orthogonal tensor, 388
orthogonality condition, 333, 388
orthotropic elastic shell deformation, 376
orthotropic plasticity, 85, 188
orthotropy coefficients, 98
overconsolidated state, 271
– heavily (dense), 278, 304
– lightly (loose), 278, 303
overconsolidation ratio, 271
overstress, 202, 218

perfect plasticity, 43, 70, 134, 219, 308
perfectly plastic material, 42, 60, 88, 182, 193
permeability coefficient, 267
Piola-Kirchhoff stress, 29, 321
pipe
– elastic-plastic deformation, 80, 155
– orthotropic, elastic-plastic deformation, 197
– thermoelastic-plastic and creep deformation, 234, 245
plane strain conditions, 370
plane strain problem, 138
plane stress conditions, 138, 187, 229, 252, 370
plastic corrector, 114
plastic modulus, 42, 132, 147, 153, 216
plastic potential, 68, 278, 315
plastic work, 54, 62, 67
plate

– elastic-plastic deformation, 174, 199
– orthotropic, 100, 102, 200
– plastic bulging, 364
– thermoplastic and creep deformation, 242
Poisson's ratio, 74, 216, 368
polar decomposition, 333
pore, 264
porosity, 185
position vector, 325
positive definite, 332
Prager's hardening rule, 68
Prandtl-Reuss equations, 53
principal (material) axes, 375
principal directions, 336
principal stress space, 55
principal stretch, 335
principle of maximum plastic dissipation, 71
principle of superposition, 374
principle of virtual work, 23, 28
proportionality coefficient
– in creep, 213
– in plasticity, 49, 59, 74, 111, 115, 133, 191, 193, 284, 306
– in viscoplasticity, 248
pseudo-time, 227

quadratic rate of convergence, 17, 179, 242

Ramberg-Osgood formula, 43, 170
rate formulations, 323
reduced effective stress, 66, 133
reference temperature, 216, 374
relaxation time, 220
return mapping algorithm, 112, 114, 125, 190, 355
rigid body rotation, 336
rolling direction, 89
rotation tensor, 335, 386

saturated, 264
scalar product, 133, 214, 382, 385
secant stiffness method, 123

shape coefficients, 86
shear modulus, 251, 286, 369
shearing of material, 270, 345, 359
sheet metals, 89
shell conditions, 138, 187, 189, 229, 252, 372
shell deformation
– elastic-plastic, 138
– elastic-plastic (Gurson model), 187
– elastic-plastic-creep, 229
– viscoplastic, 252
softening regime, 308
specific volume, 264
spin tensor, 349
static condensation, 372
stationary creep, 203
stiffness matrix, 4, 24, 32, 107
– geometrically nonlinear, 32
– linear, 32
strain
– Almansi, 338
– creep, 203, 216, 224
– deviatoric, 131
– deviatoric plastic, 287
– elastic, 40, 138
– elastic logarithmic, 353
– engineering shear components, 131, 142, 368
– generalized, 337
– Green-Lagrange, 28, 338
– inelastic, 106, 109
– infinitesimal, 339
– logarithmic, 39, 339
– mean elastic, 369
– plastic, 40, 129, 216, 350
– tensorial shear components, 368
– thermal, 216, 224, 374
– thermal (orthotropic), 378
– total, 40, 106, 138, 216, 350, 374
– trial elastic, 355
– viscoplastic, 247
– volumetric, 130, 353, 369
strain increments, 28, 33, 112, 148
– linear, 28

– nonlinear, 28
strain rate
– creep, 204, 207, 213
– plastic, 68
– viscoplastic, 218, 220
strain rate effects, 44
strain softening, 288
strain vector, 31, 130
strain vectors, 113, 247, 368
strain-displacement matrix, 31, 106
– linear, 31
– nonlinear, 31
strain-driven problem formulation, 38, 106, 252
strain-hardening, 42
stress
– at end of time step, 106
– at start of time step, 106
– deviatoric, 130, 139, 189, 224
– generalized, 340
– hydrostatic, 185
– mean, 130, 189, 369
– trial elastic, 355
– trial elastic deviatoric, 355
– trial elastic mean, 355
stress integration, 106, 109, 188, 223, 353
– for cap model, 285
– for viscoplastic models, 247
stress integration algorithm
– explicit, 124
– implicit, 109, 125, 184, 355
stress relaxation, 219
stress-plastic strain relations, 49, 87
stretch, 330
stretch tensor
– elastic, 353
– left, 336
– right, 335
successive elastic solution method, 121
successive substitution method, 15
summation convention, 382
superscript left

– end of time step, 4, 106
– start of time step, 4, 106
superscript right
– iteration counter, 106
– trial counter, 117
support reactions, 159, 247

tangent
– constitutive matrix, 109, 141
– modulus, 22, 41
– stiffness matrix, 4
tangent stiffness-radial return method, 123
Taylor series expansion, 24
tension cutoff, 283
tensor, 384
tensor notation, 384
tensor transformation
– of constitutive matrix, 378
– of strain, 378
– of stress, 378
tensorial transformation, 385
thermoelastic material
– isotropic, 374
– orthotropic, 377
thermoplasticity, 216
thin-walled tube, 77, 166, 236
three-dimensional deformation, 128, 185, 223, 250, 368
time at end of time step, 3, 24, 106, 223
time at start of time step, 3, 24, 106
time independent plasticity model, 111
time step, 3, 23, 106, 225
torsional moment, 77, 166, 236
total formulation, 323
total Lagrangian (TL) formulation, 27, 321
transformation matrix, 378, 384
Tresca yield condition, 47
trial elastic deformation gradient, 355
trial elastic state, 112, 355

trial elastic stress radius, 355
triaxial test, 273

unbalanced energy, 179, 243, 361
unbalanced force, 179
undrained conditions, 267
uniaxial
– compression, 298
– creep, 203
– creep model, 203
– elastic-plastic model, 39
– loading, 42, 48, 151
– straining, 269
– stress-strain curve, 39, 203
– viscoplastic model, 217
unit vector, 330, 384
updated Lagrangian (UL) formulation, 27
updated-Lagrangian-Hencky (ULH) formulation, 349

vector, 381
velocity strain, 340, 354
vertex yielding, 284, 290, 295
virtual displacements, 28
virtual work, 28
– external, 28, 31
– internal, 28, 29
viscoplastic deformations, 201
viscoplastic model
– general 3-D, 220
– stress integration, 247
– uniaxial, 217
viscosity
– coefficient, 218, 222
– function, 222
void, 185
void ratio, 264
volumetric plastic strain, 49, 186, 285
von Mises
– material, 47, 129, 358
– material model, 64
– yield condition, 47, 249

work conjugate, 340
work-hardening, 62

yield condition, 42, 45, 85, 112,
    185, 216, 222, 249, 300, 308, 314
yield curve, 42, 55, 97, 130, 133,
    250, 275
yield function, 45, 68, 113
yield point, 40, 44
yield stress, 22, 41, 86, 129, 216,
    218, 250
– in principal direction, 85
– initial, 22

yield surface, 56, 97, 112, 129, 315
yield surface radius (stress radius),
    64, 129, 224, 250, 257
yield surface size, 129
Young's modulus, 20, 41, 216, 368

Printing: Mercedes-Druck, Berlin
Binding: Stein+Lehmann, Berlin